ARMY EXTENSION COURSES

ARMY
EXTENSION COURSES

Special text No. 266

INFANTRY IN OFFENSIVE COMBAT

Prepared under the Direction of the
Chief of Infantry for use with the
Extension Course of the Infantry School

GOVERNMENT REPRINTS PRESS
Washington, D.C.

Ross & Perry, Inc. Publishers
216 G St., N.E.
Washington, D.C. 20002
Telephone (202) 675-8300
Facsimile (202) 675-8400
info@RossPerry.com

SAN 253-8555

Government Reprints Press Edition 2002

Government Reprints Press is an Imprint of Ross & Perry, Inc.

Library of Congress Control Number: 2001098748
http://www.GPOreprints.com

ISBN 1-931839-33-6

Book Cover designed by Sapna. sapna@rossperry.com

⊗ The paper used in this publication meets the requirements for permanence established by the American National Standard for Information Sciences "Permanence of Paper for Printed Library Materials" (ANSI Z39.48-1984).

CONTENTS

CONTENTS

CONTENTS

CHAPTER 1

GENERAL CONSIDERATIONS OF OFFENSIVE COMBAT

1. *THE ROLE OF INFANTRY.*—*a.* Decisive results are obtained only by the offensive. The infantry is the arm that carries the fight to the enemy, the arm that has the ability to hold what it gains, and the arm that has the power, mass, and stamina to crush the enemy. It is the essential arm of offensive combat.

b. In offensive combat infantry demands the highest order of training, leadership, and morale. Numerically superior forces do not necessarily insure victory. Discipline, leadership, skill in the combined use of weapons and ground, and superior resolution, are far more essential in gaining the decision. Only with these can infantry advance great distances under fire that it cannot return, take advantage of every aid of nature, overcome the enemy's fire, close with the enemy, and hold every advantage gained.

c. Infantry, moreover, consists of such a variety of elements that a high grade of intelligence is required in its handling. The leadership of no other arm needs more skill and care. Its chief element is men; and to lead them and control them in offensive combat is not a science concerned mainly with mechanical precision and accurate data. It is instead an art, which deals not only with how the infantryman fires his weapons, but how he uses their fire in conjunction with the ground on which he fights. It has been said wisely that the tactics of infantry is based "on nature and human nature, not on mechanics or geometrical perfection". The infantry attack is accordingly founded upon a fighting spirit. It requires a stout

1

aggressiveness as well as a full measure of skill in using fire, ground, and movement. This spirit of the offensive—a full determination to advance at every opportunity —must be inculcated in every individual. For it is this that heartens infantry to bear heavy losses and yet close to grips with an enemy.

2. *THE PURPOSE OF OFFENSIVE COMBAT.—a.* The purpose of offensive combat is the attainment of victory by physical disorganization of the enemy forces. The offensive alone embodies those positive qualities which seek and promise a decision. The first of these qualities is initiative. The choice of objectives, the selection of decisive directions, the concentration of the attack forces, the choice of time, and the actual creation of opportunity, all spring from initiative. With initiative goes the high morale that promotes confidence, and the vigorous action essential to successful combat.

b. The action of an army in the field is directed toward a military objective. For the purpose of securing this objective a number of secondary military objectives must be given to the subordinate commanders. The attainment of these lesser successes helps the main force in turn to succeed. Every force should so maneuver as to gain its objective with the minimum expenditure of energy and casualties. Eventually, these maneuvers will force the enemy to interpose his troops between the objective and the force striving for it. The maneuvering force must then attack. The ultimate objective then becomes the decisive defeat of the enemy in battle. Offensive combat invariably seeks victory through the disorganization of the enemy's armed forces, which is accomplished by overcoming or destroying his moral and physical strength. Tactical victory is obtained by attacking the enemy in a decisive direction with an overwhelming combination of maneuver, fire power, and shock action. Through maneuver, infantry takes the maximum advantage of the terrain of the battlefield to place its fire accurately upon the enemy. The fire inflicts losses on the enemy and reduces his effective strength and morale. The shock, or threat of it, causes the adversary to yield or to be destroyed.

3. *BASIC CONSIDERATIONS.*—The basic considerations of offensive combat apply to both large and small forces. These considerations are:

a. The plan of offensive combat must be based on the best information obtainable.

b. The plan must be simple of execution.

c. The plan must take into consideration the advantages and disadvantages of the ground over which the attack is to be made, in order to profit by the former and avoid the latter, as far as possible.

d. The preparations must be as nearly secret as possible.

e. As regards its strength, time, and direction (all three if possible), the main attack must come as a surprise to the enemy.

f. All forces must cooperate to attain the strongest possible attack.

4. *STEPS OF THE OFFENSIVE.*—The steps of offensive combat are:

a. The advance in route column.

b. The advance to contact, partly deployed.

c. Contact and complete deployment.

d. Advancing the attack (fire and movement).

e. The assault.

f. Continuing the attack (reliefs to continue).

g. Organization of the ground and redistribution of force to hold the objective attained, or—

h. Pursuit.

5. *SPECIAL ASPECTS OF THE ATTACK.*—*a. Tanks.* —Tanks are a component of the infantry arm. Their tactical use from the viewpoint of the foot soldier, is that of supporting him in the attainment of his objective by overcoming specific points of hostile resistance with which

they are designed to deal. The presence of tanks in the attacking force requires a careful coordination of their missions with that of foot troops and close coordination between the two infantry elements. Consequently, it also has a decided influence in the formulation of attack plans. This subject is covered in detail in a separate chapter of this text.

b. *The effect of limited visibility* (*darkness and woods*). —The restrictions imposed by limited visibility in darkness, in wooded areas, and in smoke and fog, have a decided influence on the conduct of all military operations. This is especially true in offensive operations because of the movement involved therein. The subjects of attack in darkness and in woods are covered in detail under separate chapters of this text.

c. *The defense in offensive combat.*—(1) The preceding introductory paragraphs rightly emphasize the importance of offensive combat. Battles and wars are only won by attacking. But they may be lost, too, by attacking. The odds against the success of an attack are sometimes too great. To rush into an attack blindly without at least a reasonable idea of what and where the enemy is; to omit considering the condition of troops; to strike at a force that is vastly bigger; in other words, to assume the offensive rashly, may lead indeed to tragedy. Boldness and risk are inevitably a part of successful war; but fortunate is the rash commander who does not meet defeat.

(2) But if a commander is not always justified in assuming or continuing offensive action, what does he do instead? He defends. Until he is stronger, or in a better position, or has less odds against him, he uses defensive warfare to hold off his opponent. It is true that by doing this he postpones his effort toward final victory, he sacrifices in large degree the ability to maneuver the elements of his force, and he may—if he sticks to the defensive too long—reduce the morale of his troops. But on the other hand he conserves the energies and numbers of his force for the time being. He makes it costly for an enemy to act against him effectively. And it may be, if the enemy

is too careless and headstrong, that he can opportunely resume the offensive once the enemy has fairly spent himself attacking a strong position, and thus lead his force to success.

(3) At least we cannot overlook the fact that the defensive has its valuable place as a temporary alternative to the offensive. A boxer, fighting an opponent of greater strength—or at any other disadvantage—soon exhausts himself. He must clinch and cover himself much of the time, and wait for his opportunity. An army, unlike a boxer, is likely to receive help eventually. Hence it has all the more reason at times for foregoing the offensive. But in the end, no matter what the preceding rounds of battle have been, it is only by attacking that a decisive victory can be gained.

(4) Defensive combat is a subject of such importance that it is treated in a separate special text. (Special Text No. 9, Army Extension Courses, Infantry in Defensive Combat.)

CHAPTER 2

THE NECESSITY FOR DEPLOYMENT

1. *WHAT DEPLOYMENT IS.—a.* Military units that are not actually in or near battle, ordinarily move from place to place in long narrow columns. Roads are very narrow compared to the full sweep of the country they run through. Hence troop formations except in and near the battlefield have always been long and narrow. But when armies, or parts of armies, get close enough to an enemy force, they have to spread out over wide distances in order to fight. This spreading out from the columns in which they usually move is called deployment. The reasons for deployment can best be shown by means of several illustrative situations.

2. *ILLUSTRATIVE SITUATION: A SMALL UNIT IN COLUMN.*—A rifle squad is marching in single file towards the enemy, as indicated below. Suddenly fire opens on the squad from the front (Figure 1). Here it is plain that the squad is not in suitable formation to fight. Only one or two men can fire on the enemy while the squad is thus in column. And in this formation, moreover, the squad is very vulnerable to hostile fire.

3. *ILLUSTRATIVE SITUATION: A SMALL UNIT IN LINE.*—Now let us consider the situation shown in Figure 2. Here we see a squad in line of skirmishers advancing towards the enemy. Now, if hostile fire suddenly opens on the squad from the front, every man can fire without interference. This formation is less vulnerable to hostile fire than that shown in Figure 1.

7

4. *ILLUSTRATIVE SITUATION: A SMALL UNIT IN AN IRREGULAR EXTENDED LINE.*—*a.* In the two figures we have just studied we could see the difference between encountering the enemy in column and in line. Certainly the line is better than the column, but before we move on to consider units larger than a squad, we

ENEMY

FIGURE 1.—A small unit in column.

should examine one more diagram in order to learn the lesson that neither a parade-ground column nor a straight close line of skirmishers is the best of all. Let us look at Figure 3.

b. This time the men in the squad form a rough line, but some are a little farther ahead than others and some are farther apart than others, and all are farther apart than in Figure 2. Why should this be? Simply because the men in the squad are taking advantage of every tiny irregularity in the ground and concealment of vegetation, and are spread out so that each man has room to do this. If they try to stay in a straight close line, as they return the enemy's fire or crawl forward to defeat him, they cannot use the ground the way a trained infantryman should. Every little hump and depression helps the in-

ENEMY

FIGURE 2.—A small unit in line.

fantryman who knows how to use it. Thus, although small units such as the squad should be in line to meet the enemy's fire and return it, rather than in column, we must remember, too, that the most practical formation, all things considered, is an irregular line of men in which every man makes the best use he can of the ground on which he fights.

c. When small units deploy from line to column there is often no intermediate stage. Thus, the men in the squad

ENEMY
♀

o   o   o   ó   o   o   o   o

FIGURE 3.—A small unit in an irregular line.

we have just studied would move directly from the formation of Figure 1 to that of Figure 3. But larger units sometimes use, as we shall see in Chapter 4, one or more intermediate formations, which divide their deployment into two or more successive stages. In the illustrative situations immediately following, we shall consider the necessity for deployment in larger units.

5. *A LARGE FORCE IN COLUMN.*—*a.* A force of all arms marching in one column with a fraction of the command in front and small groups on the flanks as security detachments is indicated in Figure 4. This force, just like

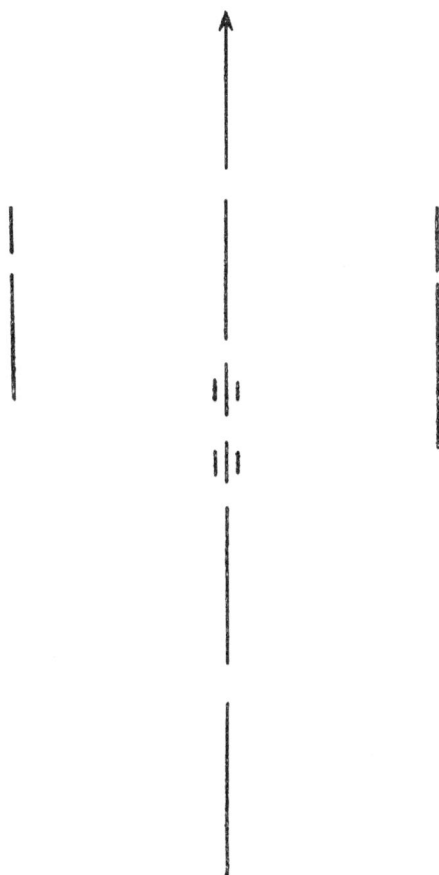

FIGURE 4.—A large force in column (diagrammatic).

the squad in Figure 1, is not suitably disposed for combat to the front, because it can bring only a small fraction of its fire power and maneuver power readily into play in that direction. The formation shown has only one advantage, that of being a good march formation—less fatiguing and easier to control than a more spread-out formation. However this advantage is not a primary consideration in combat. If a force in this long narrow marching formation does meet an enemy, it has to be extended out on a much broader front, before it can be used as a whole in battle. The rear elements must come up to their combat positions. (See Figure 5.) These elements are too far back to be readily available for the battle. They must move farther up and the units making up a considerable part of the whole force must spread out much as the men in the squad did in Figure 3, so that they can effectively fire at the enemy, and take the best advantage of the ground, and so that the commander and his subordinates can move them about as they need to, without long delays. It only takes a squad a few seconds to extend from a column to a suitable combat disposition, on any kind of ground. But to extend a force of all arms is a matter of hours, and to extend a whole army takes days.

b. A long narrow marching column like that of Figure 4 is also extremely vulnerable to hostile fire, and not only to the fire of the enemy's small-arms weapons but to his larger means of fire. The range of the enemy's rifles and machine guns is, of course, limited. Hence all parts of a force of all arms marching as in Figure 4 do not ordinarily come under hostile rifle and machine-gun fire when the enemy is met. (It is not at all safe to say that this will *never* happen because it has often happened in war, and later in this text we shall see why.) But, generally speaking, when a long marching column encounters a hostile force, only its leading elements are liable to come under the enemy's rifle and machine-gun fire at first. On the other hand, the enemy's attack aviation can cover the full length of a long narrow column in a very brief time.

And his artillery may be able to put fire along the column for several miles back.

c. Figure 6 brings out one point about artillery that we should learn here. The dispersion of artillery fire is long

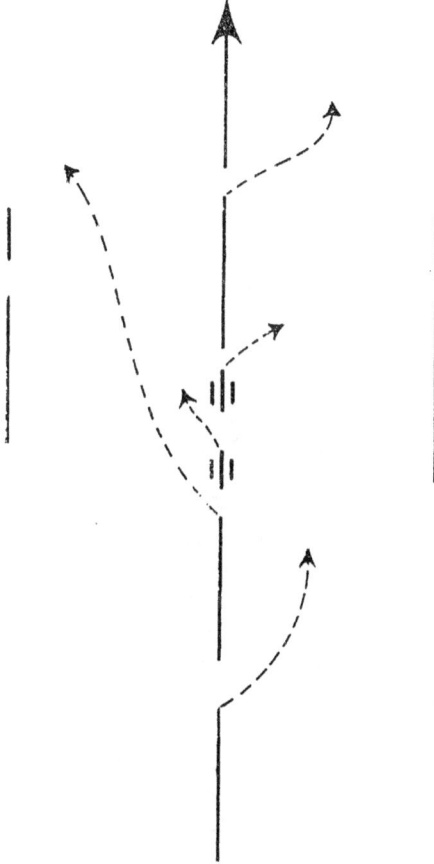

FIGURE 5.—A large force deploying from column (diagrammatic).

and narrow in the direction of fire. Consequently artillery
fire placed on a long narrow column of troops is much
more effective than on a wide shallow formation. (We
must remember that Figure 6 is only a diagram and that
neither columns nor lines are exactly straight, as shown,
with the single exception of a marching column on a long
straight road.)

6. *REASONS FOR DEPLOYMENT.*—*a.* By this time,
we should be able to see clearly why any infantry unit
needs to take up an extended (deployed) formation before
it actually comes to grips with an enemy. A command
deploys in order to—

(1) Place its fire on the enemy effectively.

(2) Assume dispositions which permit the commander
and his subordinate commanders to take advantage of
favorable ground and move their units with the least delay
wherever necessary to win the battle.

(3) Reduce the effect of hostile fire.

b. In the following paragraphs we shall see in more
detail when deployment takes place.

7. *TIME, PLACE, AND DIRECTION OF DEPLOY-
MENT.*—Deployment is not simply a matter of course, nor
merely a mechanical operation that must be accomplished
at some time before battle. Much, indeed, depends upon
a correct decision as to the time, place, and direction of
deployment. We shall not, in this chapter, go deeply into
this subject, since we need to study the subject of advance
guards as covered in Chapter 3 before we can intelligently take up the methods of deployment in detail. But
to emphasize beforehand the necessity for correct deployment, three illustrative situations follow.

8. *ILLUSTRATIVE SITUATIONS: DEPLOYMENT.*—
a. Hostile forces of approximately equal strength and mobility are moving toward each other as in Figure 7. Contact is imminent. Force *B* has a decided advantage over
Force *A* because it can bring its fire and maneuver power

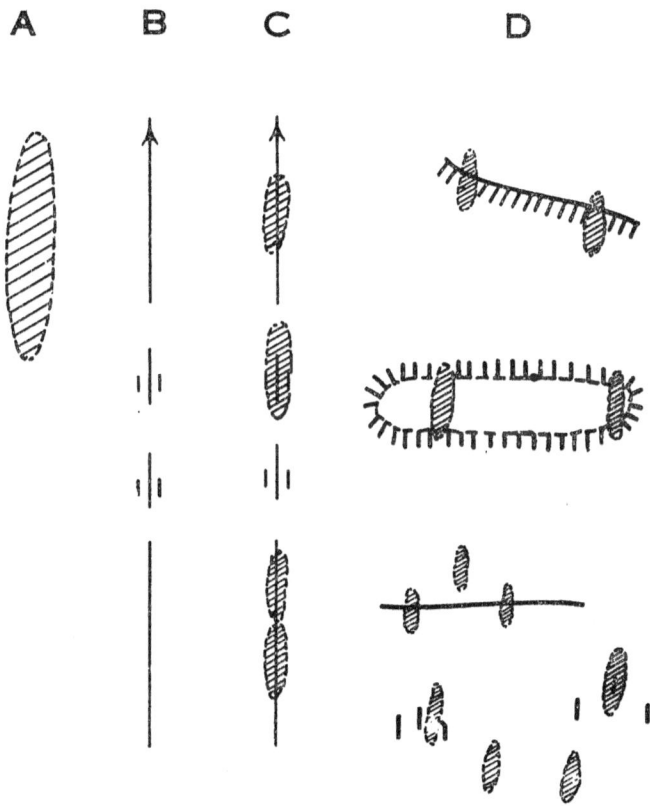

FIGURE 6.—Artillery fire and troop formations.
    A: The general shape of artillery dispersion.
    B: The general shape of a column of troops.
    C: Artillery fire on a column of troops.
    D: Artillery fire on lines, shallow areas and
           deployed formations.

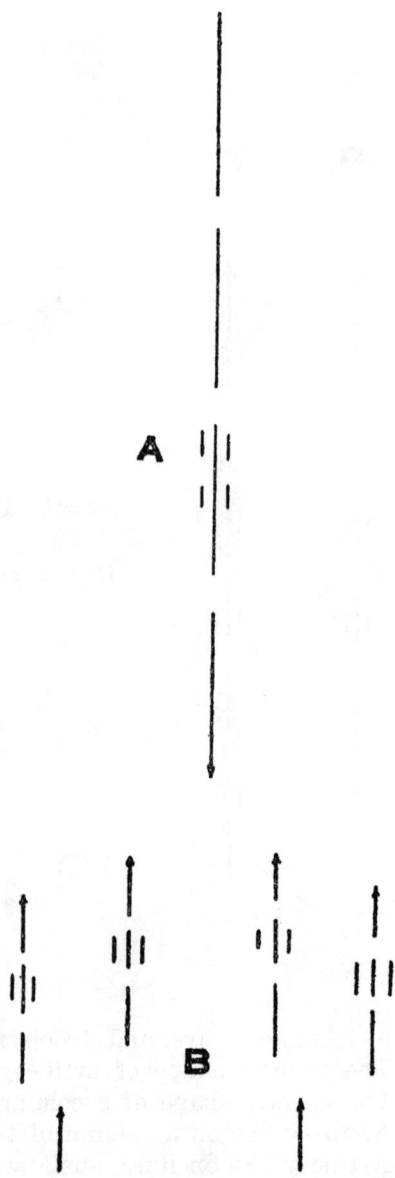

FIGURE 7.

to bear more quickly than Force *A* can, and because it is less vulnerable to fire. In other words, Force *B* is ready for battle and Force *A* is not.

b. Hostile forces of approximately equal strength and mobility are moving as shown in Figure 8. Contact is imminent. Here Force *A* has the advantage. It will take Force *B* longer, other things being equal, to change its dispositions so as to be able to bring its fire and maneuver power to bear in the direction of Force *A* than Force *A* will need to bring its fire and maneuver power to bear on Force *B*. Force *B* must wheel around to the left to fight. If Force *A* acts aggressively there is a strong probability that it will overwhelm Force *B* and prevent the latter from maneuvering, or at least punish it severely while it is maneuvering. For Force *A* has one of its elements already forward and moving in a favorable direction. The others can come up in succession, taking a good deal of time it is true, but still free to maneuver as they do come up. Force *A* would have even more advantage, if it were deployed like Force *B*. Force *B*, on the other hand, is all ready to fight, but in the wrong direction. It has deployed too early.

c. Two hostile forces of approximately equal mobility are moving as shown in Figure 9. Force *B* is only three-fourths the strength of Force *A*. Contact is imminent. Force *B* has such an initial advantage here that Force *A* would have great difficulty in changing formation so as to bring its fire and maneuver power to bear on Force *B*. Force *B*, consequently, is in a position to win decisively.

9. *THE POSSIBILITY OF UNEXPECTED CONTACTS WITH THE ENEMY.*—*a.* The three diagrams we have just examined bring out the point that deployment is just as necessary for large forces as for small ones. But the student may wonder whether hostile forces ever clash as abruptly as these simple diagrams indicate. Surely such information-gathering agencies as cavalry, motor patrols and observation aviation should prevent a force from getting into an acute situation like that of Force *A* in

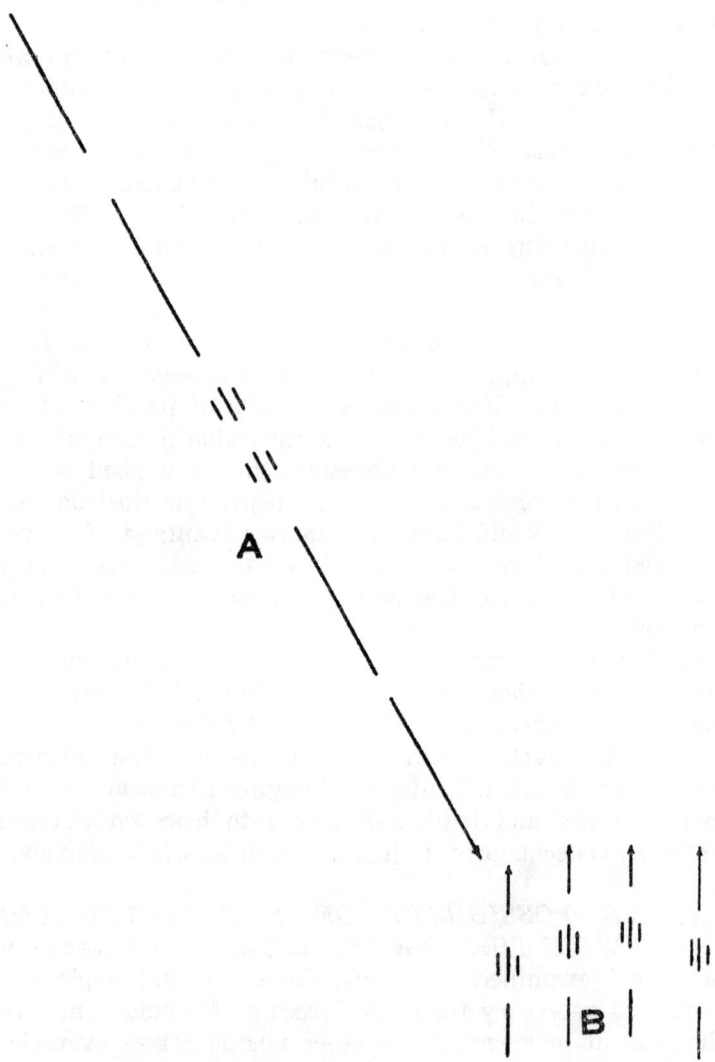

FIGURE 8.

Figure 9; or the awkward position of Force *B* in Figure 8. This point is important; and it is to our advantage to clear it up here, before we continue our discussion of deployment further.

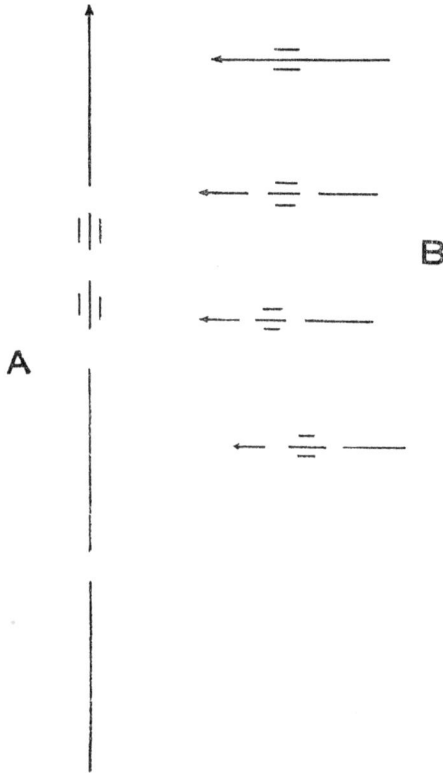

FIGURE 9.

b. Actually, such situations as shown in Figures 7, 8 and 9, have occurred in the past and there is no assurance that they will not occur again. In the early part of the World War, observation airplanes flew almost with impunity over hostile columns. Yet such clashes as these occurred even when the weather favored air observation. Today, although there have been improvements in observation aviation, there have likewise been improvements in antiaircraft measures. Consequently air observation will in all liklihood have to operate from much higher altitudes. Again, troop movements are now made largely under cover of darkness; and when troops must move by day, we exercise every practicable measure for concealment from air observation. Thus it is true that, even when the weather is favorable for flying and for observing from the air, the effectiveness of air reconnaissance is little, if any, greater now than it was in the World War. Furthermore, there are always periods when the weather prevents hostile planes from carrying out observation missions over large parts of a theater of operations, if not the whole area. And even in good weather, the ability of hostile aviation to cover large areas thoroughly is in proportion to the number of planes available. Thus we can only conclude, as far as aviation is concerned, that unexpected contacts with the enemy are still entirely possible.

c. Means and methods of making motor reconnaissances have been vastly improved since the World War. Reconnaissance is now more rapid and more distant than was possible with the horse. But motorized reconnaissance also has its limitations. It must sweep the country on all sides of an advancing force with equal thoroughness, since hostile motorized elements may attempt to approach from any direction. Thus motor reconnaissance has far more ground to cover. It may also, like reconnaissance in past wars, become lost, or be driven aside or avoided at certain points by an adroit enemy. It is not impossible for sizeable reconnaissance elements to be swallowed whole. They may slip or break through the enemy's reconnaissance or counterreconnaissance and find themselves cut off from return. Always, we must remember, the motor elements protecting the main enemy force can move as far and as fast as our own. Thus, all things considered, we cannot by any means look upon motor reconnaissance as an absolute safeguard against unexpected contacts. Its mobility promises a more effective and far-reaching screen, perhaps, but far from an infallible one.

d. It is also an error to assume that commanders will never make the mistake of providing no protection, or of providing inadequate protection for their commands. Even a superficial study of past wars shows that such an assumption is without foundation.

e. We must conclude, then, that unexpected contact with the enemy is still possible. It is especially possible of occurrence early in a war between major powers when one commander takes advantage of his opponent's lack of information and strikes at him from an unexpected direction. Thus Figures 7, 8, and 9, which we have used to illustrate the advantage of proper deployment, are not based on false assumptions.

f. As we continue the study of offensive combat throughout this text, we shall see more and more that deployment for battle is a matter of major importance and that the disadvantage of a poorly made or ill-timed deployment is often a deciding factor. In Chapter 3 we shall take up the study of advance guards, which we shall need to know about in order to understand the details of how deployment for battle is accomplished. In Chapters 4 and 5 we resume the study of forces deploying from march column upon more or less sudden contact with the enemy. And in Chapter 6, we next take up the more usual and preferable method of beginning to deploy a force when contact with the enemy is imminent but not yet actual.

CHAPTER 3

ADVANCE GUARDS AND CONDITIONS AFFECTING THEIR CONDUCT

SECTION I

INTRODUCTION

1. *WHAT ADVANCE GUARDS ARE.*—An advance guard is a security detachment that precedes and covers a force on the march and as it enters battle. If a force marches in more than one column or group, each is preceded by its own advance guard. Each advance guard operates under the commander of its own column until the situation requires the coordination of advance guards under a single head.

2. *MISSIONS OF ADVANCE GUARDS.* — Advance guards have two principal missions. For as long as possible they insure the uninterrupted advance of the main body elements of their column or columns following in rear. And then, upon contact with a hostile force of such strength that the uninterrupted advance of the main body is no longer possible, advance guards insure enough time and space for the main body elements to maneuver in accordance with the will of the commander of the whole force.

3. *SPECIFIC DUTIES OF ADVANCE GUARDS.*—In accomplishing their mission, advance guards perform a number of specific duties:

a. They remove obstacles, repair roads, and facilitate in every other practicable way the uninterrupted advance of the main body.

b. They reconnoiter to the front and flanks in order to guard against surprise, and to obtain information of the enemy.

c. They drive back small parties of the enemy in order to prevent their observing, firing upon, or delaying the main body elements.

d. They secure for their main bodies such terrain features as hills and ridges that defilade the main body elements from possible hostile fire and observation and provide good defilade and observation for friendly artillery.

e. Upon encountering the enemy in force, they check his advance long enough to give the commander of the whole force time to employ his main body elements as he desires.

f. When the enemy is encountered in force, they determine his lines and flanks, and other dispositions.

4. STRENGTH AND COMPOSITION OF ADVANCE GUARDS.—a. The strength and composition of advance guards vary with the strength of the whole column, the mission, the situation, and the terrain. A large command requires longer to prepare for action than a small one; hence, its advance guard is correspondingly larger. When the mission of a force is aggressive and there is a probability of strong hostile resistance, relatively strong advance guards are required. Greater strength is also required as the distance to the enemy decreases. The proportion of infantry varies from a small fraction to one-third or more of the infantry of a column, according to the size of the command and other attending circumstances.

b. From the nature of the mission and duties of advance guards, it is apparent that they are both information gathering agencies and combat units. Their combat functions are usually to attack, defend, or fight delaying action. Therefore, in large commands the advance guards consist of units of all arms.

c. In a small force the advance guard may consist of infantry alone.

SECTION II

CONDITIONS AFFECTING THE CONDUCT OF ADVANCE GUARDS

5. *CONDITIONS AFFECTING THE CONDUCT OF ADVANCE GUARDS.*—Before we can begin a discussion of how advance guards are conducted, we must have a clear understanding of the two different conditions under which advance guards operate. These conditions are:

    a. When contact with the enemy is not imminent.

    b. When contact with the enemy is imminent.

6. *DEFINITION OF THE TERM "IMMINENT".*—The meanings of "imminent" that here apply are: "threatening to occur immediately; near at hand; or impending", said especially of·misfortune or peril; hence, "full of danger", "threatening", "menacing", "perilous". Hence it follows that when contact is imminent, it is time to do something.

7. *WHEN CONTACT BECOMES IMMINENT.*—*a.* The question immediately arises, "*When* does contact become imminent?" It stands to reason that a commander cannot estimate an exact moment, point or line on the route of advance where the situation changes. Yet he can, however, by making the most of his reconnaissance agencies and other sources of intelligence, determine the approximate *area* within which actual contact will occur. He can, in effect, look at the country to his front as it appears on his map, and say: "From what I know of the enemy's movements and our own rate of advance, the line of probable contact is Line *A* (Figure 10). But contact with the enemy—particularly with the fires of his longest-range artillery (or with his mechanized elements, if such he has)—is possible as close as Line *B*. Therefore, contact becomes imminent when my force reaches Line *B*; and from then on, unless I receive definite information to the contrary, I will be within the area of imminent contact and assume that the enemy is near at hand, an immediate threat and menace."

b. Just how far Line *B* is from Line *A* is a matter for

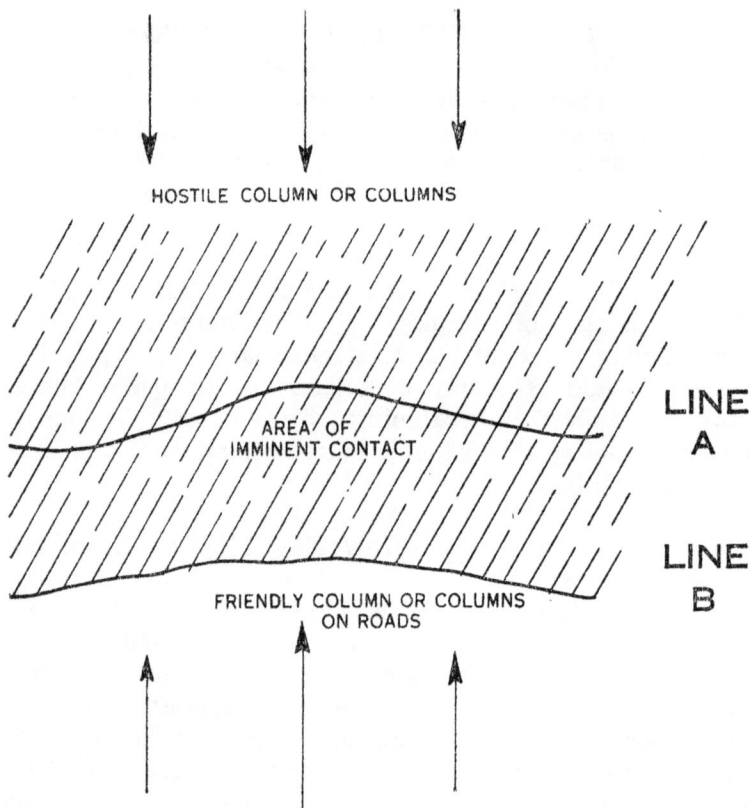

FIGURE 10.—Contact imminent.

the commander to decide in each situation. It is impracticable to lay down a specific distance. In most situations, the range of the enemy's longest-range field artillery (or the probable radius of action of his mechanized force) is the best determining guide.

8. *WHO DECIDES WHEN CONTACT IS IMMI-NENT?*—*a.* *"Who* decides when contact is imminent?" is naturally the next question. The commander of the whole force is in the best position to make this decision. It is he who issues the orders for the movement of the whole force with which the advance guards must co-operate, who decides whether the whole force is to act offensively or defensively. Also, he has before him all available information of the enemy gathered by many intelligence agencies.

b. For these reasons, it is the commander of the whole force who, having determined the lines of probable and imminent contact, and therefore the area of imminent contact, orders his forces to be redisposed, so that they *enter* this area prepared for impending combat. However, in the exceptional case of unexpected contact, which was discussed in paragraph 8, Chapter 2, an advance-guard commander acts on his own initiative.

c. It is important, of course, for a column to have an advance guard as it marches along a road when contact is not imminent, but far more important is the operation of advance guards when contact with the enemy is imminent.

d. In Chapter 4 we shall take up the study of advance guards when contact is not imminent, and in Chapter 5 we shall see how a force goes into battle when contact suddenly occurs. In Chapter 6 we shall consider the most important phase of all in which advance guards are concerned: that when contact with the enemy is imminent.

CONDUCT OF ADVANCE GUARDS WHEN CONTACT IS NOT IMMINENT

1. *GENERAL PRINCIPLES.*—When contact is not imminent, convenience and the comfort of the marching troops govern rather than tactical requirements. Therefore the command moves in column formation on roads. The whole force moves forward in one or several columns, depending upon its size and the number of roads available. Each column has its own advance guard, consisting of a point, advance party, support and reserve, which march on the road. For convenience let us refer to this as "route formation".

2. *FORMATION.*—*a.* An advance guard the size of a battalion or larger is usually divided into two major parts, the support and the reserve, which advance in the order named. The support marches in three groups: the point, the advance party, and the support proper, these also in the order given. (See Figure 11.)

b. Each of these four elements—point, advance party, support proper, and reserve—is detached and sent forward from the element of the column next in rear. Thus, the whole advance guard is detached forward from the whole force. The support is a detached part of the reserve, and the support sends forward the advance party which in turn sends forward a point. Each succeeding element, from front to rear, is larger than the preceding element.

c. The commander of the whole column designates the troops to form his advance guard. The advance-guard commander designates simply a reserve and a support.

The support commander divides the support into the support proper and the advance party, and advance-party commander sends out the point.

d. If the advance guard is less than a battalion, it usually has no reserve. Otherwise the formation is the same.

e. In detailing the advance guard and in dividing the advance guard into its parts, commanders should maintain tactical unity. For example, if the support consists of a company, the support commander should detail a whole platoon, and not four or five squads, as advance party; and the point should be a squad, not five or six men.

3. *STRENGTH OF ADVANCE-GUARD ELEMENTS.*
—The size of the advance-guard elements increases from front to rear. The advance party proper is larger than the point, and the support proper is larger than the advance party, and the reserve in turn still larger than the support. The reason for this is that a small body, which can deploy rapidly, should be first to gain contact. A squad, for example, can deploy almost instantly whereas larger units require more time in proportion to their size. If a large unit came under sudden hostile fire while marching on the road, its elements would remain under this fire without being able to return it during much if not all of the time it required to deploy. Thus we see that each element of the advance guard protects the deployment of the next succeeding element.

4. *DISTANCES.—a. Distance between advance-guard elements.*—(1) This naturally leads to the question: How far apart should the several advance-guard elements be as they proceed in route formation? None of these distances can be set to fit all marches. Like other distances, intervals, frontages, and depths in tactics, these distances vary with the conditions of each situation: the mission of the command, the size of the elements, the terrain, the proximity of the enemy, and whether it is day or night.

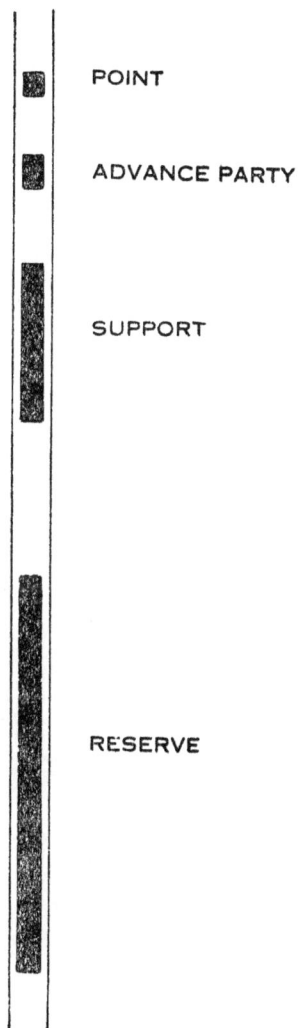

POINT

ADVANCE PARTY

SUPPORT

RESERVE

FIGURE 11.—Diagram of an advance guard in route formation.

In close or rolling country, where successive positions afford protection, the distances should be less than in flat, open country. They should be less at night than during the day.

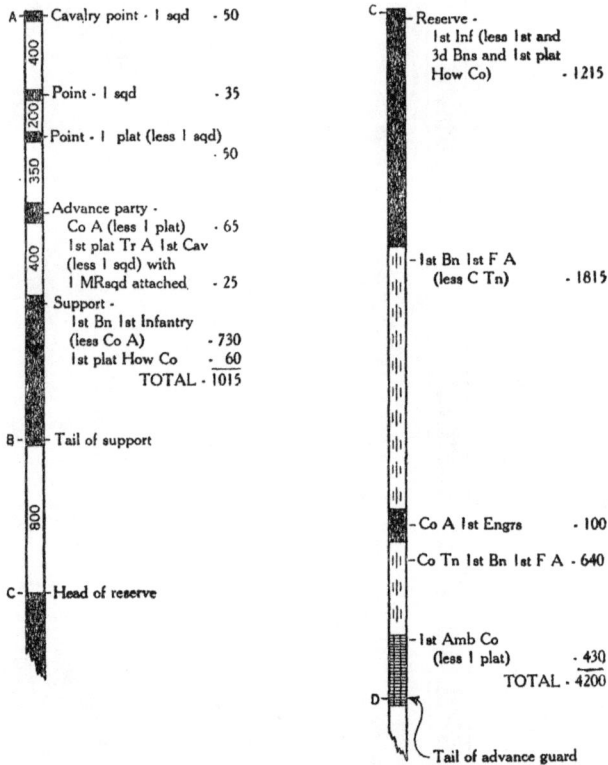

A -▨- Cavalry point · I sqd   - 50

400

▨- Point · I sqd   - 35

200

▨- Point · I plat (less I sqd)
                            - 50

350

Advance party ·
   Co A (less I plat)   · 65
   I st plat Tr A I st Cav
   (less I sqd) with
   I MRsqd attached.   - 25

400

Support ·
   I st Bn I st Infantry
   (less Co A)   - 730
   I st plat How Co   - 60
              TOTAL · 1015

B -▨- Tail of support

800

C -▨- Head of reserve

C -▨- Reserve ·
   I st Inf (less I st and
   3d Bns and I st plat
   How Co)   - 1215

- I st Bn I st F A
   (less C Tn)   - 1815

- Co A I st Engrs   - 100

- Co Tn I st Bn I st F A - 640

- I st Amb Co
   (less I plat)   - 430
              TOTAL · 4200

D -

Tail of advance guard

FIGURE 12.—Advance-guard distances; route formation.

(2) Remembering, then, that advance-guard elements increase in size from front to rear and that it takes a larger unit longer to prepare for action than a smaller one, we arrive at the conclusion that the distances between these elements should also increase progressively from front to rear. At the same time, it is very important that

each succeeding element should be in supporting distance of the next preceding one.

b. Distance between the head of an advance guard and its main body.—The distance from the point to the head of the main body should be great enough so that minor operations of the advance guard do not interfere with the progress of the main body.   (See Figure 12.)   It should also be great enough to prevent the head of the main body from coming under effective, hostile fire.   This fire, of course, may be rifle or machine-gun fire with respective effective ranges of 1000 yards and 4000 yards; or it may be the fire of light artillery, the maximum effective range of which is 8500 yards.

c. (1) In the following table, maximum and minimum road spaces and distances between elements are given for an advance guard consisting of an infantry battalion. This table is a guide only, and should not be taken to fit all situations.   At night, for instance, the distances would be much less than the minimum shown.

AN ADVANCE GUARD CONSISTING OF A BATTALION, IN ROUTE FORMATION

UNIT	Strength	Formation	Road space in yards	Distance to next succeeding element in yards
Point	Squad	Staggered	25 to 50	150 to 250
Advance Party	Platoon (less 1 squad)	Column of twos	35 to 50	300 to 450
Support	Rifle Co. (less 1 platoon)	Column of squads	60 to 75	400 to 600
Reserve	Battalion (less 1 company)	Column of squads	730	800 to 1200

(2) If we add the minimum road spaces and minimum distances we have a total of 2500 yards from the point to the head of the main body.   The total of maximum dis-

tances is 3405 yards. Comparing these totals to the
ranges given in the preceding paragraph, we see that the
head of the main body is protected from rifle fire in
either case, but that neither protects it from long range
machine-gun fire or the fire of light artillery.

d. Hence we can only conclude that in route formation
an advance guard the size of a battalion is not big enough
to protect the head of its main body from hostile artillery
or machine-gun fire. This particular point should be
noted as an emphatic reason for a force not to remain in
route formation once contact has become imminent.

5. *REGULATION OF THE RATE OF MARCH.*—*a.
Phase lines.*—(1) The advance of a whole column in route
formation is ordinarily regulated by the designation in
advance of successive phase lines. Phase lines are simply
lines that cross the route of march at important terrain
points. Thus, formulating his march orders, the commander studies the route, notes where it passes important
tactical features such as ridges, villages, woods, selects
certain of these features, and draws lines through or
along them roughly perpendicular to, and across, the
route of march. These lines, together with the hour at
which the commander directs the column to reach each
line, are given in the march order. Thus the commanders of the elements in the column receive a schedule, so to
speak, of march, and can regulate the rate of march on
these lines. Phase lines also give the column commander
definite points of reference for possible tactical use, if the
enemy should be encountered during the march. The use
and announcement of phase lines also acquaints the whole
command in advance with the important tactical localities
along the line of march.

(2) Phase lines are of still greater value, as we shall
see later in this chapter and especially in Chapters 5 and
6, to forces of more than one column and forces in any
stage of deployment.

(3) It is important to emphasize here that phase lines
are not straight lines but take the general form of the

terrain features with which they coincide. This point will also be stressed again later.

b. Regulating unit.—It is usual to designate the support of the advance guard as the regulating unit. Hence the reserve is ordered to "follow" the support, and the main body to "follow" the advance guard. The support commander, in his orders to the advance party and support proper, prescribes the distance between them and directs the advance party to precede the support by this distance.

6. *POSITION OF COMMANDERS.*—The command post of the column commander is usually at the head of the main body and that of the advance-guard commander, at the head of the reserve. These locations for command posts should not be taken to mean that commanders must continuously remain at these points. They are simply temporary headquarters points through which the commanders can always be located. A commander is free to go to any place in the column where he feels that his presence is needed. Sometimes, indeed, he should by all means be elsewhere than at his command post. If the advance guard encounters the enemy, for example, the commander of the whole column should immediately join the advance-guard commander at the head of the reserve. There, he is in the best place to learn at first hand, and as early as possible, what is going on. If the enemy proves to be so strong that his whole force may be needed in action, the commander learns this fact early by being well forward. Accordingly, he can form his plan and get his rear elements moving forward much sooner than he could if he were farther back. And he can also direct the advance-guard commander to take the action best suited to the plan for the employment of the whole force. (This does not mean, of course, that the high commander takes over in any way the direct command of the advance guard. He is simply on the spot and therefore can point out directly to the advance-guard commander what he wants. The advance-guard commander directs the disposition of

his force in accordance with the orders of his superior.)
On the other hand, if the enemy proves to be only a small
force that the advance guard or some of its elements can
quickly overcome, the high commander learns this fact as
quickly as the advance-guard commander.

7. *CONNECTING ELEMENTS.*—Contact between elements, as the column proceeds along the route of march,
is maintained by dismounted men placed in the space between the elements. These men are called "connecting
files". Connection is usually maintained from rear to
front; and hence these connecting files are sent forward
from the element in rear to maintain connection with the
next preceding unit. They are so spaced that each connecting file can remain in constant view of the next succeeding and next preceding file or unit. (See Figure 13.)
For this reason, in rough, broken, or heavily wooded terrain with winding routes, it is sometimes necessary to
have two men act as one "connecting file". Mounted men
can be used for this task. Connecting files pass forward
all orders and messages received from the rear. They
pass back no signals except "enemy in sight", "enemy in
sight in large numbers", and special signals previously
agreed upon. Their only other duty is to conform to the
movements of the unit on which their own movement is
regulated. For instance, if a connecting file is told to
follow the support proper at a specific distance, he maintains this distance throughout the march merely by halting
when the support proper halts and moving forward again
when it moves forward. This eliminates any necessity
for signals except the two emergency signals given above.

8. *THE FUNCTIONS OF ADVANCE-GUARD ELE-
MENTS.*—*a. The point.*—The point is actually a patrol at
the forefront of the entire column. It confines its activities to the axis of march. When possible, it drives back
small enemy parties in its front; but when it cannot do
so, it covers the deployment of the advance party. When
the advance party is a platoon or larger unit, it may

send out two points, one preceding the other. The double point is particularly valuable if the advance guard has no mounted or motor elements.

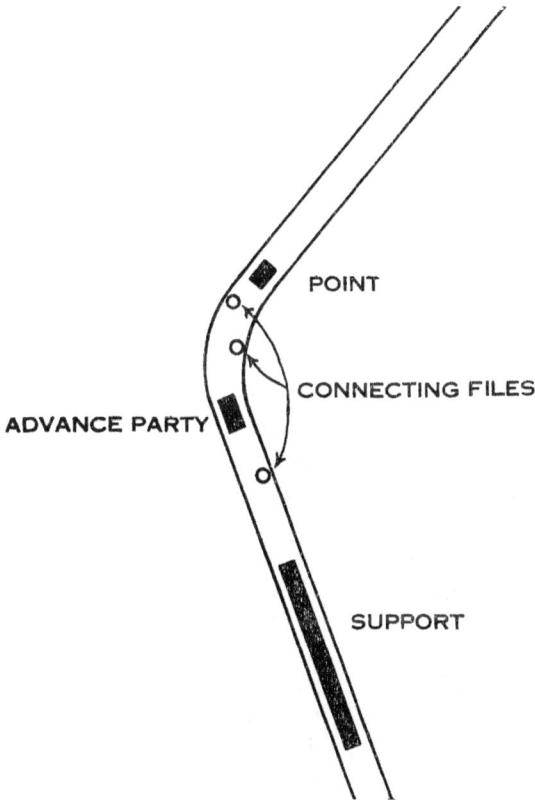

FIGURE 13.—Connecting files.

b. The advance party.—The advance party is made strong enough to provide a point, support the action of the point, and cover the deployment of the support proper.

c. The support.—The support covers the reserve by offering resistance when the enemy is encountered, thus enabling the advance-guard commander to employ the reserve in accordance with a definite plan.

d. The reserve.—The reserve is the element on which the advance-guard commander principally depends in carrying out his mission when the enemy is met in force. It is his maneuver unit.

9. *DEVELOPMENT OF ADVANCE-GUARD ACTION.*

—*a.* When a column in route formation encounters the enemy in sufficient force to indicate the employment of the entire advance guard, its commander must keep in mind that his mission is to act in furtherance of the plans of the commander of the whole force. If the force commander is not with the advance-guard commander, the latter does not delay necessary action by waiting for the force commander to come up. Instead he acts promptly as the situation requires and in accordance with the known intentions of the force commander. When these intentions are not known, the advance-guard commander must base his immediate decisions and actions on the mission of the whole force and the situation that confronts him.

b. The mission of the advance guard may be to attack, defend, or delay. One of these three missions may be given to the advance-guard commander as early as the beginning of the march. Better yet, the force commander can direct in advance the type of action he desires if the advance guard meets the enemy within specific areas along the route of march. This is by all odds preferable. But if, for any reason, the mission has not been announced in advance, the force commander gives the advance guard a specific mission as soon as possible if it unexpectedly encounters the enemy. And if the force commander is not present with the advance guard when such contact with the enemy occurs, the advance-guard commander, as stated above, must act as the situation and the mission of the whole force dictate, until he gets specific orders from the commander of the whole force.

c. But in the rapid changes that take place in a meeting

engagement, the force commander may find that he cannot carry out his original intention. For example, let us assume that the commander has previously informed the advance-guard commander that, if the enemy is encountered within a certain area, the column is to deploy to the right of the road with a view to attacking the enemy in that general direction, and has directed the advance-guard commander in this event to seize a given terrain line and cover the deployment of the main body. Let us assume further that on encountering the enemy at this point, the force commander finds that the enemy is better disposed for attack than his own force, and that instead of attacking he must take the defensive, at least temporarily. He must notify the advance guard of this change of intention at once, and, if necessary, give the advance guard a new mission, so that the advance-guard commander can act in accordance with the change of plan and the actual situation.

d. On encountering the enemy, the point, advance party, and support proper attack immediately straight to the front in order to drive back the enemy. If this is found to be impossible, these units immobilize the enemy by fire and thus form a pivot about which the reserve can be maneuvered. Rapidity of decision and action are essential. There should be no halting or hesitating on the part of leaders or units as each element of the advance guard acts to reinforce the element ahead.

e. The reserve acts as a unit. It does not fritter away its strength by piecemeal reinforcement of the support. It is employed by the advance-guard commander in accordance with the principles of attack, defense, or delaying action, whichever is directed in his mission. When the enemy is first encountered by the point, the reserve, like the other elements of the advance guard, continues to move forward at full marching speed, and does so until it receives other orders or other circumstances prevent. If the action to the front is apparently more than a mere brush with isolated hostile elements, and the terrain permits, the reserve should then be partly deployed as it continues to advance. Its units move forward along such covered routes as are available near the road. When the reserve reaches a point where the fire or proximity of the enemy makes a still less vulnerable formation desirable, the units of the reserve are further or completely deployed. The advance-guard commander forms his plan of action while the reserve is moving up.

(*NOTE.*—The term "partial deployment" is used throughout this text to mean any degree of deployment short of complete extension for combat. When a column on the march "partly deploys" it opens out on a wider front with its main subdivisions still in column formations but separated laterally by some distance. Its leading elements may or may not be completely deployed. Thus a battalion would, in partial deployment, split up into company or platoon columns with some interval between them as in Figure 14.)

f.(1) Any encounter with the enemy in force by a column proceeding in route formation is liable to be disadvantageous to that column. As this text has stated before, an alert commander is not likely to be caught thus unawares. But as we saw in paragraph 8, Chapter 2, such things do occur in warfare. And when they do occur, the force commander, the advance-guard commander, and every other leader must realize the seriousness of the situation.

(2) When the leading elements of the advance guard meet without previous warning a sizeable enemy force, this must be taken as an indication that the enemy has been more on the alert than ourselves, that his information about us is more complete than ours regarding him. He has, in other words, the jump on us. This may not be the case in actual fact. But it is far more likely than not.

(3) Accordingly, the advance guard and the whole column, once the enemy has been unexpectedly encountered in some strength, must be ready for any eventuality. The force to the front, for example, may not be the only hostile force near by. Another may strike the column at

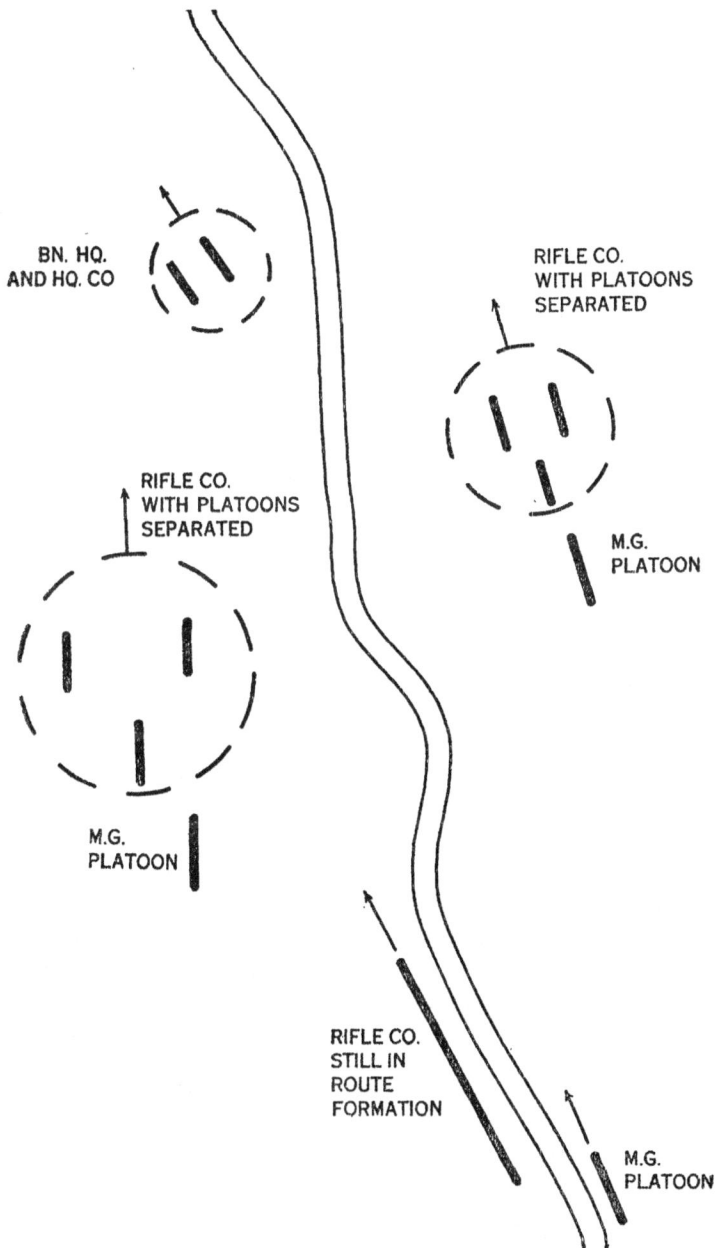

BN. HQ.
AND HQ. CO

RIFLE CO.
WITH PLATOONS
SEPARATED

RIFLE CO.
WITH PLATOONS
SEPARATED

M.G.
PLATOON

M.G.
PLATOON

RIFLE CO.
STILL IN
ROUTE
FORMATION

M.G.
PLATOON

FIGURE 14.—An infantry battalion in a partly deployed
formation.

any moment from a flank; for if reconnaissance to the front has allowed the enemy to fire on the leading units, there is little guarantee that reconnaissance to the flanks is any better.

(4) It was in exactly such situations that the most of the French columns found themselves on August 22, 1918, when the French armies advancing in parallel columns on different roads met the German armies in the Battle of the Frontiers. Here, a combination of reckless haste to advance and strike the enemy, and poor reconnaissance plus a disregard of important enemy information actually received, caused a severe defeat to the French. Some French columns were struck from the flank before their advance guards found an appreciable hostile force. Some were assaulted from both flanks and from the front simultaneously. The heads of the columns were, in effect, "bitten off" by the German forces, which had advanced in somewhat more extended order and with far more regard to careful reconnaissance and well-considered evaluation of the information obtained.

(5) At the same time, the conduct of more than one of these French columns shows conclusively that all is far from lost when a force in route formation is struck by a strong hostile force. Quick and determined action by a cool leader may not turn defeat into victory but is almost certain to make a bad situation better.

(6) The advance guard of a column can, of course, only fight that part of the enemy force it encounters. And in this chapter we must confine ourselves to that viewpoint realizing that any greater difficulties would be a matter for decision and action by the commander of the whole column. In the next chapter, however, we shall study the measures that he takes after his advance guard meets the enemy for the protection of his column against hostile forces that may advance from other directions than his immediate front. But at this point in our study of offensive combat it must be clearly realized that almost any situation may develop when a column marching in route formation blunders into a large enemy force unexpectedly.

10. *ADVANCE-GUARD ACTION WHEN THE ENE-MY WITHDRAWS.*—*a.* If a hostile force encountered by an advance guard withdraws, the commander of the column disposes his whole force in a partly deployed formation for imminent contact as described in Chapter 6, and then resumes the advance. In the usual situation of this kind, since the enemy has already engaged him once, the commander, lacking other information, must assume that contact thenceforward is imminent. Even if the hostile force that has delayed him is only an isolated delaying detachment, it will probably act again in a brief time. He may, however, receive definite information—from observation aviation, for example—that the hostile force has withdrawn by motor and that apparently there is no other force within imminent contact. In such a case, the commander would order his advance guard back on the road, to resume the march in route formation. Such a situation is very unlikely.

b. Simply because the hostile units encountered by the advance guard are very small and quickly driven back is no basis for a commander to assume that contact is not imminent and that he can resume the march in route formation with impunity.

c. The head-on collision of an advance guard with the leading element of the enemy's main force is, we may say again by way of summary, rare and exceptional in modern warfare. But such meeting engagements, and also encounters with strong hostile delaying or harassing forces, may occur when a commander fails to make thorough use of his reconnaissance agencies. The information-seeking tools of a commander have always been important but are now more important than ever, in view of the more rapid movement in the warfare of the present. In the next paragraph we shall study the methods of patrolling of a column in route formation, since this subject is highly important throughout every phase of offensive combat including the stages preliminary to battle.

11. *PATROLLING.*—*a. General.*—(1) In this chapter we are chiefly interested in the patrolling actually done by

an advance guard. We should understand first, however, that a marching force does not advance toward the enemy, relying entirely on the protection and information that its advance guard affords. In modern warfare an enemy with means of rapid transport may strike at a column from any direction—front, flanks, or rear—especially at a slow-moving column of foot troops. The enemy may not be able to move its whole force easily with great speed any more readily than we can, but it may attempt to hit a large force of ours with small, fast, motorized or mechanized units of its own for the purpose of harassing us, cutting off detached parts of our force, interfering with our routes to the rear, and similar endeavors.

(2) In the past, of course, we have used flank guards and rear guards, in conjunction with advance guards, to furnish all-around protection to a marching force. These same protective units now have even more importance, perhaps, and will continue like the advance guards to send out patrols in order to guard the main force from unexpected attack.

(3) The main force itself also sends out reconnaissance detachments. In forces of all sizes these patrols are ordinarily formed of mounted or motor units and work far out beyond the patrols of the security elements of a column (advance, flank, and rear guards). But the farther out from the column these distant reconnaissance groups go, the more there is to cover. (Even aviation, as we saw in paragraph 7*b*, Chapter 2, is limited in its ability to cover large areas.)

(4) For this reason an advancing force cannot afford to assume that nothing can unexpectedly penetrate its outer protective screen. (It was just such dependence on the advance cavalry screen that resulted in the German armies' sudden contact with the French columns in the Battle of the Frontiers referred to above. The cavalry units, for the most part, failed miserably either to afford protection or gain information of the enemy.) Therefore, we see the necessity for a column in route formation to move, in effect, as if it had to depend entirely

on its own resources for warning of the enemy and protection against it.

(5) An advance guard or other security detachment does not attempt to make distant reconnaissance unless specifically directed to do so. Ordinarily, the force commander in his advance-guard order given before the march, places a limitation on the distance that it patrols to the front and flanks of the column. In subparagraph *b* immediately following, we shall see how advance guards preceding a column in route formation send out patrols.

b. Advance-guard patrols.—(1) In considering advance-guard patrolling we must especially keep in mind the first general mission of the advance guard—to assure the uninterrupted forward movement of the main body. In fact, all patrolling by a column on the march should be carried out with this important end in view. There is no set method or system that will accomplish this for all marches, since the terrain and tactical situation are never twice exactly the same. Consequently, advance-guard patrolling must be planned for each march.

(2) Before a march begins, the advance-guard commander makes a map study of the terrain to the flanks of his route of advance and determines the localities from which hostile fire could be placed on the main body of his column. These, he assumes, are danger spots, or critical points and must be thoroughly investigated by suitable patrols before it is safe to march his force by them. Having determined these localities, he then decides upon the size of patrol to be sent to each, its type (mounted, motorized, or dismounted), how long it should remain at the dangerous area, and what element of the advance guard is to send out each patrol. In his advance-guard order he directs the patrolling thus decided upon.

(3) The commander of a small column should also supplement his preliminary map reconnaissance, when time permits, by personal reconnaissance during the march. As the march progresses, the commanders of the advance guard and its elements should constantly observe and study the terrain to the front and flanks, in conjunction with the map, in order to discover any additional danger

points or areas that may have been overlooked during the original map study. If any are found, additional orders directing patrolling thereto are issued by the advance-guard commander.

(4) Patrols may be taken from any element of an advance guard. If it is foreseen that a large patrol, or several small ones, should be sent from any particular element during the march, the initial strength of that element should be sufficient to provide them. If unexpected patrolling develops along the route, an advance-guard commander must not hesitate to order the patrolling even to the extent of reducing the strength of any advance-guard element to a point necessitating reinforcement. It may be necessary, for example, to send out the entire advance party and replace it by moving up a new one at an increased rate of march, or at the next hourly or other halt. Patrols sent out to accomplish the advance-guard mission of protecting the head of the main body should remain in the areas they are sent to investigate until the possibility of danger to the *head* of the main body no longer exists. For this reason, they may not be able to rejoin the advance guard until the end of the march.

(5) In addition to patrols sent for the purpose just discussed, it may be found necessary for advance elements to send out small patrols for the local protection of the advance guard. The advance-guard commander notifies the commanders of all the advance-guard elements regarding the patrols he has ordered sent, in order to prevent duplication of effort by commanders' sending local security patrols to the same areas. Local security patrols should rejoin their units in a short time.

(6) In general, patrols should be started out at the earliest practicable moment. Therefore, they are usually sent from the support. The support is hence made large enough originally to take care of this patrolling. If a support commander desires to send many patrols from his advance party, he, in turn, should make the advance party

large enough in the first place both to take care of this patrolling and perform its other duties. In planning his patrolling, an advance-guard commander must not, of course, lose sight of the combat mission his support must perform in case the enemy is met. His support, for example, should be a tactical unit as it enters combat. Tactical unity is best preserved for a support by detailing extra troops from the reserve to march with it in order to take care of the patrolling. For example, let us assume that a battalion is in advance guard with Company A forming the support and that the battalion commander decides that he must send ten patrols with a total strength of six squads. In that event, he should order one rifle platoon of Company B or Company C of the reserve to march with the support proper, or preferably with the advance party, to do this patrolling. This leaves Company A (less its advance-party platoon) intact for combat.

c. The use of motors in advance-guard patrolling.— Motor vehicles, even in small numbers, enable an advance guard to perform its patrolling to much better advantage than when all patrolling must be done mounted or on foot. Danger areas can be reached sooner by patrols designated to cover them. A single vehicle, for instance, can often take several small patrols to different points in a short time, and well in advance of the column. It can also pick up the patrols and bring them rapidly back to their places in column. Motor patrols are also valuable for unexpected patrolling additional to that planned before the march is begun.

12. *ARTILLERY IN ADVANCE GUARDS WHEN CONTACT IS NOT IMMINENT.*—The assignment of artillery to advance guards when contact is not imminent should be by complete tactical units; i.e. batteries and battalions. The amount of artillery assigned varies directly with the strength of an advance guard, the advance guard mission, and the length of time it alone must oppose the enemy when he is encountered. For example, from one battery to one battalion of light artillery may

be assigned to accompany the advance guard of a force
the size of a reinforced brigade in route formation while
contact is not imminent.   When the situation especially
favors its use, and distant targets such as troops passing
through defiles or at detraining or detrucking points, are
likely, medium artillery may be attached to an advance
guard.   The advance-guard artillery usually marches at
the rear of the reserve.

13. *TANKS IN ADVANCE GUARDS WHEN CON-
TACT IS NOT IMMINENT.*—*a.* Light tanks may properly be included in an advance guard when it can be foreseen that the advance guard will be required to engage in
serious fighting to accomplish its mission. When so used,
they move with the reserve. Tanks are particularly useful
when the advance guard is opposed by delaying detachments.

b. Since tanks can travel across country they can often
march off the roads parallel to the foot troop units using
the roads, so as to be prepared for quick commitment to
action.

c. Tanks moving in a column with foot troops may
move by short bounds using the interval between the tail
of the reserve and the head of the main body, particularly
when the terrain on either side of the road is unsuitable
for cross-country movement.

d. When the commander of a force desires to retain
control of tanks but still anticipates opportunities for their
use in the advance guard, he may retain them near the
head of the main body. In the event of unexpected contact with a hostile force by the advance guard, their
speed enables them to move up to the zone of the advance guard's activities within a very few minutes. Normally they are committed to action with the reserve of
the advance guard.

e. Tanks with the advance guard should be used as reconnaissance agencies only as a last resort when other
means are impractical or entirely absent.

14. *ANTITANK DEFENSE IN ADVANCE GUARDS WHEN CONTACT IS NOT IMMINENT.*—It is usual to attach a proportionate share of artillery, infantry howitzers and caliber 0.50 machine guns to the advance guard. For a battalion of infantry acting as advance guard when contact is not imminent the usual number of these weapons is: one infantry howitzer (37-mm gun), one battery of artillery (four 75-mm guns), and one platoon of caliber 0.50 machine guns (four guns). The antitank artillery usually marches at the rear of the reserve and the infantry howitzer with the reserve; but the caliber 0.50 machine guns should be distributed throughout the advance guard. This may be done by attaching a section (two guns) to the support and holding the other section with the reserve. These guns with the reserve should march one near the head and the other at or near the tail of the reserve. Other antitank measures such as slashings and demolitions or other "road blocks" carried out by patrols assisted by engineer troops at important road centers and bridges, culverts or other defiles, may be found effective as antitank means.

15. *ANTIAIRCRAFT DEFENSE IN ADVANCE GUARDS WHEN CONTACT IS NOT IMMINENT.*— There are seldom enough special antiaircraft units (coast artillery) to allow their attachment to each advance guard. Such antiaircraft artillery and combat aviation of the main force may be sent forward to protect several adjacent advance guards where their routes or zones are in close proximity to each other. But the advance-guard commander must in general depend on his own weapons and formations for his antiaircraft protection.

(*NOTE.*—Thus far this text has largely touched upon matters preliminary to offensive combat. But as we can see clearly by this time, the manner in which a force goes into combat may well be in itself half the battle. In the next chapter we shall study the method of deployment for battle from route column. We should still bear in mind that this is the exceptional case, and not the preferable

method, but nevertheless a method that is highly important when a column in route formation does meet the
enemy unexpectedly.)

CHAPTER 5

DEPLOYMENT FOR COMBAT FROM ROUTE FORMATION

1. *GENERAL.*—*a.* In the preceding chapter we saw how the advance guard of a force in route formation engages the enemy when a column runs unexpectedly into a hostile force of any size. We saw the deployment of the advance-guard elements in succession, each in support of the preceding element. In this chapter we take up the conduct in such an eventuality of the main body of a column with an aggressive mission, and see in detail how it deploys and joins battle after the advance guard has come to grips with a strong hostile force. Here again we must remember that the method of entering battle given in this chapter is the exception and not the rule—a method employed, in fact, only when it has to be employed, only when it is too late to move a force into combat in the most desirable method (already partly deployed), which we shall study in Chapter 6.

b. In beginning our study of the deployment of a force from route formation, let us first examine an illustrative situation with a view to determining when a main body should deploy under the conditions assumed.

2. *WHEN TO DEPLOY: ILLUSTRATIVE SITUATION.*—*a.* The commander of a force of all arms marching in one column (See Figure 15.) on an aggressive mission was at the head of his main body. Unexpectedly, marching hostile infantry was encountered by his leading elements. He went forward promptly to join the advance-guard commander. By 20 minutes later, all indications

led him to believe that he was opposed by a force of considerable strength. The main body is still moving forward on the road. Hostile artillery, if there is any, has

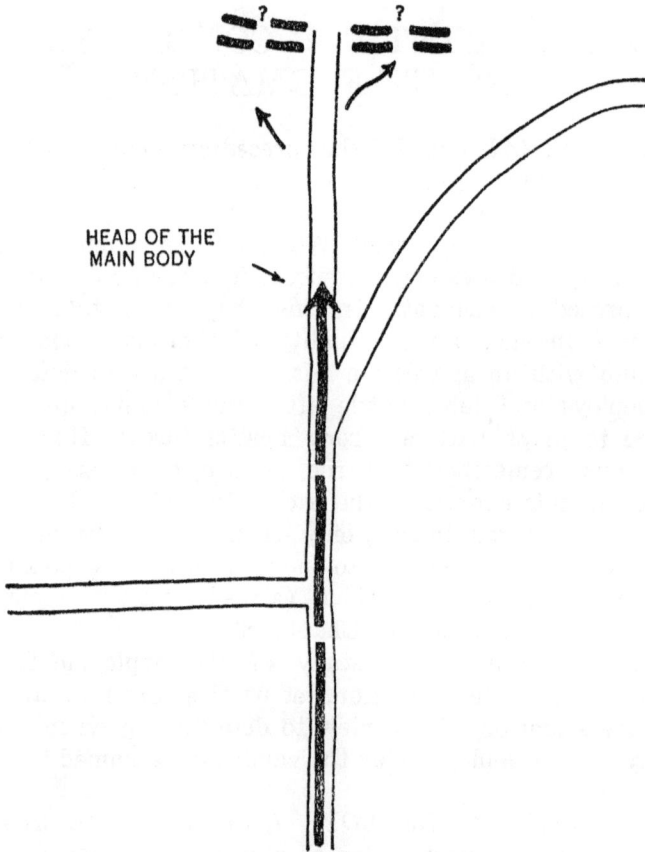

HEAD OF THE
MAIN BODY

FIGURE 15.

not yet gone into action. The commander must now decide when and where to deploy his main body.

b. In this situation the commander should begin a complete deployment immediately. If the enemy has artillery

—and he is not likely to be without it—it will probably go into action soon. When it does, if the main body is still on the road it is liable to suffer heavy casualties from hostile artillery fire. If the enemy force with which the advance guard is now engaged is only one of the leading elements of a large hostile force already partly deployed and advancing towards this column, any delay whatever in deploying may prove fatal. In a few moments, moreover, the main body will be within range of hostile machine guns to the front, which is all the more reason for beginning deployment.

c. Even when he is with the advance guard, a commander in a situation such as this will seldom know at once whether the advance guard is seriously engaged or not. But in case of doubt he should by all means err on the side of safety and begin deployment. There is seldom a good reason for halting the main body in route column for any appreciable length of time. It is, in fact, dangerous to do so.

d. Now let us consider the question of where to deploy the command in the situation given. The commander, let us assume, does not know yet whether he will attack or defend. He does not know anything of the composition, strength, or dispositions of the hostile main forces. Therefore he decides to deploy his force, generally in depth and to both sides of the road. With his infantry on both sides of the road he can shift its mass to either the right or left if developments in the situation indicate that it is desirable to attack the enemy around a flank. He can also move quickly in a direction generally along the road, to penetrate the hostile force, if that appears better. Or, he can rapidly occupy a defensive position, if the hostile force to his front proves to be so strong or have such an advantage that such a step is necessary, or if his force should be attacked by another part of the hostile force from a different direction. Moreover, he can resume march formation on the road with a minimum of delay, if the hostile force by any chance happens to be a small isolated detachment soon defeated by the advance guard.

3. *THE NECESSITY FOR EARLY DEPLOYMENT FROM ROUTE FORMATION.*—*a.* The principal lessons of the illustrative situation immediately preceding are these:

(1) When the advance guard of a force in route formation advancing on an aggressive mission encounters a hostile force of unknown size, the commander of the whole force must attempt immediately to determine whether the unexpected contact is with a strong hostile force or with an isolated delaying or harassing detachment of the enemy.

(2) As soon as there is any indication whatever that he has met the enemy in strength, rather than a small detachment, he must, as a general rule, begin the deployment of his whole force.

b. We may say in general that the distance from the enemy a force is deployed depends principally on the size of the force and on the terrain. When it is known, the effective range of the hostile artillery is a decisive factor. In general, also, the larger the force, the farther back it should deploy initially; and the flatter the terrain, the farther back it should deploy initially. Moreover, every endeavor should be made to deploy in an area not under ground observation from points held or likely to be held by the enemy while the column is breaking up into a deployed formation.

c. But when the advance guard of a force in route formation unexpectedly encounters a strong hostile force the situation suddenly changes without warning from one of contact not imminent to one of actual contact. The leading elements, at least, of the main body are already within range of the hostile artillery. Indeed, they may already be within observing distance of the enemy. And the enemy, by aggressive action in force, may drive back the advance guard so rapidly as to interfere with the deployment of the main body, if that deployment is delayed. It is by no means impossible for the enemy to succeed in accomplishing this even if the deployment of the main body is begun at once. For in such a situation, the very unexpectedness of contact, as we have seen heretofore, gives the enemy force a definite initial advantage.

d. Nevertheless, terrain may prevent immediate deployment. An unfordable stream may lie across the direction of advance, for example, preventing an extension of the main body on a broader front until it has crossed the stream. Again, the hostile force with which the advance guard becomes engaged may prove to be only one part of the enemy's force, and the commander may find himself attacked from more than one direction. In this event, he has indeed a different problem to solve, and may well have to abandon all thought of carrying out his aggressive mission in order to save his force by defensive action. These two possibilities aside, however, an aggressive mission and consideration for the safety of units in column dictate immediate steps toward the deployment of the main body under the conditions we have been discussing. This deployment as we shall see in the ensuing paragraphs is not a complete and immediate deployment for combat of all units, but a gradual process usually comprised of several stages.

4. *HOW DEPLOYMENT IS ACCOMPLISHED.—a.* When a commander deploys his force from route formation to attack an unexpectedly encountered hostile force, the method of deployment depends largely upon his plan for attacking the enemy. The need for haste may require the piecemeal entry of his force into battle, with his infantry units entering combat one after another as they come up. This is called an "uncoordinated" attack. But often a commander will desire to make his attack with the bulk of his force moving to engage the enemy at one time. This is called a "coordinated" attack and requires a slower process of deployment, since the leading elements of a main body must wait for those in rear to come up, if all or the greater part of the units of a force are to attack simultaneously. We shall not at this point discuss the merits of these two methods but simply how de-

ployment for them is accomplished.

b. The deployment for an uncoordinated attack is much simpler than that for a coordinated attack. A brief illustrative situation will serve to explain the former; and then we shall consider in detail the method of deploying a force for the simultaneous entry of its several units into combat.

5. DEPLOYMENT FOR AN UNCOORDINATED ATTACK: ILLUSTRATIVE SITUATION.—*a.* A force of all arms on an aggressive mission marched north early this morning in two columns, as indicated in Figure 16. At daylight this command unexpectedly meets a superior force of all arms marching south in one column. The ridge *CD* in Figure 16 commands the bridge *E* and the terrain beyond the bridge, which is all open country. The advance guard is already on ridge *CD*.

b. The commander of Force *A* must decide when and how he will deploy his main body. This decision should be to deploy at once. First, he should get all of his artillery into action as quickly as possible, and then feed infantry battalions into the line as rapidly as he can get them up—in other words, he must make an uncoordinated attack.

c. If he can seize ridge *CD* he will probably force that part of the enemy force to the south of the bridge back across the river. If this is accomplished, the enemy, in order to attack, must deploy his main elements north of the unfordable stream. Hence, if the commander of Force *A* acts promptly by throwing in his units one after another he has an excellent chance of seizing ridge *CD*.

d. As the successive elements of the main body advance to enter combat they move in a partly deployed formation like that described in paragraph 9*e*, Chapter 4, for the advance-guard reserve. And when they come so close to the enemy that hostile fire requires still less vulnerable formations, these elements complete their deployment for entry into battle.

e. It is important to consider here, also, the artillery elements. Usually, when a force of all arms marches on a road out of contact with the enemy, the artillery con-

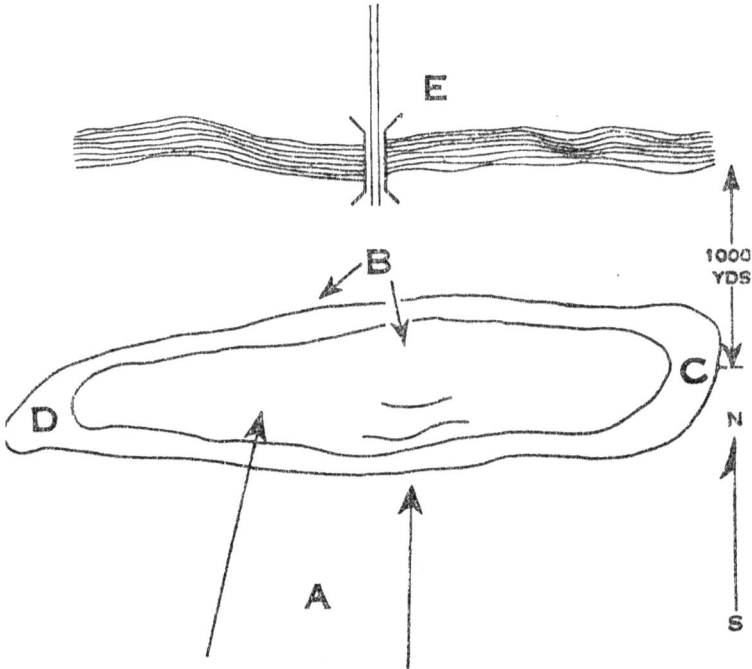

FIGURE 16.

forms to the infantry rate of march. But whenever it is desired to move the artillery forward quickly, as in an un-expected encounter with a strong hostile force, it imme-diately receives priority on roads. It then moves at the rate prescribed by the artillery commander and the in-fantry marches so as to clear the road for it.

6. *DEPLOYMENT FOR A COORDINATED ATTACK FROM ROUTE FORMATION: ASSEMBLY AREAS.—a.* When a commander decides to deploy his force for a co-ordinated attack from a marching column, he first designates areas to which the main infantry units first move and halt before finally entering combat. These are called "assembly areas". The assembly area of each unit is located roughly in rear of where it will probably enter the battle. Figure 17 shows diagrammatically how units move from column to their assembly areas.

b. The use of assembly areas permits leading units in a column to wait for those in rear to come up so that part or all may enter battle at the same time rather than in succession. It also enables a commander to keep his units assembled in more or less compact groups in definite localities so that he retains a maximum of control until he and all the units of his command are ready to attack the enemy in a coordinated manner.

7. *SELECTION OF ASSEMBLY AREAS.—a.* When time permits, it is preferable for the commander of the whole force to select the assembly areas for his subordinate units. The locations of assembly areas must be carefully considered because they have a considerable influence on the commander's future course of action. Only when it is impracticable for him to do so himself does he direct his staff to select them. When the direction of the future employment of a unit can be foreseen, its assembly areas should be located so that its movement to its line of departure is facilitated.

b. Again, the areas into which units are moved in the initial stages of deployment should, if practicable, afford concealment from air observation.

c. If the command has motorized infantry, the road net into and out of these areas must be given consideration.

8. *ROUTES TO ASSEMBLY AREAS.—a.* The route and formation of units in moving to their assembly area are governed by the possibility of enemy observation.

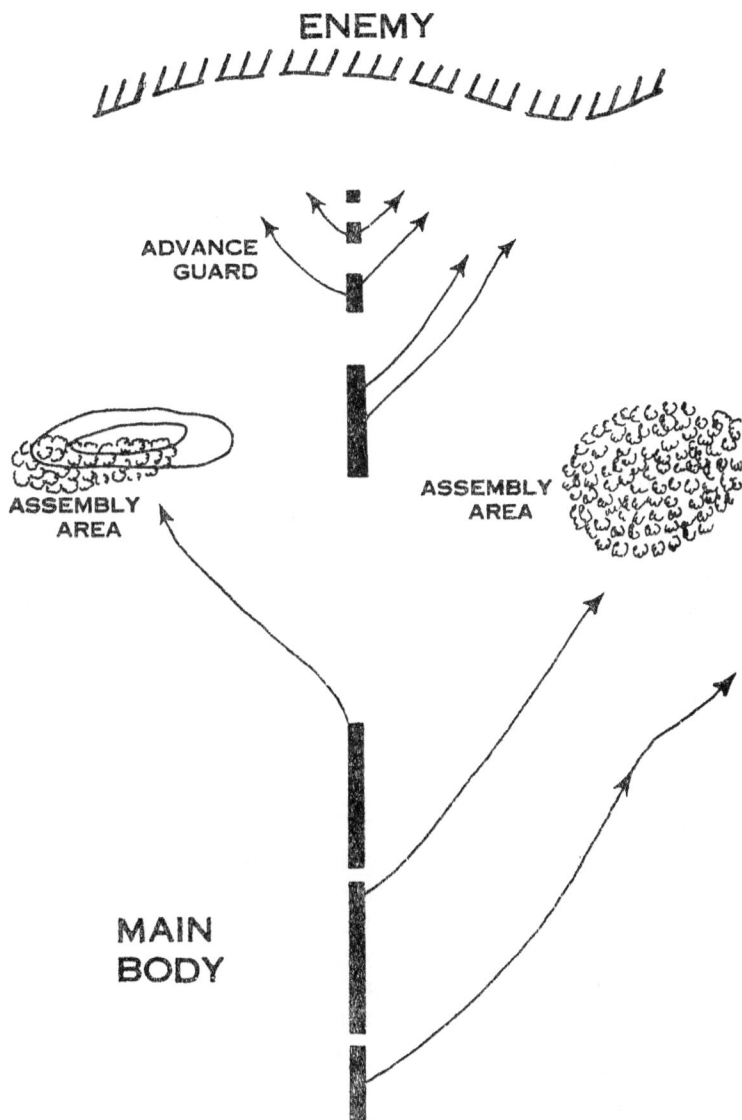

ENEMY

ADVANCE
GUARD

ASSEMBLY
AREA

ASSEMBLY
AREA

MAIN
BODY

FIGURE 17.

When time permits, routes that cannot be observed from ground in the hands of the enemy, and that offer the maximum concealment from the air should be taken. The nature of the terrain may be such, of course, that good routes are not to be found. Again, routes that interfere with the movement of other units must not be followed.

b. The formation of the unit is determined by the terrain and the likelihood of coming under hostile observation and fire. In general, the greater the chance of being observed—and hence of being fired on—, the more extended the formation should be.

9. *FORMATION IN ASSEMBLY AREAS.*—There is no prescribed formation for infantry units in assembly areas. In an assembly area in close terrain, defiladed from hostile fire and well concealed from the air, the units ordinarily remain close together while waiting to attack. In more open terrain, where they may be subjected to hostile fire or are more liable to be attacked from the air, a unit in its assembly area takes up a more extended formation.

10. *ISSUING OF EXTRA AMMUNITION.*—*a.* In assembly areas extra ammunition is usually issued and infantrymen take off their pack rolls. The pack roll is not essential to the soldier while he is fighting. Moreover, its weight reduces his mobility. The pack rolls are transported by the field or service trains until they are needed again.

b. The amount of ammunition carried by troops when they wear full packs is not enough for a combat period of any considerable duration. And once infantry units have deployed, issuing ammunition is much more difficult than when the unit is still assembled. Furthermore, it is also desirable to issue out the ammunition loads of rifle-unit combat vehicles before these units start the attack so that the vehicles can go back to bring up more ammunition as early as possible.

c. Ammunition is usually issued simultaneously within

each infantry battalion after it reaches its assembly area. Sometimes, however, it may be necessary or desirable to issue ammunition to one or more battalions somewhat earlier while they are still in route column on the road. The two examples that follow illustrate this point.

(1) A battalion is in a defile. Ammunition must be issued without delay. The surrounding terrain is such that the vehicles carrying ammunition cannot get off the road. It is apparent that if ammunition is to be issued at once pack rolls must be dropped, and extra ammunition issued, on the road.

(2) The commander of a marching column of all arms decides to effect a partial deployment by forming three columns. To do this, he plans to have the leading infantry elements of the main body halt until the other two columns have been formed and have taken their respective places in the new formation. In this situation the leading infantry elements might either go into assembly areas first, or drop pack rolls and issue extra ammunition on the road as in the situation in (1) above.

11. *OTHER STEPS TAKEN IN ASSEMBLY AREAS.* —Let us assume, now, that an infantry unit of a command has reached its assembly area and that the commander of the whole force has decided to make a coordinated attack. Before the whole command is ready to fight several important things must be done:

a. The commanders of all units, from the commander of the whole force down, must in turn reconnoiter the ground forming the approaches toward the enemy and the probable battlefield itself, formulate tactical and administrative plans for the attack, and then issue orders to their subordinate units.

b. These orders include directions for the advance of the next lower units to the lines or areas of departure from which they are to begin the attack, and for the attack itself. Hence in following out the orders, the subordinate units take up suitable formations for their respective parts in the attack.

c. Thus the infantry units of the whole command must still further deploy and extend laterally, and units in rear are brought still farther forward to their attack positions.

d. A system of communication between higher and lower units, and between adjacent units must also be established.

12. *THE PROTECTION OF A FORCE DURING PREPARATION FOR BATTLE.—a.* During this process of getting ready for battle, a force as a whole is not yet set for aggressive controlled action and is particularly sensitive, consequently, to aggressive action on the part of the enemy. Unless a command is protected during this phase, its chance of defeat, or at least of serious interference with its plans, at the hands of even an inferior but aggressive enemy, is great.

b. Let us consider first, how a force composed of all arms should protect itself during this phase before battle. Such a force obtains the protection necessary by employing the advance guard or advance guards, reinforced by artillery and perhaps by some additional infantry, to cover the front; and also by employing other security detachments to protect its flanks and rear if they are not otherwise protected.

c. If, however, a force consists, let us say, of only an infantry battalion with a howitzer platoon attached, it obtains protection during the preliminaries to battle through the use of the advance guard, reinforced perhaps by some machine guns and the howitzer platoon to cover the front, and by security detachments to protect the flanks and rear if they are not otherwise protected. Naturally, in such a small force, artillery is not ordinarily available.

13. *PROTECTION DURING DEPLOYMENT: ILLUSTRATIVE SITUATION.—a.* The advance guard of a command marching north as indicated below met a marching hostile force on the northern slopes of ridge *CD* as shown

in Figure 18.   The commander of the whole force ordered
the advance guard to hold ridge *CD* and cover the deployment of the main body.   One infantry battalion of the
main body of the column was ordered to assembly area *A*
and another to assembly area *B*.   The terrain is broken
and partly wooded.

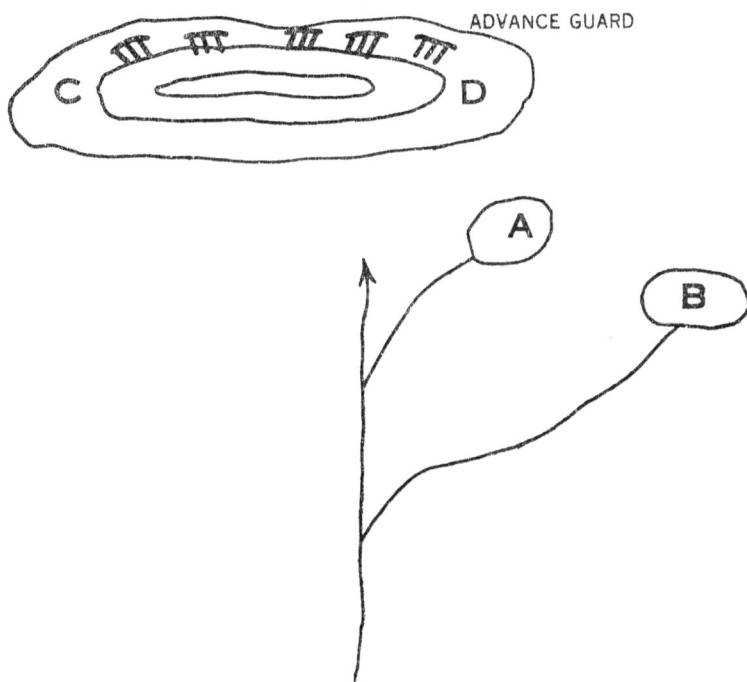

FIGURE 18.

b. Here, the commanders of the two battalions moving
to *A* and *B* should each put out at least a small advance
guard, and patrols to the flanks.   In a meeting engagement (abrupt encounter with the enemy) all units should
protect themselves while they are moving to their assembly areas, while they are in them, and while they are
moving from their assembly areas to the line of departure for beginning the attack.   In the Battle of the

Frontiers in August 1914 many French units suffered unnecessarily because their commanders failed to appreciate the need for continuous protection by elements of their own commands. They chose to assume that the cavalry of higher units was enough protection. The cavalry failed badly not only by failing to furnish protection but also by not obtaining accurate information of the enemy. Consequently many infantry units were surprised and taken at a serious disadvantage.

c. In the illustrative situation, although the advance guard was ordered to cover the deployment of the main body, the two battalion commanders moving to *A* and *B* have no assurance that it will or can furnish absolute protection for all units. A neglect to provide local protection continuously after the battalions have left the column is taking an unwarranted risk and may prove disastrous.

d. In addition both commanders should adopt suitable formations to reduce casualties from hostile fire.

14. *ARTILLERY DISPOSITIONS.*—*a.* The commander of the whole force designates general areas in which he wants the artillery to be emplaced and gives the artillery its specific missions. The artillery commander subdivides the area and assigns the subdivisions to the various artillery units. The location of the artillery is likely to have a considerable influence on the future course of action of the command. Unless its dispositions are directed by the commander of the whole force, he is liable to find his artillery unsuitably disposed for the most effective use in the execution of his plan of operations. The artillery commander is of course better qualified technically to designate the exact positions for his units once he knows the plan of the commander.

b. Assembly positions for the artillery are not designated by the commander of the whole force. Its long range makes it well suited to support the action of the advance guard, to cover the deployment of the infantry units and, in great part, to participate in the future action of the infantry—all from initial firing positions. It is desirable, therefore, to get it into action as soon as practicable. Although this may cause subsequent changes in the locations of some batteries in order to support the main attack better, the advantage of getting the artillery early into battle usually more than offsets this disadvantage.

15. *CONTROL DURING THE PERIOD PRELIMINARY TO BATTLE.*—*a.* Maintaining control over a force from the time of its initial deployment from column toward and into assembly areas, and until it is ready to attack, is also highly important. In large forces the commander designates zones of advance in which each column of his command moves forward.

b. To further insure control he also designates phase lines extending across these zones of advance. In general, one subordinate unit does not advance beyond a given phase line until all units have reached it. Thus, the progress of the whole command toward battle is kept even, and one unit cannot move ahead of, nor fall behind, the others. If phase lines were not prescribed, a unit whose zone of advance lay across open country over which troops could march easily might lose contact with other units and outdistance them. And a unit that had to cross rough ground on which progress was slow might fall behind. As we shall see in the next chapter, phase lines are not straight lines arbitrarily drawn at equal distances across the zones of advance. Instead, they are lines of irregular shape connecting important terrain features, and they lie at varying distances apart, inasmuch as large units must take the utmost advantage of terrain just as we saw that small units must, in paragraph 3 of this chapter.

c. When a command consists of only two columns as it moves forward toward battle during the stages of deployment, phase lines may not be used. Continuous liaison (keeping in touch) and supervision by the commander of the whole force suffices to insure uniformity of advance.

In small units a base unit may be designated and other units directed to guide on it.

d. Another measure a commander takes in large commands to prevent one or more columns from becoming prematurely committed to action is to prescribe the maximum forces the column commanders may put into battle without further orders.

16. *ORDERS FOR THE ATTACK.*—Complete oral orders for an attack rather than fragmentary oral orders, are usually issued to subordinate commanders or their representatives and to certain members of the staff, assembled preferably at a point from which at least part, if not all, of the terrain to be passed over during the attack can be seen. As soon as the plan of attack has been formed, those concerned are assembled so that the orders can be issued.

17. *INFORMATION OF THE ENEMY.*—The chief sources of information of the enemy while a command is deploying are security detachments such as the advance guard, and such purely reconnaissance agencies as motor patrols, mounted and dismounted patrols, and observation aviation. A commander cannot wait until he has obtained complete information of the enemy before making his attack plan. In fact, it is seldom possible to obtain complete information of the enemy. And if he waits too long before formulating a definite plan of action, he gives the initiative to the enemy, as we have seen, and runs the risk of being defeated before his command is ready to fight. On the other hand, if he issues orders prematurely they may have to be modified or changed entirely before the action begins. The commander's whole object must be to reach a decision, formulate a plan, and issue orders in such a way as to cause no unnecessary delay in deploying his command and launching the attack.

18. *ROUTES FROM ASSEMBLY AREAS FORWARD.* —*a.* The route and formation of units from their assem-

bly areas to the line of departure are also governed by
the possibility of enemy observation.   Routes that cannot
be observed by the enemy, and that offer the maximum
concealment from the air, should be taken, if time permits.
Good routes may not be available, of course; and routes
that interfere with the movements of other units must
not be used.

b. The formation of the unit is determined by the terrain and the likelihood of coming under hostile observation and fire.   In general, as always in approaching battle, the greater the chance of hostile observation and fire,
the more extended the formation.

CHAPTER 6

THE ADVANCE TO CONTACT (ADVANCE GUARDS AND DEPLOYMENT WHEN CONTACT IS IMMINENT)

1. *GENERAL.*—*a.* In Section II of Chapter 3 the conditions affecting the conduct of advance guards were given as:

(1) When contact with the enemy is not imminent.

(2) When contact with the enemy is imminent.

b. Everything we have studied thus far has dealt with the first of these conditions and what occurs when a force advancing in route formation—used only when contact is not imminent—suddenly encounters the enemy. In this chapter we take up the conduct of advance guards when contact with the enemy is imminent and see how they, and the main body elements they precede, enter battle.

c. The method of advancing to meet a hostile force that is described in this chapter is by all odds preferable. Through its use a command does not meet the enemy while it is still in long vulnerable columns, unready for battle, but in a partly deployed, fan-shaped formation, with its advance guards ready to fight as soon as they strike the enemy forces. The difference between this method and the method described in Chapters 4 and 5 is essentially the difference brought out in Chapter 2 on The Necessity For Deployment. It is a matter of preparing for combat in advance, of making those preparations before a force gets too close to an enemy force for comfort.

d. The use of the method covered in this chapter depends

on only one thing—information of the enemy. A commander has to know that he is soon to meet the enemy in
force before he can decide that contact is imminent and
take action accordingly. If this information does not come
to him, he finds himself, more often than not, forced to
enter combat at a disadvantage in the manner we have
studied in Chapters 4 and 5. But if his sources of information are active, alert, and diligent, he can move to battle prepared for it, in the way we are about to consider.

2. *ZONES AND BOUNDARIES.*—*a.* Once a commander
decides that contact with the enemy must soon be considered imminent (as described in Chapter 3), he then
must decide upon the details of the partial deployment of
his force. His command as a whole is to advance in several
adjacent zones as soon as contact becomes imminent. Hence
boundaries between these zones must be established. And
in the selection of these boundaries we see again the inevitable importance of terrain.

b. The boundaries of these zones are drawn, for the
most part, through natural and artificial terrain features
which separate terrain corridors and compartments. Thus
the zones of advance correspond, as far as the ground
permits, to adjacent strips of terrain, each of which is defiladed by terrain features from all direct hostile fires
not within the strip. A simple illustration is a small
valley extending in the direction of advance (Figure 19).
This valley is made the zone of a unit of suitable size,
and the boundaries follow the ridges on each side of the
valley.

c. Corridors such as that in Figure 19 are, of course,
ideal, and exceptional. Adjacent valleys with prominent
terrain walls lying between them are seldom found extending in the direction of advance. But a study of any
terrain that forms an area of imminent contact discloses
features which, although they may not be as prominent as
ridge lines, nevertheless divide the area into corridors.

d. Figure 20 illustrates such a corridor. Here the

features through which the boundaries are drawn do not continuously protect the flanks of the zone yet defilade it to a large extent. In such cases, where distinct natural corridors are not present, the boundaries should be drawn

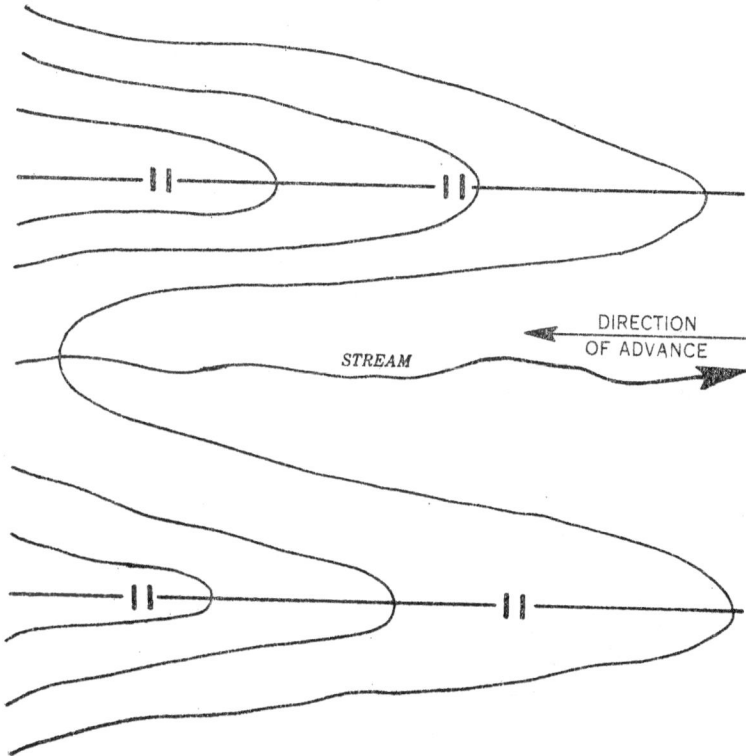

FIGURE 19.

over the higher parts of the terrain and through villages and woods. Thus the zone is defiladed to the maximum from fire delivered from an adjacent zone.

e. If the boundary is drawn as shown by the broken line, B, in Figure 20, troops moving forward in zone II may be stopped by fire from hill A, and may be unable to advance until troops in zone I drive out the resistance on hill *A*. If the troops in zone I are unable to do this, then much time is lost in coordinating the actions of the two advance guards against hill *A*. But by drawing the boundary as shown by the solid line, this coordination is obtained initially, and each advance guard is authorized to place fire upon or otherwise drive the enemy off of that part of hill *A* lying within its zone.

f. Resistance along the boundary can often be neutralized by fire, and then forced to withdraw by passing the bulk of the advance guard by it and threatening complete envelopment. With the boundaries drawn as shown by the solid lines, Figure 20, such action may be promptly executed, for each advance guard is free to fire on that part of the terrain along the flanks of its zone without the necessity of further coordination with its neighbor.

g. On the other hand, if hill *A* is an important tactical locality which cannot be passed by, but must be captured by actually occupying it, or secured for observation or other purposes, it is best to run the boundaries so that hill *A* lies wholly within the zone of a single unit big enough to deal with it. Its capture then becomes a task for that unit only. (This is especially important along phase lines, which are discussed later.) It is exceptional, however, in assigning initial zones of advance across the area of imminent contact, to run boundaries around high ground, villages, and woods, instead of through them.

h. Such a change of boundaries rarely happens, in fact, until the advance guards meet such strong hostile resistance that the commander of the whole force must issue new orders to cover the future action.

i. Furthermore, boundaries between adjacent elements in the area of imminent contact are not straight azimuth lines, nor are they continuously parallel, because the width of each zone is influenced by the terrain features over which its boundaries pass. In order to take advantage of terrain, a zone may vary considerably in width.

3. *ORDERS FOR THE PERIOD OF IMMINENT CON-TACT.*—*a.* As we have already seen, it is the commander of the whole force who estimates when contact will become imminent, basing this prediction upon all enemy information at hand. He must then forewarn the elements

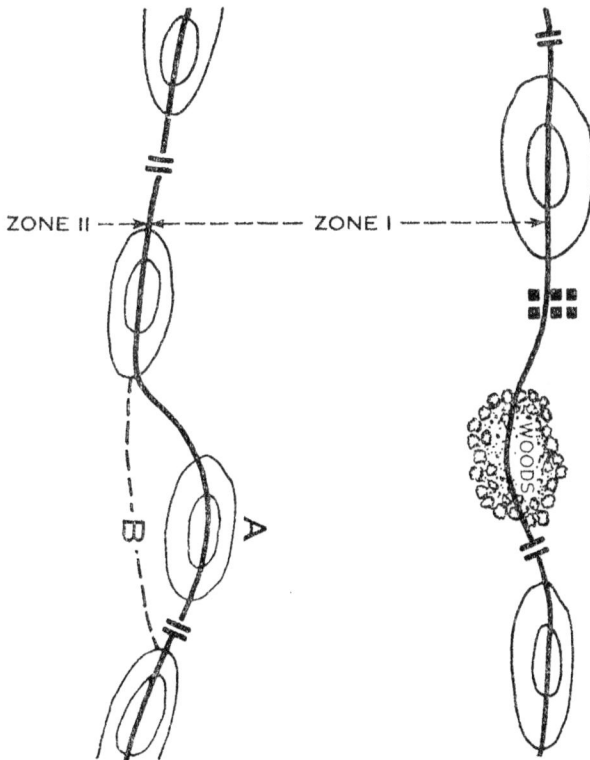

FIGURE 20.

of his force, and issue the order for their redisposition. The order must reach the smallest units in time for them to comply with it fully before they enter the area of imminent contact. If, after the commander has determined the area of imminent contact, some time is likely to elapse before the actual issuing of the order, he should send out a warning order. Later information may, of course, change his forecast of the area of imminent contact, and correspondingly, the place and hour of partial deployment.

b. On the other hand, there may be too little time, instead of too much. The commander, it is possible, may find that a move by the enemy threatens him unexpectedly. In this day of movement by motors, conditions change rapidly, although in an alert command, active reconnais-

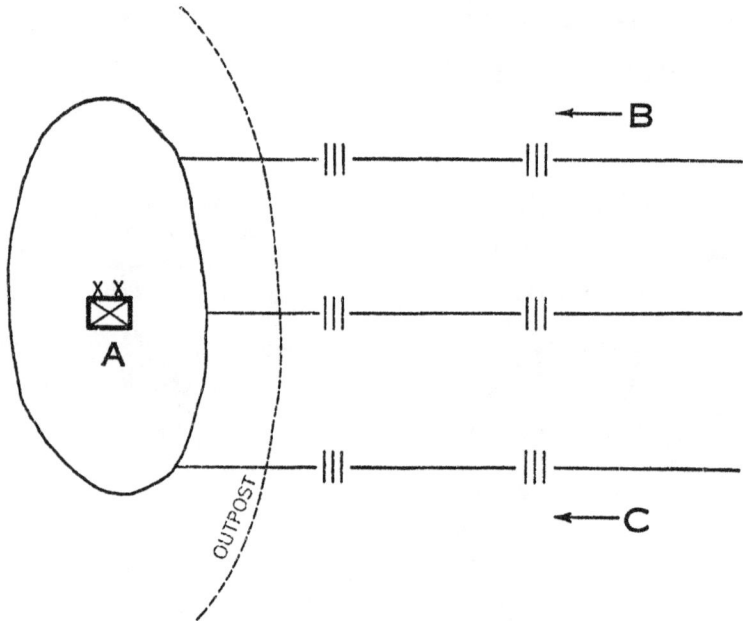

FIGURE 21.

sance should never permit the force to be caught unawares.

c. Nevertheless, contact may become imminent at any time, and find the force variously disposed. The force may be in bivouac, for example; and because of a hostile movement, contact with the enemy may be imminent as soon as the force starts its advance from bivouac.

4. *CONTACT IMMINENT FROM A BIVOUAC: IL-LUSTRATIVE SITUATION.*—Assume that the 1st Division halts for the night in the bivouac, *A*, (Figure 21), with outpost as shown protecting the division during the

night; and that contact with the enemy is not considered imminent even for the next day's march. Then further suppose cavalry patrols report that a hostile force, estimated to be an infantry brigade with artillery, detrucked at points B and C at midnight and started moving toward

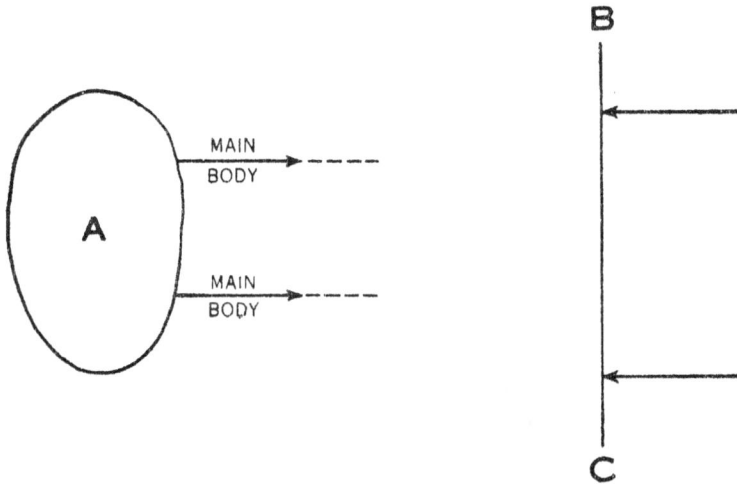

FIGURE 22.

A in two columns, so that the 1st Division might expect to meet the hostile force anywhere between its outpost and the line BC. In this situation the advance guards should be ordered to take up the partly deployed formation as they advance from the bivouac—certainly no later than passage of the outpost lines.

5. *CONTACT IMMINENT FROM ROUTE FORMATION: ILLUSTRATIVE SITUATION.*—On the other hand contact with the enemy may become imminent during the progress of a march. For example, let us again assume that the 1st Division began a march from bivouac A expecting to reach the line BC without meeting the enemy. (Figure 22.) During the march, while the di-

vision is in the formation shown, or any similar formation,
we shall further assume the receipt of information that
a hostile reinforced brigade, in two columns, is approaching the 1st Division with the heads of hostile columns on
the line *BC*. If the distance to line *BC* is, say, only four or
five miles, then the commander of the 1st Division should
reason that contact has become imminent and therefore

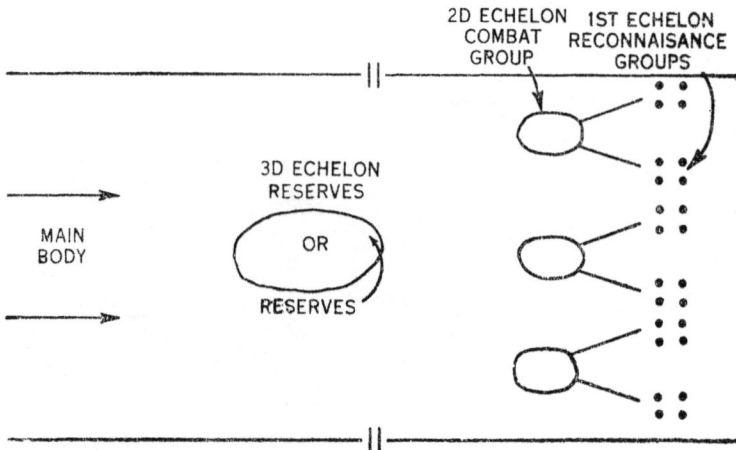

FIGURE 23.

immediately issue an order for the change from route
formation to a partly-deployed formation.

6. *FORMATIONS.—a.* The order of each commander
in turn, from the high commander down, prescribes the
zones of advance to be covered by the next lower units,
upon entering the area of imminent contact. The advance
guard in each zone partly deploys and continues thereafter to move forward on a wide front until contact is
gained or the situation otherwise changes from one of
imminent contact.

b. Each advance guard advances in three echelons.
(Figure 23.) The leading echelon is composed of small

reconnaissance groups covering the entire width of the zone. The second echelon is made up of combat groups from which the reconnaissance groups are sent, and which act in close support of the forward echelon. The remainder of the advance guard, formed in one or more reserves, composes the third echelon.

c. The exact formation varies considerably, of course, and depends on a number of factors; the width of the zone of advance, the size and composition of the advance guard, the location of expected resistance, and the anticipated employment of reserves. The forward echelon of reconnaissance groups, however, must cover the entire width of the zone. But this does not mean that there is a continuous thin line of skirmishers stretched across the zone. This echelon is made up, rather, of an irregular line of reconnaissance groups sent forward from the echelon of combat groups. They may be half squads, or squads, acting in the manner of reconnaissance patrols. At any stage of the advance, their number varies according to the width of the zone and the nature of the terrain as it effects visibility and good patrolling.

d. But the width of the zone varies according to the nature of the terrain. A regimental zone with a battalion in advance guard, may be at one place only 1000 yards wide, and at another, 2000, according to the varying width of the corridor. In open country in the narrower parts of the zone, it may be possible for one company to cover the entire regimental front, and yet permit its commander to retain adequate control of the company. But in the wider parts, the company commander might not be able to direct his company to the best advantage. Therefore the disposition of the units of the advance guard depends primarily on control, and control depends to a large extent on the nature of the terrain.

e. A battalion acting as an advance guard in a regimental zone, or a narrow brigade zone, may be disposed in various formations, of which the following are typical and suitable:

(1) *One rifle company composing the first two echelons* *(reconnaissance and combat groups), and the other two* *rifle companies in one or more reserves.*—(Figure 24.) This formation is rarely used and is possible only in a narrow zone of advance in open country where visibility is excellent. Otherwise control of the two leading echelons is too difficult.

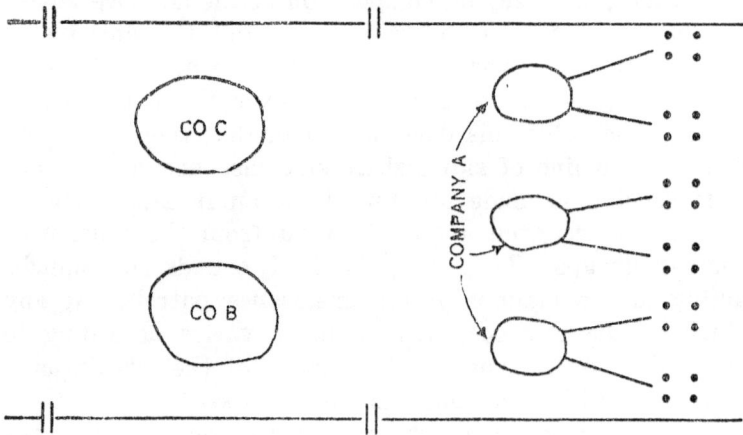

FIGURE 24.

(2) *Two rifle companies composing the two forward* *echelons and one rifle company in reserve.*—In this case the entire zone is subdivided into two company zones. (Figure 25.) This formation is subject to variations. For example: Figure 26.

f. If the advance-guard commander expects resistance to be uniform over his entire front he may adopt the formation shown in Figure 25. But if he has reasons to believe that he will need greater force in the right of his zone he may use the formation shown in Figure 26. Some variation of this formation will find greater application than that with one company leading (Figure 24), or the formation shown in Figure 27. Here all three companies are in line, each making up a part of all three echelons.

This formation has two distinct advantages: (1) easy lateral control, (2) strong driving power past isolated hostile resistance—the best method of reducing isolated hostile resistance is for part of the advance guard to move on beyond it, which usually causes the resistance to fall. This formation has, however, the disadvantage of a scattered reserve, a disadvantage that can be partly overcome

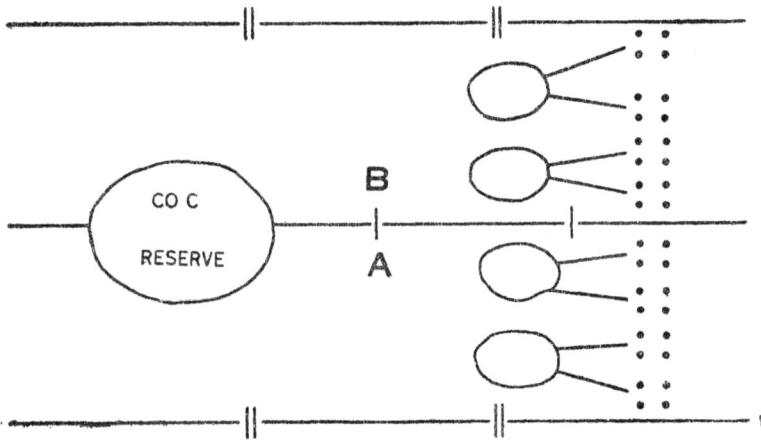

FIGURE 25.

by advancing the reserves of the outer companies close to the interior boundaries and by the advance-guard commander's prohibiting their use without his authorization.

  g. The distribution of the units in advance guards is a function of each advance-guard commander. He is told by higher authority that contact is imminent and is assigned a zone. This indicates to him that he is to adopt the partly deployed formation. The higher commander depends on him to cover the entire zone, and it is his responsibility to dispose his advance guard to gain this end.

  7. COORDINATION AND CONTROL.—a. Coordination of the advance of adjacent advance guards, and between each advance guard and its main body, is obtained

by designating successive phase lines and prescribing a
method of controlling the further advance after each phase
line has been secured. There is no set method that applies
in all cases. There are many methods, some of which are:

(1) To prescribe a time schedule, directing all advance
guards to clear each phase line at a certain time.

(2) To direct each advance guard to move forward
from a phase line when those next adjacent arrive on the
phase line; or

(3) To prohibit advance from each phase line until
such movement is ordered by higher authority.

b. The system based on a time schedule is impractical
for an advance over several phase lines. Owing to the
variations of the terrain and differences in resistance
encountered in the several zones, some advance guards
are usually unable to comply with the schedule. Of the
three methods listed above, a combination of (2) and (3)
is generally the best. However, any sound method of
coordination and control must comply with the principle
that each echelon in rear of the first remains continuously
within supporting distance of the next preceding one.
And this means that it is also in communication therewith. A main body commander should not allow his advance guard to get so far ahead of an adjacent advance
guard that its safety is jeopardized. To prevent a flank
from thus being uncovered an advance guard does not
ordinarily move beyond a phase line until adjacent advance guards have arrived thereon.

c. The fact that control and coordination are always
difficult to maintain requires especial effort on the part
of all leaders. Constant patrolling between adjacent advance guards and adjacent main bodies is necessary, and
visits of staff officers to units on the right and left must
be frequent. Not only the transmission of information
from front to rear but the transmission of information
and instructions from rear to front should be prompt and
vigorously followed up. Commanders, indeed, must make

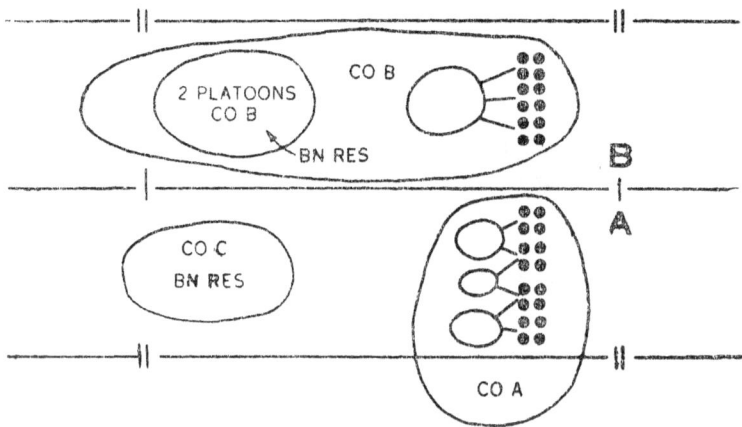

FIGURE 26.

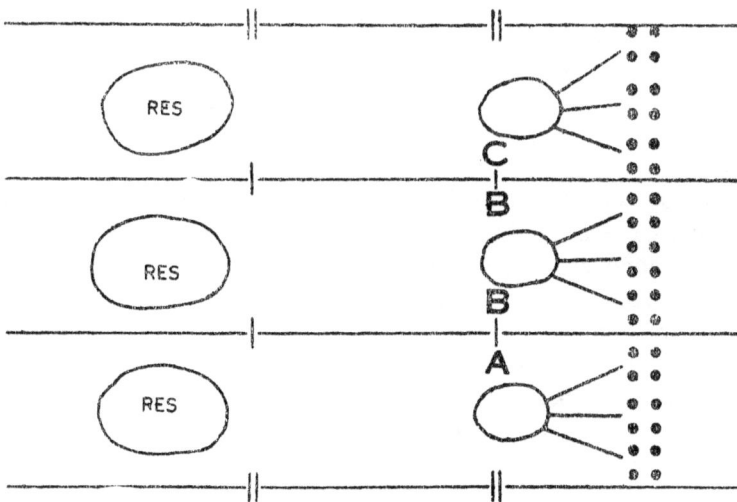

FIGURE 27.

the most of all means at their disposal in order to insure
control and coordination.

d. Furthermore, control and coordination are enhanced
by a proper and intelligent selection of phase lines. Where
practicable, phase lines are prescribed so that the main
body is not required to pass beyond one phase line before
the advance guard secures the next succeeding one. This,
of course, is not always practicable. The terrain may
prevent it, and the principle must then be observed that
the distance between advance guard and main body
should never be so great as to allow the defeat of the
advance guard before the main body can support it.

e. It is evident, therefore, that no fixed distances are
prescribed between elements of the advance guard or between the main body and the advance guard. These distances vary according to the distances between phase lines
and in accordance with the principle governing the conduct of the advance—that each succeeding echelon should
at all times be in supporting distance of the preceding one.

8. *THE IMPORTANCE OF TERRAIN COMPART-
MENTS TO ADVANCE GUARDS.*—*a.* The movement of
an advance guard is controlled laterally by the boundary
lines of its zone, and its forward movement is regulated
by phase lines traversing the zone more or less perpendicular to the boundary lines. Subject to certain variations
discussed above, the lateral boundary lines of an advance
guard zone should, as we saw in paragraph 2, pass over
the higher parts of the terrain and through woods and
villages. Thus the zone is defiladed as far as the terrain
permits from fire delivered from adjacent zones.

b. Phase lines likewise should pass along the high terrain and through woods and villages so that the area in
rear of each phase line is defiladed from direct fire and (as
soon as the line is occupied) from observed indirect fire
delivered from positions forward of the phase line. The
reason for this, of course, is that such a line is usually

the best position both for organizing a defense and delivering an attack.

c. Thus, the zone boundary lines and any two successive phase lines (*AA'* and *BB'*, Figure 28), form a terrain
compartment inclosed by high ground, woods, or villages,
or other terrain features that defilade the compartment

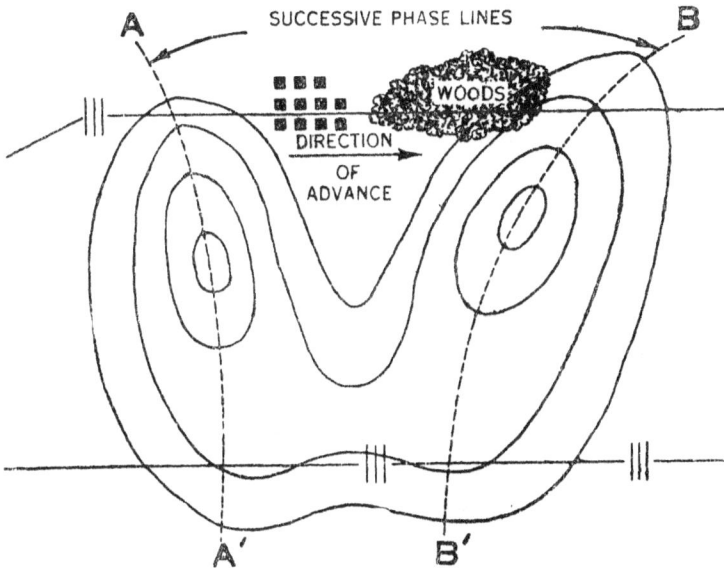

FIGURE 28.

in all directions. Common sense tells us that we will seldom
find this ideal. But, in any situation, a careful study of
the terrain must be made and these lines so designated
that the ideal is approached to the fullest extent possible.

9. *DEVELOPMENT OF ADVANCE-GUARD ACTION.*
—*a.* Now we must consider the important question: How
does an advance guard act when it encounters enemy forces
of various sizes? Before answering this let us recall to
mind the general mission of the advance guards. This
mission is twofold. For as long as possible the advance

guards insure the uninterrupted advance of the main body. And then, upon contact with a hostile force of such strength that the uninterrupted advance of the main body is no longer possible, an advance guard insures enough time and space for its main body to maneuver in accordance with the will of its commander. In accomplishing this dual mission, an advance guard performs the specific duties given in paragraph 3, Chapter 3.

b. In performing these duties an advance guard must continuously guard against surprise. This it does by adopting the partly deployed formation, and exploring all places likely to contain the enemy in any force, and also by regulating the advance so that in general the main body does not advance beyond one phase line until the next succeeding one is secured by the advance guards.

c. As it encounters small parties of the enemy each echelon of an advance guard drives them back to the extent of its power. The leading echelon (reconnaissance groups) should be able to cope with small enemy patrols, which are likely to be the first hostile elements encountered. But it may be necessary for the second echelon (combat groups) to assist the leading one to drive back some small groups offering particularly stubborn resistance. After these initial contacts occur, it can be expected that as the advance continues, the bodies of hostile troops encountered will be of increasing size, until finally contact with the hostile main force is gained. And then, as the resistance increases it eventually becomes necessary to employ the reserves.

d. But at some time prior to the commitment of advance-guard reserves, the commander of the whole force must tell the advance-guard commanders what he intends to do with the whole force, and what he wants the advance guards to do. This step should be taken as soon as the commander arrives at his decision. It may be taken even at the beginning of the advance, or at any time during the advance, and in any case must be taken before it becomes necessary for advance-guard reserves to be used.

The earlier the better, of course, so that advance guards can act without delay and with intelligent cooperation in furthering the contemplated action of the command as a whole.

e. The commander of the whole force may decide to defend, or to attack. If the decision is to defend, he orders the advance guards to secure a certain line. Each advance guard then acts according to the situation. If the enemy advance element is holding this line, the advance guards attack and drive it back. If there are no hostile troops on the line, the advance guards merely continue forward far enough to hold the enemy beyond small-arms fire so that the main body can occupy the line.

f. If the decision as regards the whole force is to attack, the advance guards are usually given the mission of driving in the hostile covering forces and continuing to advance until they are stopped by the enemy in force. If the enemy is found in a defensive position, they locate his strong and weak points, and his flanks. If the enemy is advancing, the advance guards seek to delay him sufficiently for the main body to make deliberate preparation to carry out its intention.

g. All of this action by advance guards, it should be remembered, is for the benefit of the main body or bodies. An advance guard does not fight independent actions, but must, instead, closely coordinate its action with that contemplated by its main body.

h. Whether the advance guards attack or defend, they continue to carry out their mission of gaining information of the enemy. Often the information that they gain forms the chief, if not the sole, clear knowledge a commander has of the hostile force opposing him.

10. *CONDUCT OF ADVANCE GUARDS WITH AN ATTACK MISSION.*—*a.* If an advance guard is given a mission requiring it to attack, it employs, in general, the principles of attack for a unit of its size. Naturally this attack must be over a wider front than is normal for the

unit. It, therefore, attacks with shallow depth, and does
not hesitate to commit all reserves, even initially, when
necessary to accomplish its mission. (In Chapter 7 we
shall see how the action of the advance guards verges
into part, at least, of the main battle.)

b. The flanks of the zone are usually protected from
hostile observation and fire except from within the compartment they enclose. Extending across the zone lies the
next phase line, usually on high ground. This phase line
is the objective of the advance-guard attack. It is along
the protected flanks that the attack has its best chances
of advance to its objective. Therefore, unless the center
of the zone offers an excellent avenue of approach, the
attack should be stepped forward on the flanks of the
zone.

c. Small parties of the enemy are likely to be found
along the flanks of the zone. To advance up the center of
a zone, then, is to invite flanking fire from such groups.
Therefore the action of an advance guard should be directed principally along the flanks of a zone. One good
method, for example, that may be used is to take out
initially small hostile groups simultaneously, as at A and
B in Figure 29. Then, according to the resistance uncovered, to concentrate all fires of the advance guard on
C until that position is taken; then on D until that
position is captured, and continuing in like manner until
F and G are taken, and then to place all fires on E until
that resistance is captured or neutralized. By this time
the entire advance guard of this zone is close to its objective and prepared to attack it.

d. This alternate "shouldering forward", with all fires,
including advance-guard artillery, concentrated successively as just described, produces a more successful, less
costly, and in the end, more rapid advance than does action
contemplating a simultaneous advance over the entire zone.
But it is not the only method. For example, stubborn resistance may be uncovered at D and weak resistance on
the left flank. In that event, it is probably better to push

the attack successively on *C*, *F*, and *E*, thus bringing about the downfall of *D* and *G*.

11. *COORDINATED ADVANCE-GUARD ATTACK AND TACTICAL INTEGRITY.*—*a.* At some stage of the advance the commander of the entire force may anti-

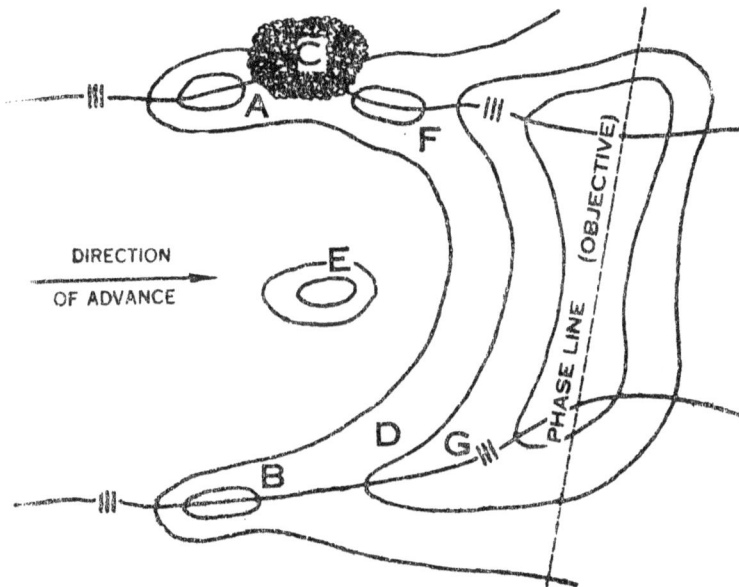

FIGURE 29.

cipate the necessity for a close coordination of the attack of the several advance guards. When this is the case, he assumes direct control over all the advance guards, or places them under a single commander responsible to him. When this occurs, they pass, of course, from the control of the column or zone commanders. The commander of the whole force forms plans and issues orders to the now consolidated advance guards. Nevertheless, in this case the actions of the units that comprised the several formerly separate advance guards are essentially the same as explained heretofore.

b. After such a consolidation, a battalion in the advance guard will find itself acting under a new commander and may even find that it is no longer preceding its own regiment. For example, Figure 30 shows that since the

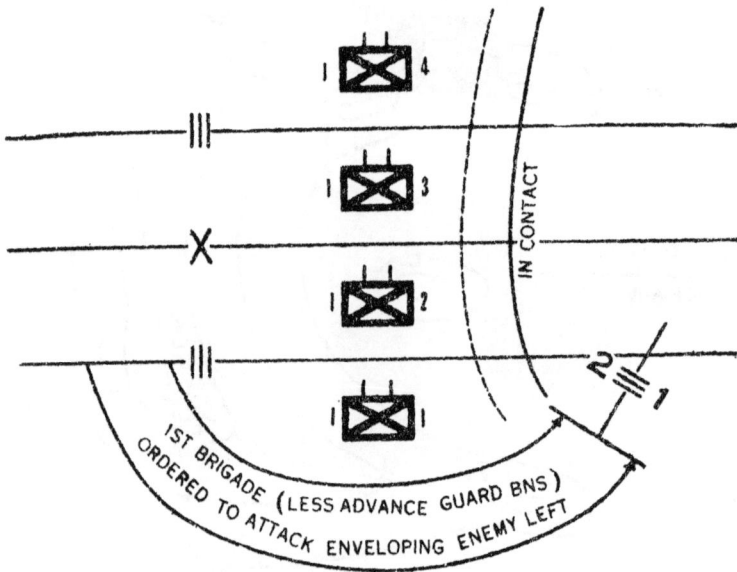

FIGURE 30.

1st Battalion 1st Infantry and 1st Battalion 2d Infantry are engaged with the enemy, the tactical unity of these two regiments is broken up and will propably remain so until the attack is over.

12. *A SMALL FORCE OPERATING ALONE.*—*a.* This discussion thus far has been based primarily on the advance of a large force across the area of imminent contact. A large force is generally opposed by a hostile force of roughly comparable size. By the time such a force has partly deployed upon entering the area of imminent

contact, the opposing enemy force is then too close to change direction without dangerously exposing its flank. But let us consider also a small force, such as a regiment or battalion, acting alone.

b. In the first place a regiment or battalion seldom operates on an entirely independent mission. Such units, however, are often called upon to move alone, wholly or partly uncovered, on semi-independent missions such as a flank guard, or the enveloping force in an attack. In such cases, they may not have within themselves sufficient advance intelligence agencies to determine that contact is imminent. But they are usually operating in conjunction with a larger force whose commander should supply the smaller unit with all possible information pertinent to the determination of imminent contact. Through these means and its own reconnoitering, a small force endeavors, and is often able, to determine when contact is imminent. Having once determined that, its advance guard, for the same reasons that govern those of a large command, should thereafter move forward in a partly deployed formation ready for battle.

c. Small independent advance guards modify, of course, the formations suggested above for large units. Advance along a terrain corridor may not be feasible. Instead, they can, for example, move, partly deployed, astride the route of advance covering some distance to its flanks. Having no adjacent advance guards, such an advance guard must patrol to its flanks.

d. Whatever the size of a unit, it is responsible for its own security. And when there is reason to believe that contact with the enemy is imminent, it is inexcusable for a unit of any size to continue its advance without taking the precaution of preparing itself for combat, by partly deploying at least its covering force—the advance guard.

13. *ADVANCE IN CLOSE OR BROKEN TERRAIN.—* Even when contact with the enemy is imminent, there

will be cases, in rare instances, when the advance guard
cannot be moved forward in the manner herein described.
In very close or broken terrain, with limited routes, the
advance guard may temporarily be force to revert to
column fomation with considerable distances between units.
This would require patrolling to cover the zone and maintain contact with adjacent advance guards.

14. *EFFECT OF OTHER RECONNAISSANCE AGEN-
CIES.*—When advance guards are preceded by reconnaissance agencies other than infantry, the necessity for reconnaissance to the front on the part of the infantry
components is reduced. But this does not relieve them,
however, from the responsibility of maintaining their own
security. As we saw in paragraphs 7b and c, Chapter 2,
it cannot be expected that aviation, cavalry, or mechanized
or motorized reconnaissance units will be able to sweep
the entire area around an advancing force or the intervals between columns marching on roads. Therefore it
is essential that patrolling across the entire zone of advance, or to the flanks between partly deployed columns,
be carried on by advance guards, in order to detect and
drive off small parties of the enemy that may have filtered
through the more distant reconnaissance screen.

15. *ADVANCE-GUARD ARTILLERY WHEN CON-
TACT IS IMMINENT.*—a. When contact is imminent,
the advance-guard artillery moves generally across country, taking advantage of covered routes, in rear of the
infantry of the advance guards. It moves by bounds from
phase line to phase line. A part or all of it is placed in
position prepared to support the advance from one phase
line to the next. If the advance guard attacks, its artillery, from positions well forward, supports the attack by
fire. This fire, in the attack, usually consists of:
   (1) Fire on enemy elements that oppose the greatest
resistance to the infantry advance.
   (2) Fire on hostile artillery.

(3) Fire on hostile elements to force their early deployment.

b. Likewise, if the mission of an advance guard requires it to hold defensively, its artillery, from positions echeloned in depth (if more than one battery is present) supports the defense by fire. This fire is usually delivered to—

(1) Force early hostile deployment.

(2) Interfere with hostile deployment.

(3) Break up the hostile attack formations.

(4) Support counterattacks by our advance guard.

(5) Cover its withdrawal, if the advance guard is forced to withdraw.

16. *TANKS WITH THE ADVANCE GUARD WHEN CONTACT BECOMES IMMINENT.*—*a.* In his order directing the partial deployment of his whole force when contact becomes imminent, the high commander designates the disposition of such tank units as may already be attached to the advance guard preceding the force in route formation. The tanks may be divided up among the advance guards moving forward in each zone or attached to one or a part of the several advance guards and not to all. They move, ordinarily, with the reserve echelons of the advance guards of which they form a part.

b. If the commander decides to take over the command of the advance guards as a coordinated whole, he may, instead of attaching them to one or more of the advance guards, hold them out instead, as a single reserve element to be used later in a particular zone or zones of advance when the need for them arises.

17. *ANTIAIRCRAFT DEFENSE AFTER CONTACT BECOMES IMMINENT.*—After contact becomes imminent, infantry units, in the same manner as before, must depend on their own weapons and measures for protection against hostile aircraft.

18. *LATE DISCOVERY THAT CONTACT IS IMMIN-ENT.—a.* A commander may first discover that contact is imminent long after it has actually become so but before his route formation advance guard has actually encountered the enemy. In this event he must decide whether to take up a partly-deployed formation or attack direct from route formation as described in **Chapter 5.**

b. If the hostile force is very close, he should, in general, attack at once, provided of course that his mission is aggressive and the enemy's preponderance of numbers does not make a temporary adoption of the defensive advisable. But if the hostile leading elements are still some distance off, the commander should immediately take up the formation of imminent contact and continue his advance partly deployed.

19. *THE COMPLETION OF DEPLOYMENT.—a.* The completion of deployment from the partly deployed formation of imminent contact is simpler than that from route formation since the force is already disposed in considerable width. Once the advance guards have come to grips with the enemy, the commander of the whole force should soon acquire enough information of the enemy's dispositions to form a basis for his plan of attack, assuming, of course, that his mission is aggressive.

b. In the usual case, the commander will find it necessary to assign assembly areas for units in rear of the advance guards, although he may have them move directly into battle in making an uncoordinated attack. An uncoordinated attack, however, from the partly deployed formation of imminent contact, is much less piecemeal and much more rapid than from route column.

c. The issue of extra ammunition to rifle companies is made in assembly areas or beforehand, according to the situation.

20. *CONCLUSION.—a.* Large forces are unwieldly. Once they are in movement, it is a difficult and slow task

for them to change direction even while they are still in
route formation, but much more so after they have partly
deployed in the area of imminent contact. A commander
does not thus partly deploy his force until he comes so
close to the enemy that this precaution is necessary. Then,
however, it becomes imperative to change from a vulnerable column formation to dispositions that are more convenient for a rapid entry into battle.

b. By the time partial deployment is necessary, the
distance separating the opposing forces is generally too
small for the hostile commander to risk exposing the flank
of his force by attempting to change its direction of movement. He must in turn continue on—with a part of his
force, at least—to meet the threat of our partial deployment and advance, although he remains as free as his
opponent to maneuver his rear elements. (The presence
of a natural terrain barrier or the possession of greater
mobility, of course, may create exceptional situations.)

c. For the most part, then, a commander loses nothing
and gains many advantages by a timely determination of
when and where contact with the enemy becomes imminent,
and taking measures accordingly as described in this
chapter. To do this he needs only one thing—information
of the enemy.

CHAPTER 7

THE ATTACK AND ITS FORMS

SECTION I

INTRODUCTION

1. *INTRODUCTION.*—Up to this point our studies have been directed almost entirely upon the steps of offensive combat preliminary to battle. There may be, of course, and usually is, sharp actual fighting during these preliminary stages. Advance guards, as we have seen, must often attack, and attack vigorously, to accomplish their missions. But although these attacks are conducted in a manner similar to attacks by a whole force, they nevertheless comprise only the introduction to battle. What we have so far covered regarding offensive combat are the first three steps as given in paragraph 4, Chapter 1. In the remainder of this text we are concerned almost entirely with the stages of battle in which the whole force takes part—in other words, the battle proper rather than its inceptive stages.

2. *PURPOSE AND DOCTRINE OF ATTACK.*—The purpose of attacking the enemy is naturally indistinguishable from that of offensive combat—the attainment of victory by a physical disorganization of the enemy forces. The doctrine or guiding principle of the attack is to strike unexpectedly with the maximum available force at a decisive point and in a decisive direction.

95

SECTION II

OFFENSIVE SITUATIONS AND OTHER GENERAL CONSIDERATIONS

3. *SITUATIONS IN WHICH THE ATTACK MAY BE USED.*—*a. General.*—While a force is gaining contact with a hostile force, and finding out what the enemy consists of, and how his forces are disposed, the commander estimates the situation, arrives at a decision, and forms his plans. If his mission is aggressive and the enemy has not surprised him at too great disadvantage, his decision in the great majority of situations must be to attack. Even when his mission is not aggressive, he may find it best to attack; for example, when he has a temporary advantage of numbers, terrain, or position, over the enemy opposing him, or can effectively surprise the enemy.

b. Attacks classified with regard to the enemy situation. —From the viewpoint of the enemy's readiness and expectancy of attack, attacks fall into five general classifications:

(1) Surprise attacks.
(2) Meeting engagements.
(3) Attack against an enemy deployed for defense.
(4) Attack against an enemy in position.
(5) Attack against a defensive zone.

The division between these classifications is not sharp. Moreover, any of the last four types may (and should) contain a large degree of surprise.

4. *SURPRISE ATTACKS.*—*a.* In surprise attacks the enemy is caught unawares by a sudden onset against him. His force may be in route column, bivouac, or any other formation unsuitable for opposing the attack, but is not necessarily so. Ambuscades, aerial attacks from low altitudes, raids, and night attacks are specific types of surprise attacks. The unexpected attack by a strong force upon an advance guard in route formation is another

example. (From our own viewpoint, the possibility of such attacks was discussed in paragraph 7, Chapter 2, and in Chapter 3.)   Suddenness and rapidity characterize surprise attacks. In local efforts (efforts by certain parts only of a force) the moral effect of surprise on the enemy is often so great as to prevent organized counter-measures. Insufficient reconnaissance and observation, or faulty security disposition on the part of the force attacked, are the chief contributing factors that make surprise attacks possible.

b. *Degree of surprise.*—For the forces that are first engaged by a surprise attack the surprise is usually complete and immediate, although there may be a brief warning. But considering the size of the whole force engaged and the whole force attacked, the degree of surprise possible in ground operations is roughly in proportion to the size of the forces involved, other things being equal. Thus the wide swing through Belgium by the German forces in 1914 is classed as a surprise attack. It was unexpected, was not detected immediately, and its exact nature was not, in fact, ascertained by the Allies until several weeks after it had begun. Nevertheless, once the general direction of the operation was known to the Allies, the Allied forces had many days in which to dispose their forces to meet it. On the other hand, Washington's crossing of the Delaware and surprise attack upon the British forces in Trenton, on Christmas Eve, 1776, found the Hessians either drunk or abed, or both. The surprise was complete and immediate from the viewpoint of all members of the small enemy force involved (about 1000 men).

5. *MEETING ENGAGEMENTS.*—a. A meeting engagement is a collision of two forces, neither of which is fully deployed for combat. The whole tendency of modern infantry tactics is, as we saw in the first six chapters of this text, to avoid unexpected meeting engagements.

b. When contact is imminent, a force takes up a partly deployed formation to be all the more ready when contact actually takes place. In paragraph 7, Chapter 2, the possibility of unexpected meeting was discussed at some length; but in that place it was indicated that one of two forces thus meeting would in all probability be better disposed for combat than the other. The possibility of a meeting engagement between two forces neither of which is aware of the other's presence is considerably more remote, although such encounters, too, are by no means impossible. The term "meeting engagement," however, is broad enough to apply to any encounter between two forces when neither has yet determined upon a specific course of action with regard to the other.

c. Meeting engagements require a prompt decision by a commander as to his general plan of action. Developments occur so rapidly that delay in forming a plan and putting it into effect is liable to be the cause of failure in the ensuing battle.

6. *ATTACK AGAINST AN ENEMY DEPLOYED FOR DEFENSE.*—An enemy deployed for defense has had time to decide upon his course of action—to defend—but has not had time enough to organize in close coordination his troops and the ground on which he intends to make his defense. With every hour that passes, however, it is to be assumed that the enemy will achieve a greater degree of defensive preparation. Hence here too, rapidity of action is important. But inasmuch as the enemy has surrendered the initiative and foregone to a large degree his possibilities of maneuver by setting about the defense of a specific area, a coordinated plan of attack, with a definite time and direction, should be employed against him. Such an attack may nevertheless contain the element of surprise. The enemy expects to be attacked or he would not take up the defense. But the hour and direction of attack can be kept a surprise; which is effected by rapidity or secrecy of preparation, or both.

7. *ATTACK AGAINST AN ENEMY IN AN ORGANIZED POSITION.*—a. In an organized defensive position, the disposition of the enemy's units with their various

weapons, the natural and artificial obstacles that the enemy is using to assist his defense, as well as all other terrain features of which he is making use whether for cover, concealment, or observation, are arranged to complement and supplement each other to the highest degree. The effectiveness of the defense is further increased by such means as the digging of trenches and emplacements. In other words, the enemy fortifies the area he defends with every means in his power.

b. Before an attack can be made against a position so held, a thorough study of the position, in order to discover its points of weakness, is almost imperative. By thorough reconnaissance, both ground and air, the precise location of all defensive means within the position, and those with which the position may be strengthened in case it is attacked, must be ascertained. The necessity for careful planning and the desirability of surprise, in attacks against such positions, are both obvious.

8. *ATTACK AGAINST A DEFENSIVE ZONE.*—A defensive zone is simply a large area containing several or many organized defensive positions, organized in depth so that if the enemy is driven out of one, he can quickly fall back upon, occupy, and defend another to the rear. Offensive operations against such zones usually require many successive attacks by large forces and the treatment of such operations as a whole is, of course, beyond the scope of this text. Each step in an operation against a defensive zone is, however, in no wise different from the attack against an organized battle position which is fully discussed herein.

9. *PURPOSE AND DOCTRINE REMAIN UN-CHANGED.*—Whatever the enemy's situation may be when we decide to attack him, the purpose and doctrine of the attack as given in paragraph 2 above remain unchanged. But to these, we may add one or two other general considerations before we turn to the study of the forms in which attacks are made.

10. *COORDINATED AND UNCOORDINATED AT-TACKS.*—Coordinated and uncoordinated attacks have already been described in paragraphs 4-6, Chapter 3. The limitations of an uncoordinated attack in which the elements of a force are put into combat piecemeal should now be clearer in the light of the preceding discussion on the five types of situation in which the attack may be used. Piecemeal attacks against organized defensive sectors were found to be entirely unavailing during the early years of the World War. Yet when once a force detects a big advantage over another in the beginning stages of combat, a piecemeal attack may be the only way of exploiting the advantage without losing too much time.

11. *THE IMPORTANCE OF GROUND.*—Again, regardless of the type of situation, the use made of ground by the attackers is of the utmost importance. It was not until the true use of terrain was found that the deadlock of the World War was broken. Not only is advantageous employment of the larger variations of terrain essential to the commander's plan; but the use of the very ground over which he crawls and fights is a primary requisite to the successful attack of the individual soldier. When, in 1918, the German General Staff decided that attacks could only be carried through by troops that crept and crawled along the ground, using every irregularity of its surface to conceal their movements and reduce their exposure to fire as they advanced, only then were great gains made by attack after nearly four years of warfare. The fire power of even a hasty defense, as when the enemy has deployed but not organized his defensive position, is so great in modern warfare, that unless the vital value of ground and the way to use it are thoroughly realized by attacking troops, an attack has small chance of success.

12. *THE TIME OF DAY AS AFFECTING ATTACKS.*—*a.* At this point, too, we should consider briefly the time of day as it affects attacks. This factor has had a vital

effect on many battles in the past when night has fallen too soon and the attacking force, hampered by darkness, has been unable to press home an advantage gained late in the day. This was true of Jackson's famous flank attack at Chancellorsville. The flanking force successfully moved to its attack position by night, and fighting through the next day, had the enemy in full retreat by nightfall. But in the night the Union forces slipped away. The victory was not complete.

b. On the other hand, dawn grew to be so much the accustomed time for attacks during the World War, that when a unit on one front attacked at midday the effect was a complete surprise. Early morning is nevertheless the best time for a daytime attack from the viewpoint of daylight hours available in which to fight.

c. For the most part, however, the demands of immediate action to avoid delay may require that an attack be made at any time of day or night. But we may say, in general, that it is inadvisable to begin an attack late in the day against an organized defensive position.

d. This factor naturally appears from time to time throughout the remainder of this text. Its effect on the continuation of the advance guards' initial fighting is discussed later on in this chapter. An entire chapter (Chapter 13) is devoted to night attacks.

SECTION III

FORMS OF THE ATTACK

13. *GENERAL.*—*a.* A commander who decides to attack must also decide upon one of several possible methods of attack. Whatever method he chooses should have three aims:

(1) It should contain or fix the enemy. In other words it must keep the enemy's efforts and attention centered, thus holding him immobile.

(2) It should initially direct a decisive blow of maximum force at the point or area where the enemy will be most vitally affected.

(3) It should exert continuous pressure and the maneuver of a superior force against the vital point or area.

b. In carrying out these aims, especially the second and third, secrecy and rapidity of movement are essential. The main striking force should have depth of formation. Its elements, deployed for combat, must be arranged so that advantages gained by the leading elements can be increased and exploited by other elements following in rapid succession. Thus the effort toward the vital point or area can be sustained because the main blow has weight. The direction of attack, furthermore, should be selected so that it carries the striking force to the vital area in the shortest practicable time. Once the direction has been decided upon, provision must be made for maintaining its direction by the maneuver of the striking force itself and of the reserve.

c. To realize these aims in full a commander must have a thorough knowledge of the enemy—his strength, his dispositions, and his weaknesses—so that he can decide what the enemy's vital point or area is and then plan his attack as above outlined. But more often than not a commander has only partial or even fragmentary knowledge of the enemy at the time he desires to attack. And moreover, he can seldom hope to know all he would like to know. His knowledge of the enemy's actual dispositions may, in fact, be so slight that he may be strongly inclined to delay attacking until he gets more information. For example, a commander, let us say, has fairly complete information regarding the flanks of a hostile force opposing him but very little concerning its center. It is highly important for him to fill this gap of information since, if the enemy center is weak, it would offer him an opportunity to drive through it, split the enemy in two, and defeat half of his forces at a time, or at least place the enemy in an awkward position. On the other hand, if the center is as strong as the two flanks an attack around one or both flanks may be indicated.

a. This possible predicament of a commander illustrates

the exceeding importance of reconnaissance, not only before contact with the enemy but afterward. The World War recounts more than a few lost opportunities of exactly the same type we have just used in illustration. Only through an unceasing flow of information back to him from every possible source can a commander plan and act intelligently.

e. At the same time, we must not take this to mean that nothing at all can be done when information is meager—that a commander must wait until his knowledge of the situation is complete before he can act at all. In the first place, if enemy intelligence is not forthcoming, he can, at the very least, put every means possible into effect to get it. In the second place, he can form a tentative plan of action based on what seems most probable, and also tentative alternate plans. He can begin to put his tentative plan into effect. He can even risk driving such a plan through vigorously, provided the reasonably possible differences in the situation from what it appears to be do not offer too great a threat. He can also attack with part of his force with the end in view of finding out what lies before him. This is one sure way of gaining information. Calculated risks are a part of war. There is little success possible without them. In short, a commander can, and by all means should, do something, rather than nothing at all. Inertia has lost far more battles than boldness.

14. *CLASSIFICATION OF ATTACK FORMS.—a.* Forms of attack are classified, according to the direction and movement of the forces that make the main blow, as envelopments and penetrations. (Turning movements, a form of envelopment by a widely-acting independent force, can be classified under the general term "envelopment".) According to the mission and the means and methods of execution of specific forces making an attack, both envelopments and penetrations are divided into the main attack and the secondary, or holding, attack. The form of attack to be used in a given situation depends as

we have already seen, upon the tactical situation and the mission of the whole force.

b. It is easy to see that the best plan for capturing or destroying an enemy is to completely surround him, provided we have a large enough force. Consequently, the general effect of both envelopments and penetrations is toward surrounding, if not all, then part of, the hostile force. This tendency is plainly evident in envelopments, and becomes apparent in penetrations once the attacking force has broken through the hostile resistance. From this we see that the vital point or area toward which all attacks are directed (paragraph 1) is almost always toward the rear of the enemy dispositions, or actually in rear of them, rather than toward the front. In other words, the principal effort of any attack is not the defeat by direct onslaught of the hostile elements nearest the attackers—the first elements of the hostile defense. It is the defeat of the whole enemy force in the most expeditious manner. And the most expeditious manner, in the great majority of cases, is to strike at his vital area, which is almost never in or near the enemy's front but toward or at his rear.

c. In all attacks, but especially in penetrations, it is ordinarily necessary to overcome or push aside the first defenses of the enemy to reach his vital area. Nevertheless, this purely secondary aspect of the attack should never be confused with its chief purpose.

d. The study of forms of the attack does not lend itself well to distinct subdivisions. In both kinds of attacks—penetrations and envelopments—we strike at the enemy at more than one place in order to deceive him. In both, there is a main attack and a secondary attack. The form of the attack takes its name from the nature of the main attack; yet almost every penetration contains some aspect of the envelopment, and nearly every envelopment possesses some aspect of penetration. For this reason both forms of the attack should be studied with a view to a clear comprehension of how they are interrelated.

SECTION IV

THE PENETRATION OR PIERCING ATTACK

15. *DEFINITION*.—A penetration is an attack made to pierce the hostile defense. The purpose of a penetration is not simply to bend in the enemy's defenses but to burst through them in order to get at his vital areas in rear. A penetration by a large force also contemplates piercing the enemy's force or defensive position on a front wide enough so that the resulting rupture of the hostile force or position can be followed by an attack of one or both of the flanks created, as well as the hostile rear. But for small infantry units such as the battalion and the rifle company, a penetration is more in the nature of the first step in a "shouldering process" as described earlier in this text. A small unit thrusts far enough forward to occupy an advantageous position from which fire can be delivered, or an effective assault made, against the flank of a hostile defensive element which is holding up the advance of the next larger infantry combat unit. Thus, small infantry units acting as parts of larger units, combine a penetration with a prompt use of a flanking attack.

16. *THE NECESSITY FOR DEPTH*.—Inasmuch as a penetration must be deep in order to break through a deep hostile defense, the attack itself must be organized in depth. More often than not a hostile defense is so organized that the attacker must capture a series of organized localities rather than a single line of defenses. Hence a penetration must consist of a series of impulses if the attack is to succeed. The leading assault units work their way forward, small units employing the "shouldering process" already described, and are followed by other deployed units. These fresher succeeding units are used, as soon as it is necessary, to make the succeeding impulses, and especially to exploit the successes of the leading assault elements.

17. *WHERE PENETRATIONS ARE MADE.*—*a.* Penetrations are made against weak points in the enemy's dispositions or position rather than against strong points. This applies not only to large forces but also to small units whether acting alone or as part of a large force. But this does not mean that a commander who has the choice between making a penetration or an envelopment (See Section V), should decide upon a penetration simply because a weak point in the hostile disposition or defensive position exists. The goal to be gained by the penetration is always the primary consideration. Consequently, if a penetration through a weak spot leads to no decisive objective, there is, in general, no good reason to attempt it.

b. Direct frontal assaults against strong hostile defenses.—Modern infantry weapons well organized for defense in combination with favorable terrain possess a formidable strength. Direct frontal assaults against such defenses are not only exceedingly costly but seldom effective or successful. Strong areas of enemy resistance are far more effectively subdued by working around the resistance in order to threaten it with isolation or attack from the rear. In Chapter 9, on The Conduct of The Attack, the methods by which this is accomplished are discussed at length. For the present it will suffice to caution the student that true penetrations are not to be confused with direct frontal onslaughts made over a wide front in the manner generally employed by all armies during the first three years of the World War.

18. *MAIN AND SECONDARY ATTACKS IN PENETRATIONS.*—In a penetration there is a main attack and one or more secondary or holding attacks. The purpose of the secondary attack is to fix the enemy units in position and thus prevent their movement to oppose the main attack; to deceive the enemy as to the place, strength, and time of the main blow; and to draw the enemy's fire and reserves away from the main attack. Secondary or holding attacks are treated in detail in Section VI of this chapter.

19. *THE DISADVANTAGE OF THE PENETRATION.* —The chief disadvantage of the penetration is that it gives the enemy a chance to concentrate its fire power in more than one direction against the penetrating force. (See Figure 31.) It also permits the enemy to strike the flanks of the penetration, especially in penetrations

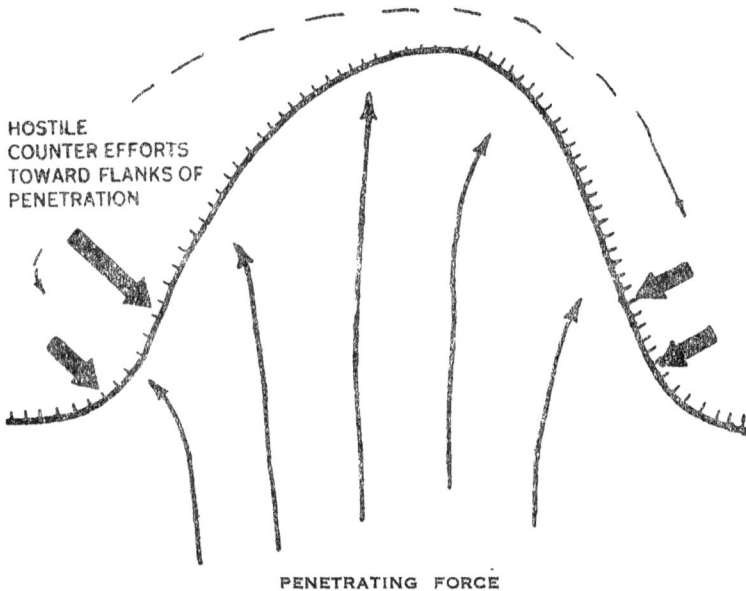

HOSTILE
COUNTER EFFORTS
TOWARD FLANKS OF
PENETRATION

PENETRATING FORCE

Figure 31.—Disadvantage of the penetration.

made by small forces on a narrow front. Furthermore, in a penetration, the attack ordinarily meets the enemy on ground of his own choosing, with his fire power already oriented toward the direction from which the attack comes.

20. *CONDITIONS FAVORING PENETRATIONS.*—A penetration may be more favorable than an envelopment—

(1) When, early after contact with the enemy, or even before contact, a rapid penetration through a gap between

hostile elements is likely to split the hostile force into two parts.

(2) When the enemy's defensive position is too extensive for the size of the force holding it. (In this case, the enemy is said to be "over-extended" and the position weakly held.)

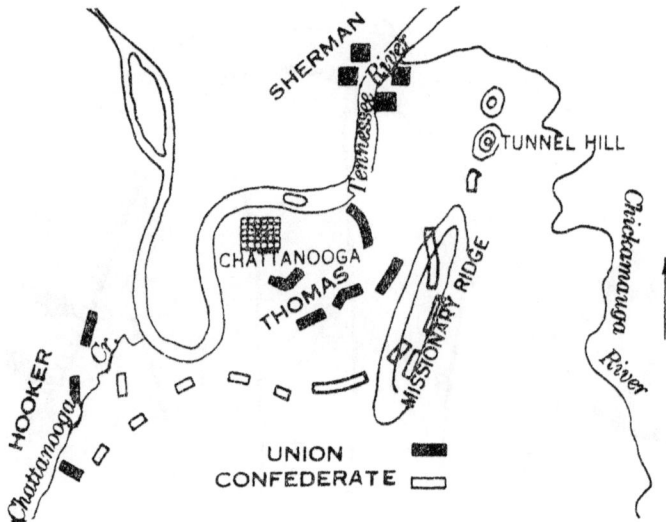

FIGURE 32.—Chattanooga, 1863.

(3) When the flanks of the hostile dispositions or position rest on strong or impassable obstacles, making an envelopment impossible.

(4) When there are well-defiladed approaches or well-defined terrain corridors leading into the hostile position, but no such approaches around its flanks.

(5) When the attacker's mission is such that no time is available for executing an envelopment.

21. *ILLUSTRATION OF A PENETRATION: CHATTANOOGA.*—At 8:00 AM 24 November, 1863, the Union and Confederate forces were disposed as shown in Figure 32. Grant's plan of attack was for Sherman to attack the

Confederate right (north) flank, Thomas to hold in the center, and Hooker to threaten the Confederate left (south) flank. Sherman was unable to take and hold Tunnel Hill, and Hooker was delayed in crossing Chattanooga Creek; so, early in the afternoon, Grant gave Thomas orders to attack the positions in his front. The attack was launched at 3:00 PM, Missionary Ridge was carried, and one Division (Sheridan's) pursued the Confederate toward Chickamauga River.

22. *HISTORICAL EXAMPLE OF A PENETRATION: AMIENS, 8 AUGUST, 1918.*—*a.* The surprise attack by the British Fourth Army against the Amiens salient in the German line on 8 August, 1918 was supported by approximately 450 fighting tanks, most of which were Mark V and Mark V, 8-ton.

b. Every precaution was taken to maintain secrecy in the preparations for this penetration. Deception figured largely in making the thrust a complete surprise. Squadron No. 8 of the Royal Air Force provided a noise barrage for the tanks during the approach march and later assisted the tank operation by reporting the location of these units to headquarters. The artillery also provided a noise barrage, but there was no preliminary artillery preparation since tanks were to be featured in the attack.

c. At 4:45 AM the tanks started forward with the various divisions. The cooperation between the infantry and tanks was good and the advance, with few exceptions, went forward according to schedule. With the exception of the III Corps objectives on the north, all objectives were captured on 8 August, and more than 16,000 prisoners and 200 guns were taken. The penetration on the 8th at the farthest point was 7½ miles. The prisoners reported that the advance of the tanks had been so rapid that resistance was useless.

d. The advance continued on 9, 10, and 11 August, when the tanks remaining in action were withdrawn. A great advance 12 miles deep had been made, 22,000 prisoners and 400 guns had been captured, and, in aiding in this accomplishment, the tanks had prevented great losses among the infantry.

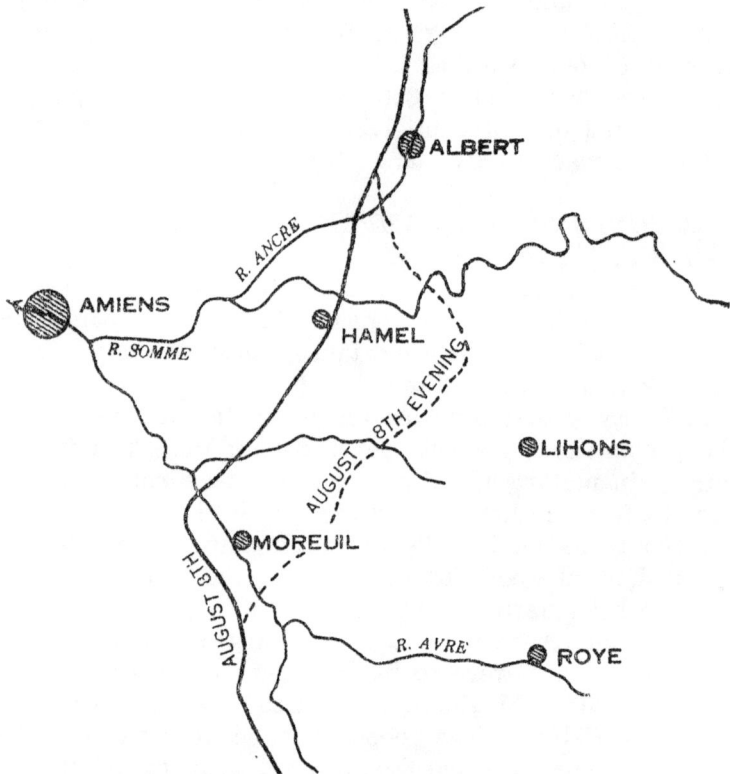

FIGURE 33.—Amiens, 8 August, 1918.

23. *HISTORICAL EXAMPLE OF A PENETRATION: ST. MIHIEL, 12 SEPTEMBER 1918.*—*a.* On September 12, 1918, and the two succeeding days, the American First Army fought and won its first battle as an army, biting off the salient, sixteen miles deep, which for four years had galled France and menaced her communications to Verdun and Nancy. (See Figure 34.) Various factors prevented the operation from being perfectly executed but

it furnishes us an example of a successful double penetration employed in a suitable situation.

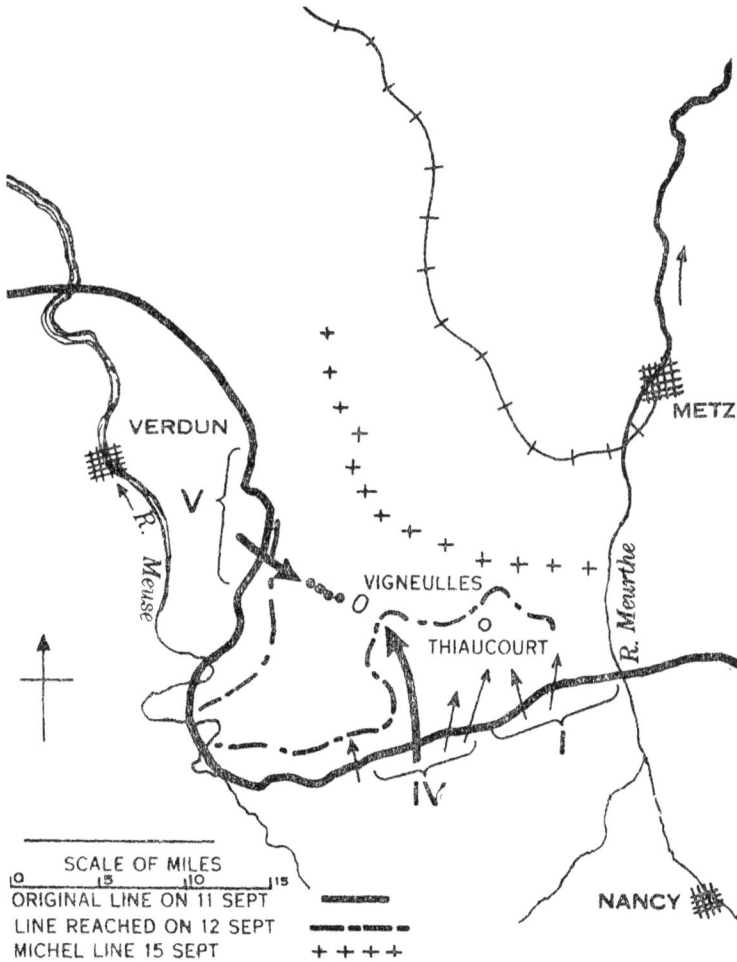

FIGURE 34.—St. Mihiel, 12-14 September, 1918.

b. The American First Army with its French supporting corps outnumbered the Germans about eight to one. French opposition to the plan had pared down the left wing so that the bulk of the force was attacking the eastern face of the salient.   The attack on the right by the I and IV Corps was to start at 5:00 AM, on the left at 8:00 AM, and all the while the French were to keep up pressure against the nose of the salient.

c. The Germans had contemplated a withdrawal and had actually begun their retreat when the attack was launched.   The swift onrush of the American 2d and 42d Divisions upset the methodical arrangements for the German withdrawal.   The I Corps reached its final objectives before midday and, soon after, its second day's objectives on the high ground north of Thiaucourt.   The other two corps fared almost as well.   Dazed and unsupported by their own artillery, a part of which had already withdrawn, the Germans made practically no resistance.

d. During 13 and 14 September, the entire line wheeled up into alignment with the I Corps facing the Michel Line. Then and there the battle was broken off.   The only serious fighting had been borne by the I Corps, which had met with counterattacks owing to the menacing direction of its advance—the enemy was willing to evacuate the salient but had no intention of allowing his base line to be crossed.

e. 15,000 prisoners and 443 guns were captured at a cost of less than 8000 casualties. The scheme of maneuver and plan of attack were successful even though the initial success was not exploited to the fullest extent.

SECTION V

ENVELOPMENTS

24. *KINDS OF ENVELOPMENTS.*—There are two kinds of envelopments, single envelopments and double envelopments, depending upon whether part of the attack moves around one or both of the enemy's flanks.

25. *THE SINGLE ENVELOPMENT.*—In a single envelopment there is a simultaneous attack against the hostile front and against one of the hostile flanks.   The

frontal portion of the attack is the secondary or holding attack. The purposes of the holding attack in an envelopment are the same as those in a penetration. Holding attacks are discussed in detail in Section VI.

26. *CONDITIONS FAVORING THE SINGLE ENVELOPMENT.—a. General.*—It cannot be assumed that a hostile force which has already taken up the defensive, or may be expected to do so immediately if strongly attacked, will neglect to defend its flanks and rear. But it is generally true that the hostile front is likely to be better prepared at any stage short of a completely organized defensive position than any other part of the perimeter of the area in which the hostile force is disposed. Two foes, whether single men or whole armies, naturally face each other and consider the position of the opponent as the most likely source of immediate danger.

b. Frontal attack least desirable.—(1) It is therefore reasonable to assume that any attempt to defeat the enemy by a direct frontal blow alone would, even if successful, be accomplished only with heavy losses. Unless time is of preeminent importance, or the situation permits only a frontal approach, an attacker should seek more advantageous conditions and avoid attacking those areas where the defender is likely to be strongest, or those that offer him the greater advantages for defense. In fine, frontal penetrating attacks are to be avoided for a decisive effort.

(2) Again we can turn to the history of the World War to illustrate this point. Throughout the war on the Western Front frontal attacks which gained little or nothing cost enormous losses. Since the two opposing lines extended for hundreds of miles, no other tactics than the frontal attack were possible. As we have noted earlier in this text, the true tactics of penetration, vastly different from the direct frontal onslaught, were put into effect in 1918. But even after effective methods of penetration were finally adopted, their employment still required tremendous effort. Even the "successful" pene-

trations of the World War were, except in one or two cases, only half as deep as they were wide. In less stabilized warfare, however, a well-conducted penetrating effort may have more chance of success.

c. Favorable conditions for the use of a single envelopment.—There are a number of specific conditions that point toward the employment of a single envelopment rather than a penetration. A single envelopment is ordinarily indicated—

(1) When the position of the hostile flank units or defenses are known or have been logically deduced.

(2) When the enemy's force is weak on one flank or when his defensive position is weakly organized on one flank (as when the flank of the position has little depth).

(3) When there are well-defiladed approaches or well-defined terrain corridors running toward or around the hostile flank (especially when terrain is better suited to attack in this direction than in the direction a penetration would have to take).

(4) When the terrain is otherwise advantageous toward the hostile flank (as when it offers good positions from which supporting weapons and arms can place effective fires upon the hostile elements or position, or affords the enemy poor fields of fire—especially with regard to his machine guns—for repelling an attack).

(5) When there is poor observation within the hostile position for the direction of the fires of the hostile supporting weapons against an envelopment.

(6) When the terrain leading toward and into the hostile flank is favorable for the use of tanks and the cross-country movement of motor vehicles.

(7) When numerous good roads for the use of motor vehicles (a good road net) lead generally toward and into the hostile flank.

(8) When one or more important terrain features lying within the enemy's dispositions or position is more likely to be captured from a flank.

(9) When, early after contact with the enemy, or even before contact, the relative positions of the enemy's force and our own, and suitable terrain approaches also, favor

a blow struck at his flank or rear.

(10) When, for any reason, the hostile force is slower of movement than our own.

d. There may never be a situation, it is true, when all of these conditions obtain. On the other hand, a single one of them may make an envelopment around one flank preferable to a penetration. One point that we should note—a point that can hardly be made too often in modern tactics—is the major consequence of terrain. Here, as in every other aspect of tactics, we can never divorce the consideration of the forces on a battlefield from that of the battlefield itself.

27. *THE DOUBLE ENVELOPMENT.*—*a.* In a double envelopment the hostile front is contained by a holding attack while both hostile flanks are attacked. The demoralizing psychological effect of a double envelopment is greater than that of any other form of attack. The enemy commander, his force struck simultaneously on three sides, immediately feels the threat of being completely surrounded.

b. This form of attack, however, requires far greater superiority in at least one of several factors: morale, numbers, mobility, auxiliary arms, and terrain, on the part of the attacking force than a single envelopment or pentration. It also requires more information of the enemy, and more efficient communication for control and coordination since the dispersion of the attacking force is greater. The risk of this dispersion and the difficulties of control to which it gives rise tend to make it unusual for a commander to adopt this form of attack even for large units. When the double envelopment is adopted as a method of attack, it requires a more detailed and carefully laid plan than any other type.

28. *TURNING MOVEMENTS.*—Turning movements are operations involving the separation of a command into two forces, one of which engages and holds the enemy while the other, operating beyond supporting distance, makes a detour so as to strike the hostile flank or rear.

(See Figure 35.)   The turning movement proper can and should be made over terrain favoring the attacker but ad-

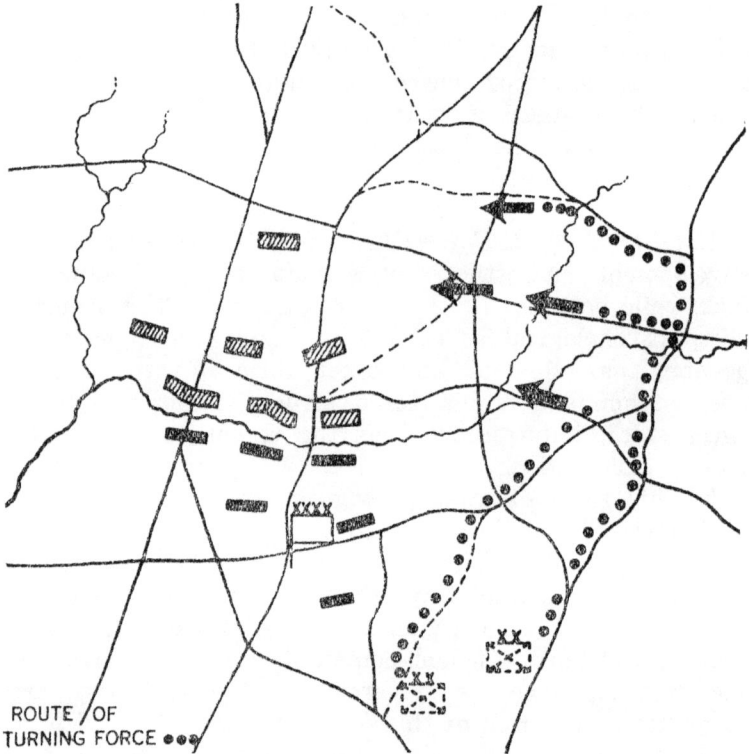

FIGURE 35.—A turning movement.

verse to the defender.   However, the enemy thus em-barrassed may counter by interposing himself between the two attacking elements, and defeat them in detail.   And moreover, although turning movements threaten the enemy's arteries of communication and line of retreat, they may, in turn, uncover the attacker's rear. Turning movements should seldom be undertaken by any but large forces, and generally only where they are superior to the enemy in numbers, mobility, or fire power.

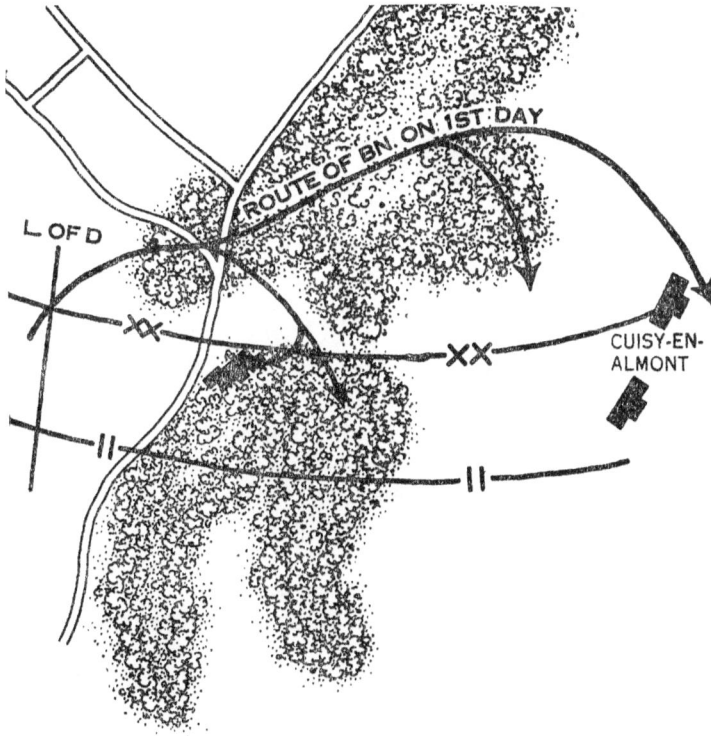

FIGURE 36.—Attack of 4th Bn., 365th Inf. (French), 20 August 1918.

29. *HISTORICAL EXAMPLE OF AN ENVELOPMENT: 4TH BN., 365TH INF. (FRENCH).*—*a.* On August 20, 1918, the 4th Battalion, 365th French Infantry, attacked toward the east in conjunction with other troops. Its zone of action was as indicated in Figure 36.

Its mission was to take Cuisy-en-Almont.

b. Just beyond the road and immediately in front of the 4th Battalion was a fortified work, which had been reduced by the artillery. The battalion commander believed that machine-gun fire from this fortified work might stop the attack of the battalion from even starting. He therefore adopted the following plan:

(1) To have the 15th Company, with one platoon of machine guns attached, the only unit of the battalion to move out at H hour. This company was to move into the zone of the unit on the north, utilizing the cover of the woods and ravines, then turn to the south as soon as the fortified work had been passed and strike it from the rear.

(2) Upon the capture of the fortified work, the rest of the battalion would attack. The 14th Company, following the route of the 15th Company, would advance via the wooded slopes that border the northwestern portion of the Cuisy-en-Almont plateau, and attack Cuisy from the north.

(3) The 13th Company was to maintain contact with the unit on the right and outflank Cuisy on the south.

(4) The battalion commander with the machine-gun company (less one platoon) would move straight toward Cuisy between the 13th and 14th companies.

c. The attack was carried out as planned and was highly successful. The battalion captured 530 prisoners and 24 machine guns.

30. *HISTORICAL EXAMPLE OF AN ENVELOP-
MENT: 1ST BN., 22D INF. (GERMAN.)*—*a.* On 7 April, 1916, during the Verdun offensive, the 1st Battalion 22d German Infantry attacked from the vicinity of Harcourt to capture a French strongpoint on a hill southeast of that city. (See Figure 37.)

b. The 1st Company was to attack to the southeast. The 3d Company was to attack towards point 288, then turn west striking the strongpoint from the rear. The 156th Infantry was to attack on the left of the 3d Company, and the 10th Reserve Company was to attack on the right of the 1st Company.

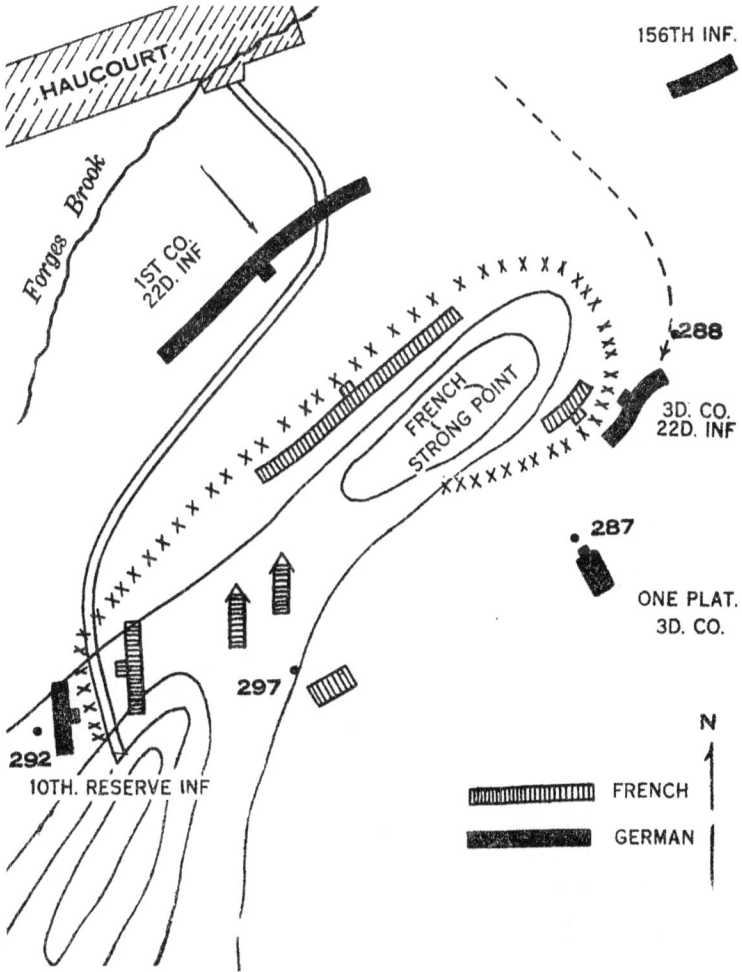

FIGURE 37.—Attack of 1st Bn., 22d Inf. (German),
7 April, 1916.

c. The 3d Company overcame enemy resistance in the vicinity of point 288, and then faced west in order to take the strongpoint in flank and rear. The company commander sent one platoon to the vicinity of point 287 with the mission of protecting the left flank of the company.

d. Upon arrival at 287 the platoon leader was confronted with the following situation:

(1) Heavy firing was heard just over the crest of the ridge towards Harcourt. Evidently the 1st Company was hotly engaged.

(2) He observed that the 3d Company was confronted by the French, who seemed to be preparing for a counterattack.

(3) Near point 292 the 10th Reserve Infantry was engaged in a fight with the French and seemed to be making no progress.

(4) Just north of point 297 he observed two columns of French troops moving north towards the strongpoint, and just south of the same point was a comparatively large number of French troops resting in reserve. His platoon had not been seen by the French.

e. The platoon leader moved his platoon to the southwest, attacked and surprised the French just south of point 297, where it captured a French colonel, two captains, and 150 men. The platoon was reorganized rapidly, and then attacked the French opposing the 10th Reserve Infantry. The attack was successful and several hundred additional prisoners were taken.

31. *HISTORICAL EXAMPLE OF AN ENVELOPMENT:* BATTLE OF WADI EL HESI (SAUSAGE RIDGE).—*a.* On 8 November 1917, the 155th and 157th Infantry Brigades (British) attacked the strong Turkish position on Sausage Ridge, six miles north of Gaza in Palestine. (See Figure 38.) The Turkish forces had occupied the ridge for several days and had strongly entrenched themselves. They had excellent fields of fire down the gradual open slopes to the west and south as shown in Figure 35.

b. Early in the morning the commander of the 155th

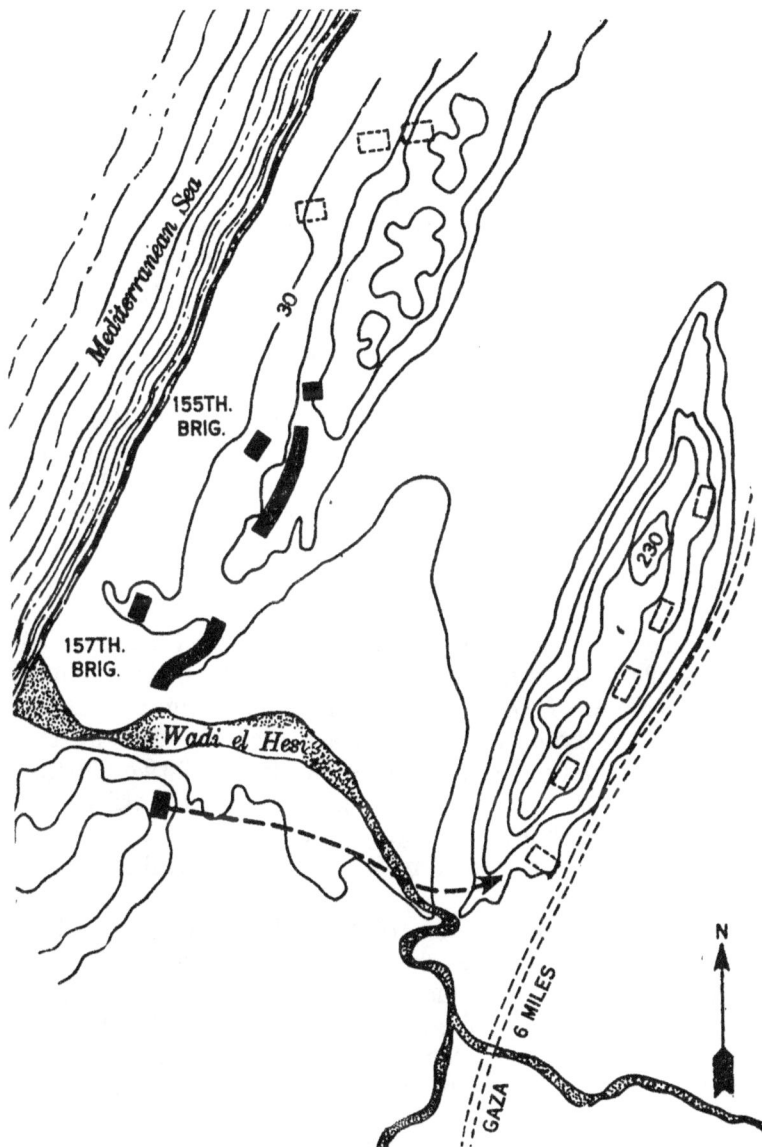

FIGURE 38.—Wadi El Hesi (Sausage Ridge),
8-9 November 1917.

Brigade received orders to attack the upper end of the ridge from the west. His troops were still some distance from the low ridge that was to form their line of departure (the position shown for the 155th Brigade on the map). While they were moving up, he went forward to observe Sausage Ridge and the ground over which his brigade would have to attack. A brief look at the ground showed him that the Turkish position was much stronger than the only available maps had indicated. He reported this at once to the division commander.

c. The division commander then ordered the 157th Brigade to support the attack of the 155th. The whole of the 157th was in column below the Wadi el Hesi. At 1:30 PM the division commander decided to have the 157th Brigade form on the right of the 155th, and attack alongside of it from the position shown, instead of supporting it. The 6th Battalion Highland Light Infantry was ordered to envelop the Turkish left in the direction shown.

d. The 155th attacked at 2:30 PM, reached a point well up the slope of Sausage Ridge and was then driven back in the course of several hours' hard fighting roughly to the positions from which its attack started. During this time its flank was threatened by an attack of two Turkish battalions from the north but this was staved off by the desperate effort of a single infantry company of the 155th Brigade (a company of the 5th Battalion Royal Scots Fusiliers).

e. In the meantime the 157th Brigade (less the battalion that was to make the envelopment below the Wadi el Hesi) launched its attack at 4:00 PM as shown. These troops found hard going, too, but reached the foot of Sausage Ridge about dark. They then charged up the hill to the top but were soon driven back. Three more assaults were made but all three were unsuccessful in holding the position. The 157th Brigade commander finally decided to put in his reserves before he tried again.

f. The flanking unit (6th Battalion Highland Light Infantry) had been the last unit to come up from the south.

It had been forced to move slowly by the heavy sand, and as darkness fell, by the necessity of maintaining contact, especially at the point where it changed direction to the left. But about 7:45 PM, just as the 157th was getting ready for a fifth try at the lower end of Sausage Ridge from the west, the 6th Battalion hit the ridge full force from the south. The Turks were not ready and were apparently completely surprised. The 6th Battalion's attack swept rapidly along the ridge; the whole line of the 157th and 155th then attacked again from the west and carried nearly every position with the bayonet. Sausage Ridge was completely in the hands of the British by 3:30 AM.

32. *HISTORICAL EXAMPLE OF AN ENVELOPMENT:* VON FRANCOIS' I CORPS (GERMAN) IN THE BATTLE OF GUMBINNEN.—(The account that follows describes only a part of a battle, and is one of the best examples of an envelopment afforded by the World War. Strangely, Von Francois' operation, although its execution left little to be desired, was not at all conclusive because, as regards the whole German force, it was done a day too soon. This aspect, however, we shall omit from discussion, in order to stress the envelopment as such.)

a. At noon, 19 August 1914, Von Francois' I Corps opposed the Russian XX Corps as shown in Figure 39. Other units extended the two lines of the two armies roughly to the south, but to Von Francois' left and the Russian right, there were only cavalry units at some distance.

b. Von Francois reported to his superior that he had a good chance to attack because the Russian dispositions were carelessly made, and received permission to do so. On the night of 19/20 August the German left division made a 10-mile night march as indicated. Then, about 8:00 AM, the whole I Corps moved to the attack as shown in Figure 40, the left division heading around the Russian flank in an envelopment, while the right division moved to a frontal attack.

c. The commander of the Russian XX Corps was unable

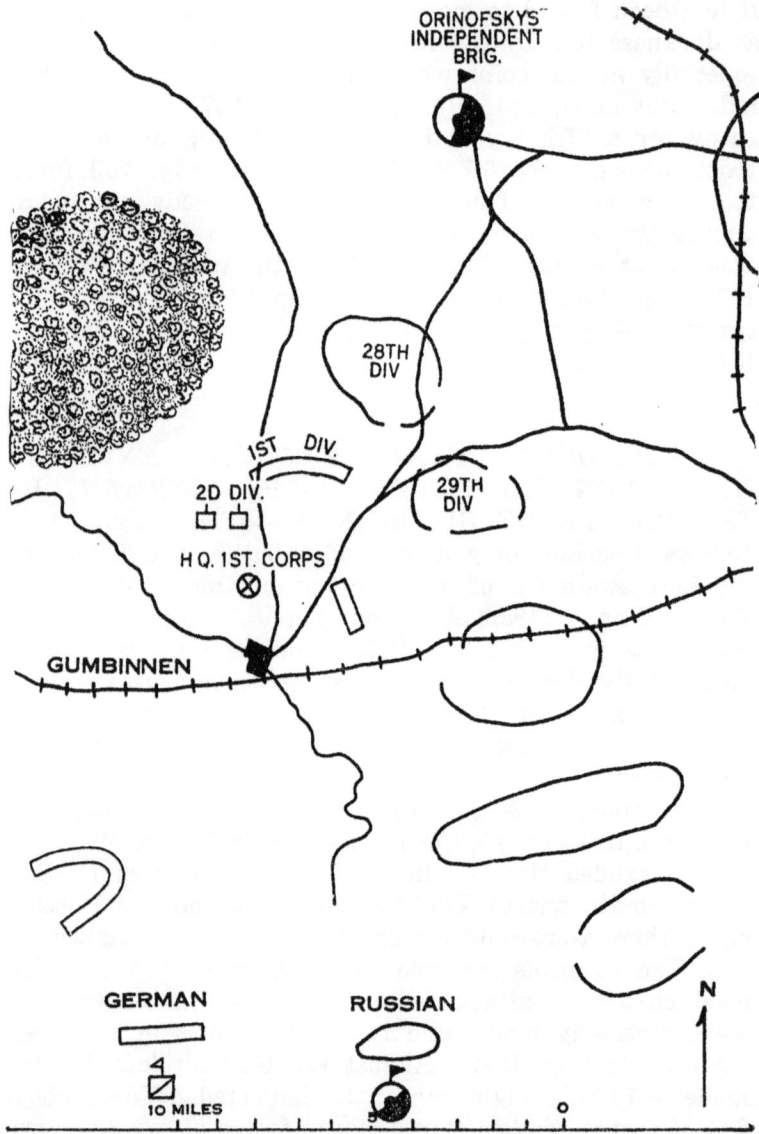

FIGURE 39.—Gumbinnen, 19 August 1914.

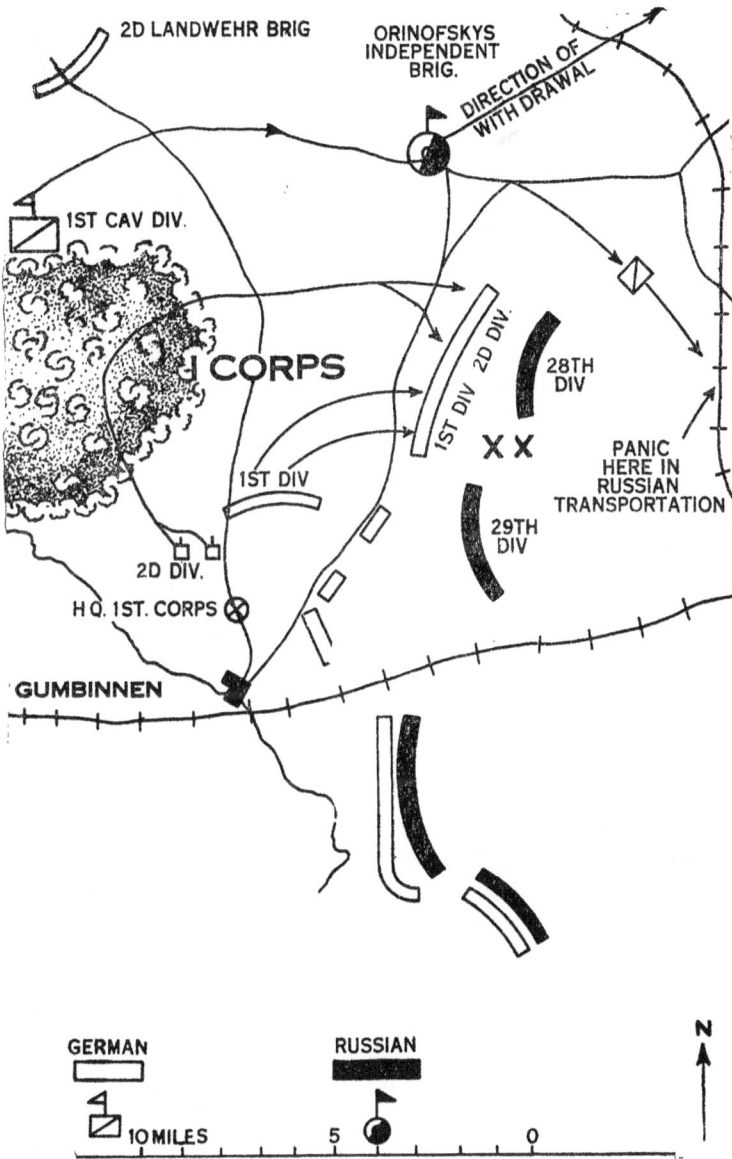

2D LANDWEHR BRIG

ORINOFSKYS INDEPENDENT BRIG.

DIRECTION OF WITHDRAWAL

1ST CAV DIV.

I CORPS

1ST DIV 2D DIV

28TH DIV

PANIC HERE IN RUSSIAN TRANSPORTATION

1ST DIV

29TH DIV

2D DIV.

H.Q. 1ST. CORPS

GUMBINNEN

GERMAN

RUSSIAN

N

10 MILES   5   0

FIGURE 40.—Gumbinnen, 20 August 1914.

to extend his force to the right fast enough to interfere with the enveloping movement, and by noon, the simultaneous attack from the west and the north had caused the Russian right division to retire in such disorder that only with difficulty was its commander able to reform his force on a position several miles to the rear.

d. The enveloping force ran a considerable risk of attack by the Russian cavalry brigade, especially since the German cavalry, whose job it was to protect the German left, was out of touch with the I Corps when Von Francois attacked. The Russian cavalry actually did make a move to attack the enveloping force, but was driven off by artillery. Having no artillery himself, the Russian cavalry commander retired from the battlefield.

e. The commander of the German cavalry, however, having no orders, but hearing the sound of battle in the distance, estimated the situation correctly and made a still wider envelopment around the Russian flank, which measurably increased the success of the operation.

SECTION VI

HOLDING ATTACKS

33. *PURPOSES OF THE HOLDING ATTACK.*—The purposes of the holding or secondary attack are the same in penetrations and envelopments:

a. To hold or fix the enemy in position while the main attack delivers the decisive blow.

b. To create uncertainty in the mind of the hostile commander as to the strength and location of the main decisive attack. This uncertainty contributes directly to the first purpose since a commander naturally cannot shift his dispositions to meet a strong attack until he knows where that attack is coming from. This especially applies to the disposition of the hostile reserves, which form the chief means of the hostile commander for meeting an attack from an unknown direction.

c. To occupy the attention of the hostile forces and thus contribute to the concealment of the main attack.

d. To force the strength and character of his dispositions, and the location of his artillery.

34. *THE RELATIONSHIP OF HOLDING ATTACKS TO THE ACTION OF ADVANCE GUARDS.*—*a.* In a single envelopment, for example, there is a simultaneous attack against the hostile front and against one of the hostile flanks. The frontal portion of the attack is often a continuation of the initial combat of the advance elements of the force. As described in Chapter 6, the advance guards, when they first gain contact with a hostile force, engage the enemy and endeavor to secure favorable ground for the future operations of the whole force. If the commander then decides to make an envelopment, the advance guards, already fighting, continue their attack while a large part of the main body elements are used in an attack around an enemy flank. If he decides to make a penetration, part of the advance guards' initial effort is continued as the holding attack and part of it is replaced by the main effort.

b. If the main attack is to be made the same day, there is no cessation of the advance guard action, which does, however, increase its efforts at the hour the main attack is launched. If the main attack is to be made the next day, the advance guards' efforts are usually slowed up, pressure being maintained by patrols until the hour of the main attack.

c. The holding attack, therefore, is not expected to make material advances, since it is through the main attack that the defeat of the enemy is primarily sought. Nevertheless, the holding attack is sometimes more successful than the main attack. When this happens a transfer of effort is usually made so that the original holding attack becomes the main attack and the original main attack becomes the holding attack. It is also possible, too, of course, for both to be successful.

35. *CHARACTERISTICS OF THE HOLDING ATTACK.*—*a.* Secondary or holding attacks are not made by fire alone but are actual attacks in which an attempt is made to move forward. There is an attack as well as a "holding" of the enemy in place. (The usual meaning of the term "to hold," implying as it does a defensive attitude, is not applicable to this offensive combat term.) The holding attack is ordinarily an attack on a broad front with comparatively weak forces.

b. Although a holding force must make an actual effort to advance, it is too much to expect a simultaneous advance along its whole front. Yet all parts of the front must be attacked by either fire or movement or a combination of both. A holding attack, therefore, consists of a simultaneous principal effort and one or more secondary efforts.

c. The principal effort of a holding force.—The principal effort of the holding force should be made, if possible, in a terrain compartment which can be shut off from the supporting fires of enemy troops in adjacent compartments. The compartment should be of a width suitable for the tactical unit making the principal effort. The principal effort is the only part of the holding attack that is likely to have artillery support. In general, it is desirable to make the principal effort of the holding attack as distant as practicable from the main attack of the enveloping force, so that the enemy will shift his reserves to meet the holding attack; or, if the hostile reserves are already better disposed to meet the holding attack than the main attack, cause the enemy to keep them where they are, or at least hesitate or delay shifting them any considerable distance to meet an attack from another direction.

d. The secondary efforts of a holding force.—In addition to the principal effort of the holding force, each other assault unit down to include battalions should make a secondary effort. Often there is little or no artillery support available; hence, machine-guns and mortars may have to furnish the bulk of the fire support. The portions of the enemy's position not included in these secondary efforts should be covered by fire to pin the enemy to the ground.

36. *LAUNCHING THE HOLDING ATTACK.*—In order to keep the enemy from determining which is the main attack, the holding attack and the main attack must be coordinated. Hence, the holding attack is ordinarily launched before or coincident with the main attack, rather than after it. When the main attack is to begin early in the morning, it may be best to begin the holding attack while it is still night. If the holding attack is launched before the main attack, the two should be so coordinated that the maximum effort of the holding attack is in progress when the main blow falls. Consequently, the holding attack must not be launched so early as to lose its impetus by the time the main attack strikes, or so early as to give the enemy opportunity to use his reserves against the holding and main attacks in succession.

37. *HISTORICAL EXAMPLE OF A HOLDING ATTACK: MEGIDDO, 1918.*—*a.* One of the best examples in modern warfare of the use of the holding attack is that of Megiddo. The broad conception of Allenby's plan of campaign against the Turks at Megiddo in the fall of 1918 was based on the mobility of his Desert Mounted Corps, Lawrence's Arabs, and his air service. His actual plan, however, involved an envelopment following a penetration of the Turkish trench system in front of the XXI Corps. (See Figure 41.)

b. Allenby's plan, as executed, was to move all but a small portion of his force from the area north and east of Jerusalem, and to concentrate the bulk of his force in the coastal plain of Sharon, north of Jaffa. While Feisal's Arabs under Colonel Lawrence blew up the railroad north, south, and east of Deraa, and the air service drove the Turkish planes from the air, the infantry and artillery of the XXI Corps affected a penetration of the Turkish front which allowed the 4th and 5th Cavalry

Divisions and the Australian Mounted Division to initiate the planned envelopment against Megiddo, Nazareth, Jenin, El Afule, and Beisan.

c. The XX Corps, on the right, executed the holding attack and, although faced with a stubborn Turkish resistance, fixed the enemy in place and formed the base upon which the XXI Corps, turning east after their penetration south of Tul Keram, was able to smash the remnants of two Turkish armies and turn that force into a mass of scattered fugitives.

d. A remarkable aspect of the whole campaign was the absolute secrecy with which troops were moved to and concentrated on the left of the line near the sea in preparation for the main effort. When all was ready, on the night of 18 September, the holding force on the extreme right began a vigorous attack. And a few hours later, early the next morning, when the main effort struck the Turks at a point nearly 40 miles away, the surprise was complete. The Turkish high command had no way of telling which was the main drive until the full force of the attack along the sea was evident.

e. This historical example also illustrates clearly how one general form of the attack is seldom clear cut in its distinction from the other general form. The main effort in this campaign had first to penetrate and then envelop.

SECTION VII

PROBABLE ENEMY REACTIONS TO THE ATTACK

38. *GENERAL.*—Outguessing the enemy has more than once led directly to success in wars of the past. But a guess is, at best, still a guess, and a hazardous basis for warfare. After a war has endured long enough for the commander of one army to learn the habitual reactions of his opponent to a given set of circumstances, assumptions as to what the enemy will do next increase in importance. Far more important, however, at all stages of

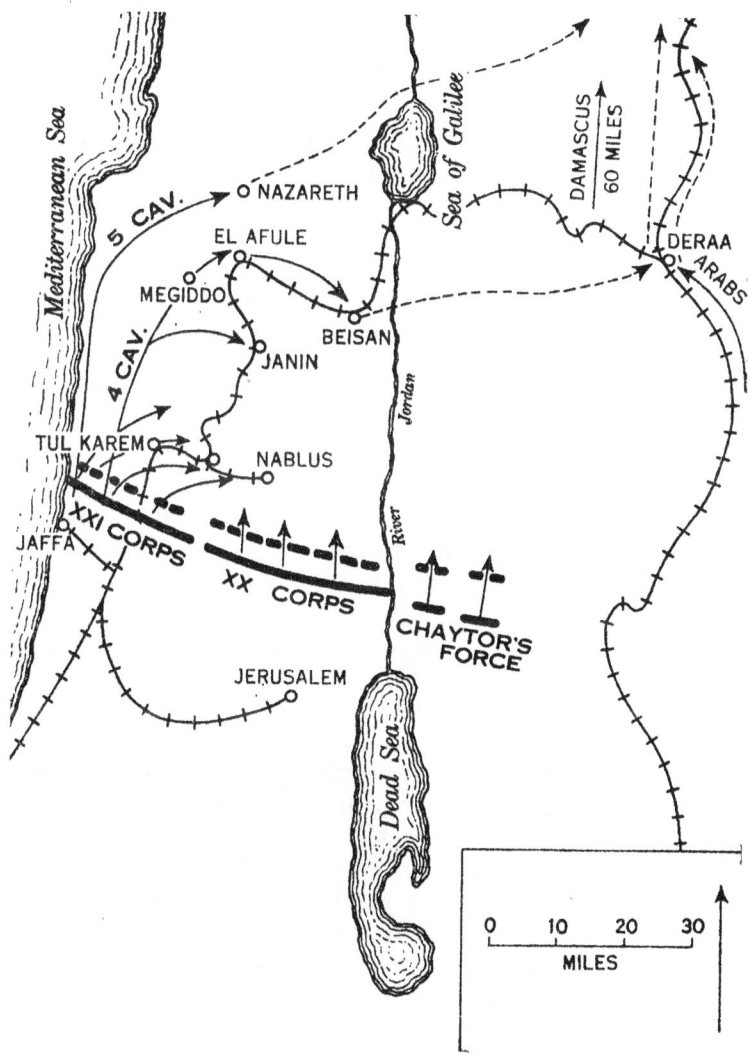

FIGURE 41.—Megiddo, 1918.

warfare is the weighing of all the possibilities open to the
enemy, and the framing of plans to meet not only what
appears to be the enemy's most likely step, but to meet
all possible steps he may take.

39. *THE DANGER OF THE PRECONCEIVED IDEA.*

—*a.* Most of all to be guarded against is the preconceived
idea.   The desire of a commander to believe that the
enemy will do a certain thing in a certain situation sometimes becomes so strong that his enthusiasm carries him
away and he begins to base his actions far too much on
his hopes.   This attitude of mind is doubly dangerous in
that it leads a commander into disregarding all other possibilities except the one desired.   A preconceived idea may,
indeed, become so strong that facts indicating that the
idea is wrong may be discounted, or even disregarded entirely.   The enthusiasm for the preconception reaches such
a degree that the commander believes nothing that he
does not want to believe.   In the words of a French military critic, the brain of the commander becomes "con-
creted".

b. The history of warfare contains many an example
of the disaster that attends the adoption of this frame of
mind.   In the Battle of the Frontiers, August 1914, the
commander of the Third French Army stubbornly disregarded indications of the proximity of the enemy, and
even announced to his subordinate commanders that the
enemy would not be encountered during the day's march
of August 22.   Without exception the French columns, as
we have recounted before in this text, were careless in
their security measures and were unexpectedly struck by
the enemy before noon.   Another illustration of the tenacity with which a commander can cling to a preconception is that of Sir John French, commander of the
British Army in 1914, who at one period consistently refused to credit adverse reports, and even issued orders
to "continue the advance" after his army had been
desperately holding, and in part falling back, for several
days.

c. Far from allowing himself to set his mind, a commander must possess elasticity of thought, or at least enough imagination to conceive which of several moves by the enemy is most likely, without discarding any possibility. He must, moreover, continually and methodically examine the situation in order to determine what the enemy can and may do. (The correct method of estimating the situation is not covered in detail in this text but is found in Special Text No. 10, Army Extension Courses, Solution of Map Problems.)

d. The estimate of the situation has its greatest importance to a commander in determining his own next step. But when he has decided upon that step he must also look forward toward what may happen afterward. In other words, he must consider the probable enemy reactions. "If I attack, what will the enemy do?" is the question a commander must ask himself repeatedly in offensive warfare, not so much from the viewpoint of caution as from a desire to consider every eventuality and to look two steps ahead instead of one.

40. *PROBABLE ENEMY REACTIONS TO A PENE-TRATION.*—*a.* The enemy reaction to a penetration may be to hold, counterattack, or withdraw. His foremost units of defense may be expected to hold without making any coordinated counterattacks. As soon as the attack actually penetrates a battle position, local attempts to drive the attack out are to be expected. Farther back, however, he may launch counterattacks with all the force of his reserves and supporting fires. Once important areas in the hostile position are seized the enemy will, in all likelihood, place artillery fire upon them. As the penetration deepens, the enemy is liable to make counterattacks toward either flank of the penetration. (See Figure 42.) and may even attempt to isolate it completely (Figure 43.) In order that such attempts may be met promptly, penetrations must be made, as we have seen before, with the penetrating force distributed in depth.

b. If the enemy withdraws before the penetration, the danger to the flanks of the penetrating force may still be

present.   The withdrawal may, indeed, be calculated to
lead the penetrating forces deep into the hostile dispositions in the hope that in the excitement of success there
will be a disregard of flank protection.   These tactics have

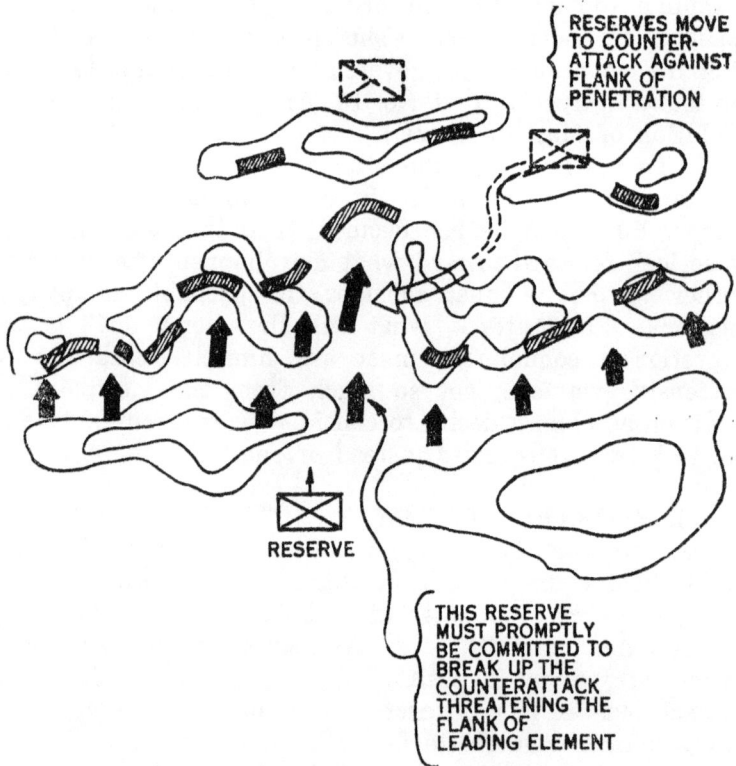

RESERVES MOVE
TO COUNTER-
ATTACK AGAINST
FLANK OF
PENETRATION

RESERVE

THIS RESERVE
MUST PROMPTLY
BE COMMITTED TO
BREAK UP THE
COUNTERATTACK
THREATENING THE
FLANK OF
LEADING ELEMENT

FIGURE 42.

been famous in warfare from ancient times.   They contributed measurably to the double envelopment and defeat
of the Russian army at Tannenburg, in which 92,000
prisoners were captured when the pincers closed and cut
off their retreat.

  c. In an attack against a defensive zone, the enemy may
hold his first defenses only briefly, making his main stand

at positions farther back. This elastic method of defense was often employed during the latter part of the World War, especially by the Germans during the last few months.

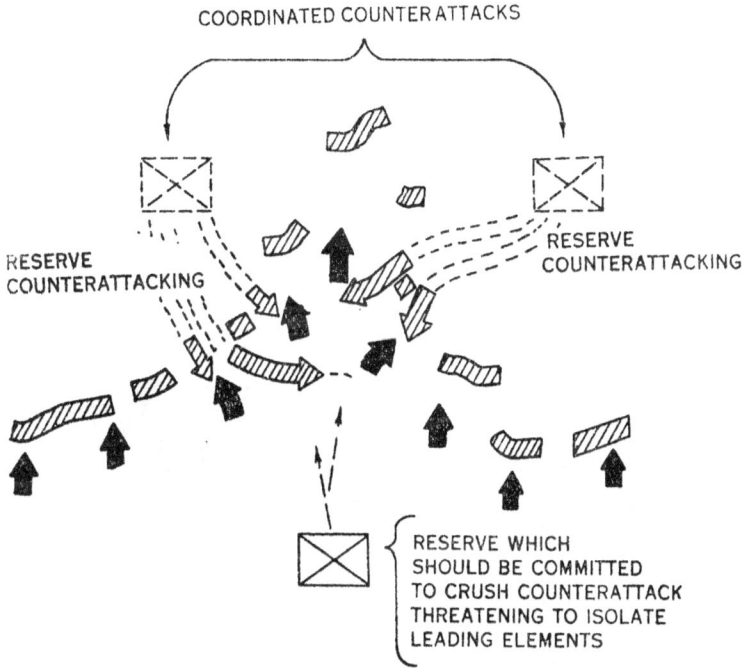

FIGURE 43.

d. At any time during an attack, moreover, the enemy may counterattack. A detailed consideration of this and the other eventualities of the attack are considered later in this text.

41. *PROBABLE ENEMY REACTION TO AN ENVELOPMENT.*—*a.* When the enemy discovers a move-

ment of our forces to effect an envelopment of his flank,
he may meet the threat by—

(1) Extending his front.
(2) Refusing his flank.
(3) Counterattacking.
(4) Withdrawing.

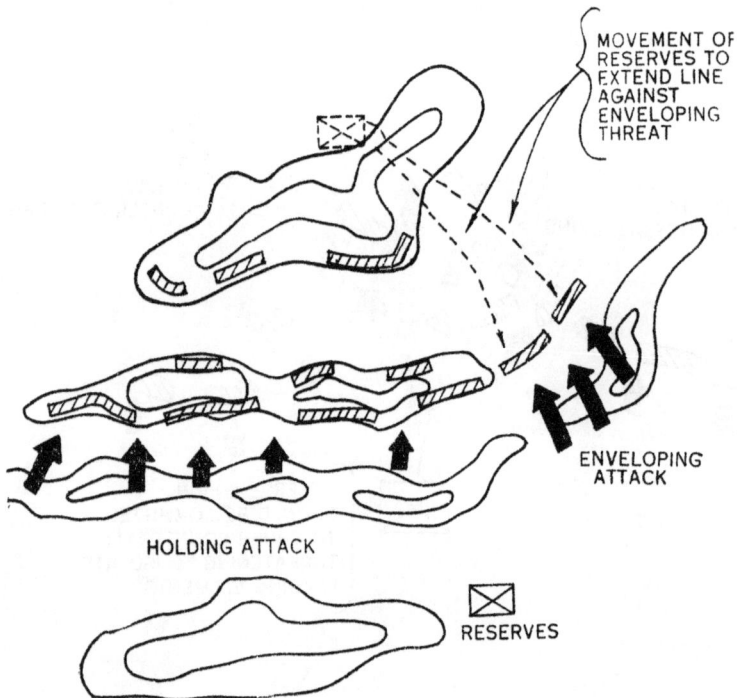

FIGURE 44.

b. The enemy extends his front by placing additional
units in position as shown in Figure 44, thus enlarging
the perimeter of his defenses in the direction of the envelopment with a view to meeting and obstructing it, or,
if the envelopment is wide, cutting it off, and separating

the enveloping and the holding force by a vigorous counterattack.

c. When the enemy refuses his flank, he simply bends
his front back at the threatened flank, or strengthens that

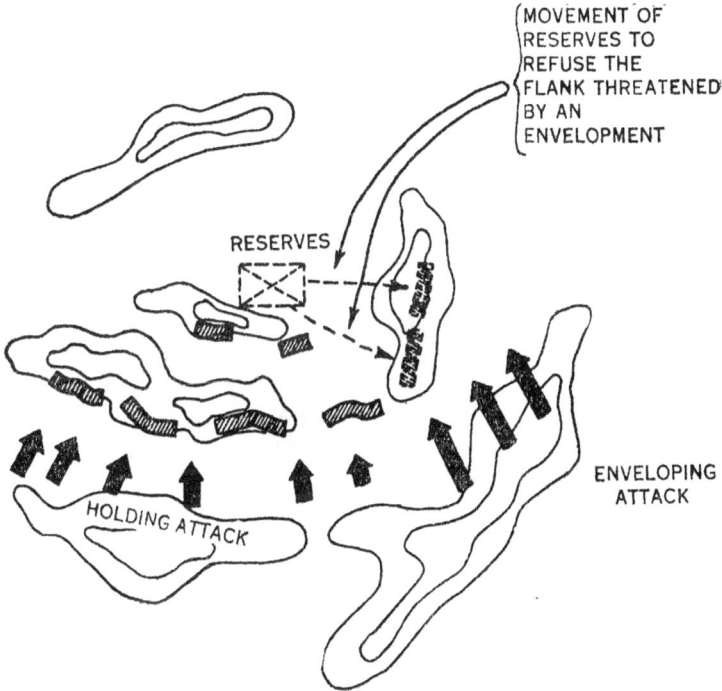

MOVEMENT OF
RESERVES TO
REFUSE THE
FLANK THREATENED
BY AN
ENVELOPMENT

RESERVES

ENVELOPING
ATTACK

HOLDING ATTACK

FIGURE 45.

flank so that his defenses are as stout to an attack from
that direction as from the front.

42. *HISTORICAL EXAMPLE OF AN ATTACK
MEETING A REFUSED FLANK: THE RUSSIAN
109TH INFANTRY AT GRAND KANNAPINNEN, 19
AUGUST 1914.*—*a.* On 19 August 1914 the advance guard

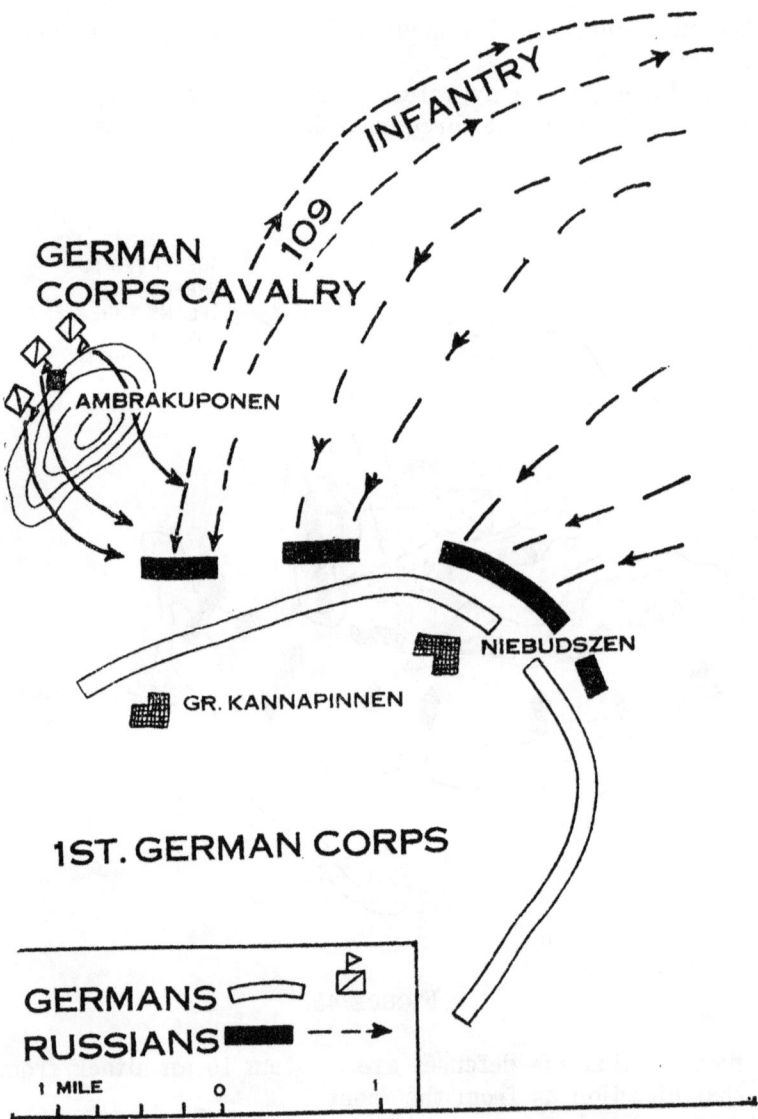

GERMAN
CORPS CAVALRY

109

INFANTRY

AMBRAKUPONEN

NIEBUDSZEN

GR. KANNAPINNEN

1ST. GERMAN CORPS

GERMANS
RUSSIANS

1 MILE   O   1

FIGURE 46.—Grand Kannapinnen, 19 August 1914

of the Russian XX Corps encountered that of the German
I Corps. The German forces in the East were generally
on the defensive, hoping to hold the Russian invasion at
bay until the war was won in France. The German advance guard fell back to a prepared position on strong
ground as shown in Figure 46.

b. The Russians attacked this postion in the direction
shown. The 109th Russian Infantry found what appeared to be a flank of the German position at Niebudszen, and the units north of this corner advanced rapidly
west because there was nothing to the front. On receiving fire from the south, however, they began to turn in
that direction, and by the time the leading elements of
the 109th were opposite Grand Kannapinnen, they were
actually attacking south instead of west. Moreover, the
whole regiment was so badly extended along the refused
German flank that, opposite the corner at Niebudszen, it
had become extremely thin. Before the German infantry
could take advantage of this fact, however, the German
Cavalry Corps, which had been observing the extension of
the Russian regiment from behind the hill at Anttsra-
kuponen, charged the 109th Regiment in flank and rear,
driving it back in disorder. The rout was completed by
heavy fire from the infantry on the south flank of the
Russian regiment.

CHAPTER 8

THE PLAN OF ATTACK

1. *GENERAL.—a.* A plan of attack is a statement of the decision of the commander and a detailed statement of his scheme for the use of each element of his command in carrying out that decision. The plan is usually given in paragraphs 2, 3, 4, and 5 of a formal attack order or in the corresponding paragraphs of the oral or dictated order. Paragraph 2 contains the decision and certain coordinating features, paragraph 3 contains detailed instructions for the conduct of each combatant element of the command, paragraph 4 contains instructions relative to supply and evacuation, and paragraph 5 contains instructions relative to signal communications and command posts.

b. In his own mind, as we have seen, an alert commander may tentatively plan his actions far ahead, or make numerous plans. But when he has decided on a specific course of action, he limits the announcement of his plan only to the immediate future, and does not issue detailed instructions farther in advance than the action reasonably can be foreseen. The announced plan for an attack, accordingly, contains specific directions to all units that are to take an active part initially, and usually a mere statement of the fact that other units are to be held in reserve. When it is time, later on, to use these units, the commander issues, in the usual case, another order and plan covering their action. He also announces, in new orders, fresh plans for future action to units already fighting, whenever this is necessary. For small units, such as a company or battalion, a new plan is often

141

necessary for each successive objective, especially in a
warfare of rapid movement. Orders covering such plans
are oftentimes simply brief oral instructions or messages.

c. A good tactical plan must be based on the latest
enemy information obtainable. The plan should provide
for secret preparation of the attack, surprise in its delivery, and skill and vigor in its execution. All attack
plans, as explained in the preceding chapter, provide for
holding part of the enemy with minimum forces, while
concentrating the greatest possible force at the point of
decisive action.

d. An attack plan should by all means be simple and
easy of execution, and once put into effect, adhered to
with tenacity. A complicated plan, involving a difficult
scheme of maneuver, offers too many chances for mistakes
and errors. After a force has once begun its attack, major
changes in plan, especially those involving the units already engaged, are not only difficult, but often dangerous
to effect because of their disruption of control and coordination. Such changes, therefore, are to be avoided. Sometimes, however, unexpected turns of the battle necessitate
partial or even complete changes in a commander's original plan of attack.

e. A plan for attack should permit a reasonable degree
of latitude and initiative to subordinate commanders in
the execution of their missions. A commander does not
prescribe exactly how his subordinates will do their parts,
unless other factors of importance hinge on a specific
performance by a certain unit, as is sometimes the case.
Nor does a commander, except when he deems the matter of great importance, include in his plan detailed orders
for units below the next subordinate units. A division
commander, for example, would seldom give specific orders
involving infantry battalions.

2. *SCHEME OF MANEUVER.*—a. An attack plan includes a definite scheme of maneuver, or statement of
how troops are to operate to obtain a given tactical effect.

The scheme of maneuver, based in large part on the best use of terrain possible, assigns tasks and prescribes the movements of elements of the command in order to secure the tactical effect desired. It is the fire power and shock action of the units primarily that gain this effect.

b. As we saw in the preceding chapter, a commander must study the terrain occupied by the hostile force and also the terrain over which his own troops must pass in order to attack. He must, in fact, analyze the whole ground of the battlefield and its approaches to determine how he can use it best from the viewpoints of observation, fields of fire, concealment, defilade, movement, and communication. His plan of attack should then be developed so that all possible features of the terrain are used to the advantage of his own force and to the disadvantage of the enemy.

c. A plan of attack also gives each immediately subordinate unit an objective or battle task. When necessary each such unit also receives definite instructions as to the assistance it is to render neighboring units in the attack. All specific tasks assigned to subordinate units should contribute to the success of the plan of the next higher unit.

3. *ZONES OF ACTION.*—*a.* In an attack plan units are usually assigned zones of action, which are primarily areas of responsibility. Zones of action are not arbitrarily assigned by measuring the width of front of the attack and then dividing it equally, or otherwise, among the units making the attack. They are based instead upon a number of factors, chief of which are the terrain the attack is to pass over, the effective strength of units, the dispositions of the hostile forces, and the scheme of maneuver of the higher commander. This scheme—and hence the assignment of zones of action—should bring the preponderance of his strength to bear against the weakest points of the hostile defense.

b. Ordinarily, a defending enemy occupies high ground, and covers the valleys between by observation, fire and obstacles, rather than by an actual disposition of troops in them. Therefore, the best avenue of approach for the attack is in general along low ground. Woods, villages and the like must also be placed in the same category as high ground since they can be utilized by the enemy as positions for weapons that place effective fire across the attacker's advance.

c. In the same way that a defending enemy does not actually occupy the whole perimeter of his defensive with troops, an attacking unit, although responsible for the full width of its zone of action, does not necessarily physically occupy the whole width at any given time.

4. *TERRAIN CORRIDORS AND ZONES OF ACTION.* —The factors mentioned in paragraph 3*a*—the hostile dispositions, the scheme of maneuver of the higher commander, the terrain, the effective strength—make it highly desirable for a commander to assign zones of action which correspond as far as possible to terrain corridors running in the direction of attack. When this can be done, a unit advancing along one corridor is not subjected to hostile fire placed upon its advance by direct laying from outside its own zone of action. It is desirable, also, for zones of action for the attack to include, where possible, not only the terrain corridors themselves but as much of the terrain features forming the sides of the corridors as may afford positions to the enemy from which he can fire across the path of the attack. Thus zones of action generally include the crests of ridges or hills bordering a corridor (See Figure 47.), and the adjacent edges of woods and villages, where it is possible to include them. At times, however, it may be essential to give the mission of capturing a certain terrain feature—a hill, a wood, a village—to a single unit. In that case, the zone of action assigned to that unit must include the particular terrain feature in its entirety.

5. *TERRAIN COMPARTMENTS AND ZONES OF ACTION.*—In assigning zones of action, a commander al-

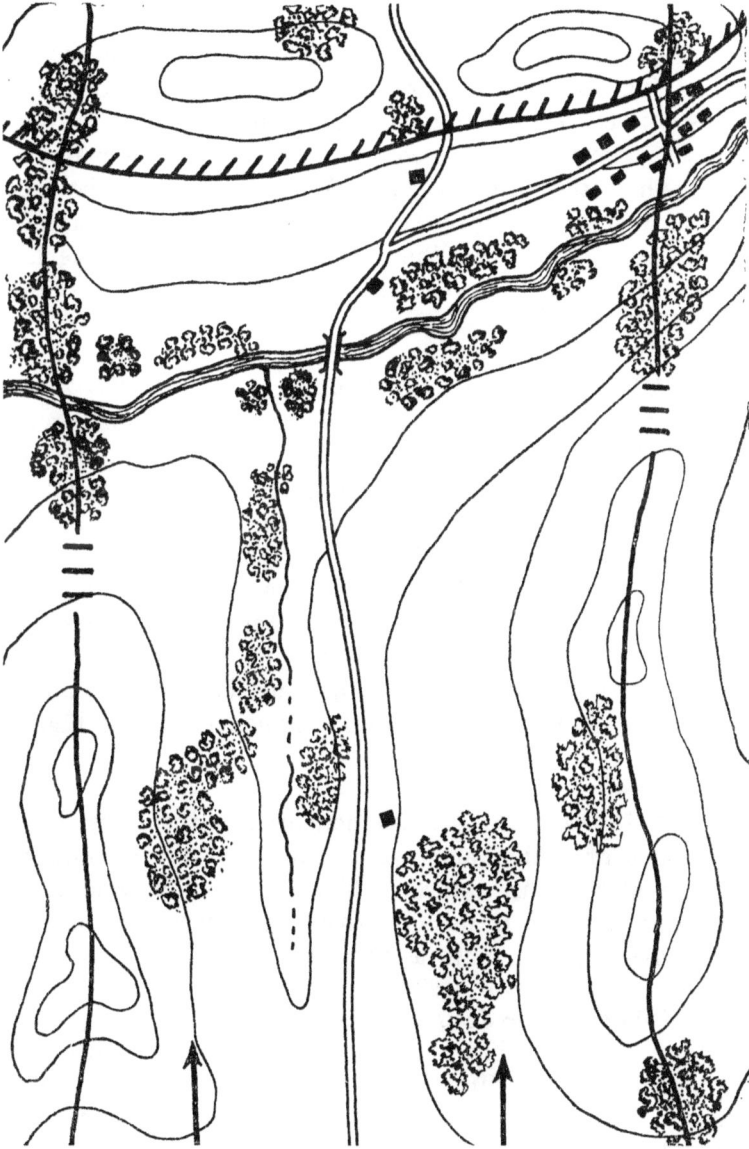

FIGURE 47.—Boundaries; zones of action.

so considers the possible uses he may be able to make of the terrain compartments within the sector, or area, over which his command must attack. The high ground, woods, etc., lying across a zone of action and forming the divisions between compartments, often afford defilade initially for the movement of troops to the lines of departure, from which they begin the attack. They may also be used as phase lines or objectives upon which the attack may be directed. Moreover, they afford defilade from the fire of hostile supporting weapons, or give cover for reorganization for a renewal of the attack.

6. *PERFECT GROUND FOR AN ATTACK.*—Zones of action for the elements of his command coinciding with perfect terrain corridors, and perfect compartments of terrain to aid the attack, are never found. A commander does the best he can with the terrain he has. He ordinarily has a considerable variation of terrain and hence a considerable choice as to how he will use and assign it. Frequently he will find, by close study, terrain corridors and compartments that fairly suit his needs. In whatever use he makes of terrain, of course, he must be governed above everything else by the scheme of maneuver of the higher commander.

7. *BOUNDARIES.*—*a.* Boundaries are the lines defining zones of action and are specified in a commander's plan of attack. An exterior (flank) unit of several units attacking abreast is usually given no exterior boundary. This gives such a unit a greater freedom of maneuver than that of interior units.

b. A commander designates boundaries by naming visible and unmistakable terrain features (Figure 48)* or by announcing the direction of attack from a given point, line, or area (Figure 49). When such terrain features are present and the boundaries of attack are parallel and

*The method of designating each type of boundary illustrated is given in the legend of the illustration.

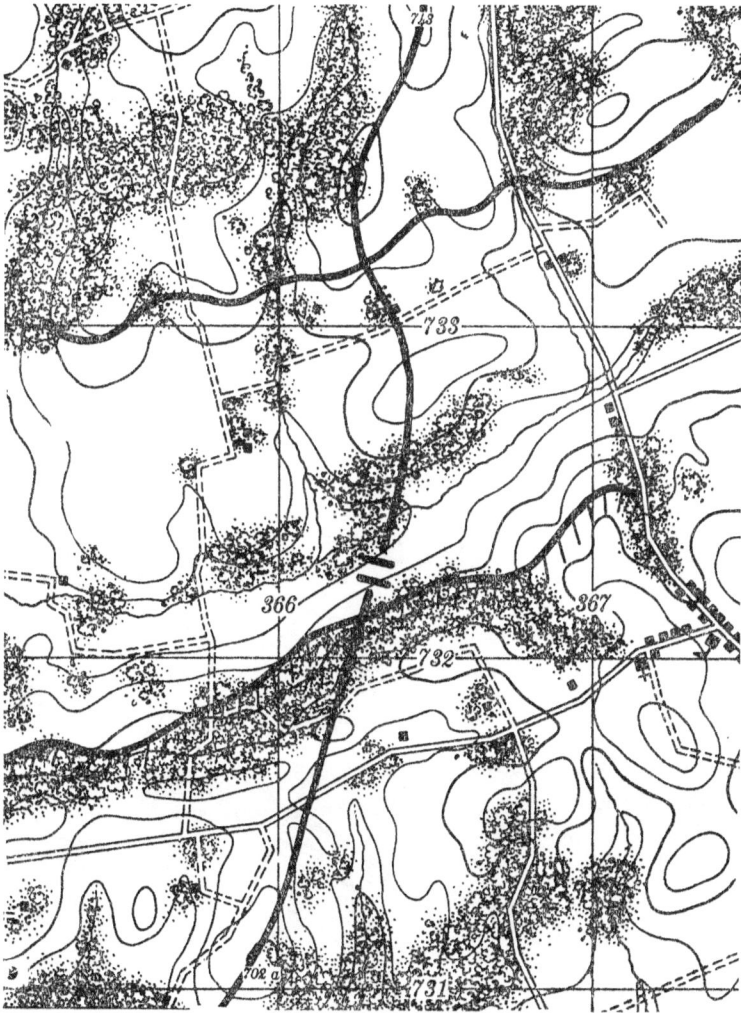

FIGURE 48.—Designation of a boundary.
Boundary between battalions: hill *743*—hill 500
yards southwest of hill *743*—house at (366.4-733.1)
—hill at (366.2-731.8)—hill *702-a*.

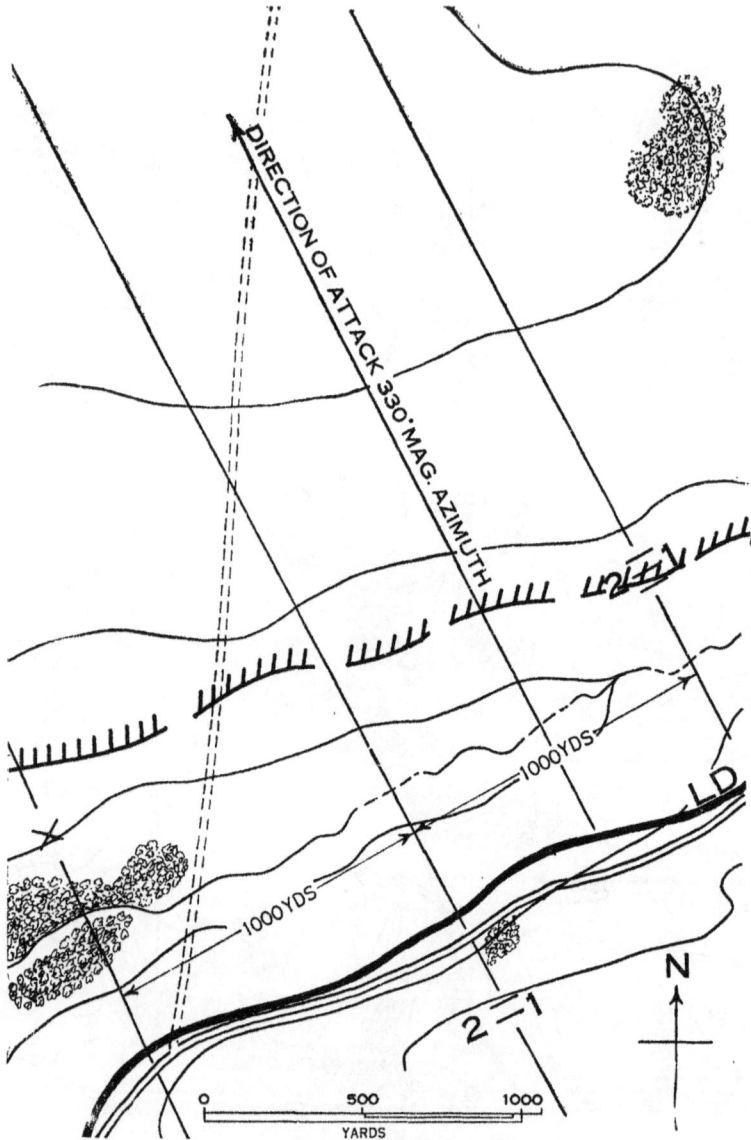

FIGURE 49.—Boundary lacking distinguishable
terrain features.

Boundary between battalions: Line through clump of trees on
line of departure; magnetic azimuth, 330°, 1st Battalion to attack
on a frontage of 1000 yards to the right (east) of initial point.
2d Battalion to attack on a frontage of 1100 yards to the left
(west) of initial point. Direction of attack, 330° magnetic
azimuth.

straight, he gives the direction of attack in addition to the boundaries, in his plan of attack. When boundaries cannot be indicated by readily apparent features of the ground, the commander designates the point on the line of departure (See the next paragraph.) where the right or left of the attacking unit is to rest, and gives the unit a frontage and direction of attack. If neither apparent terrain features nor frontage and specific direction of attack can be given, then the commander indicates a point or area from which the unit is to attack and designates an objective for the unit (Figure 50).

c. Boundaries are not simply designated forward from the line of departure into the enemy's teritory. They are prescribed far enough to the rear of the line of departure to include all combat elements of the attacking units and as far forward into the enemy's dispositions or battle position as the commander can foresee the development of the action.

d. As was inferred above, it is usually desirable for boundaries to coincide with the sides of terrain corridors leading toward and into the hostile position. But they should not divide important tactical localities held by the enemy when the capture of such a locality can best be accomplished by a single unit of the size of those for which a commander is designating zones of action. Thus in Figure 51, the division commander is careful to include hill A within the zone of a single brigade because he thinks it will be a task for a brigade to handle, rather than for his whole division. The brigade commander in turn may decide that a single regiment can do the job, in which event he would place hill A entirely within a regimental zone as shown in the figure. But if he believes both of his regiments may be needed, he may run the boundary between regiments directly through the important locality, thus indicating that the regiments will cooperate in taking hill A. Or he may give hill A to one regiment, and specifically direct the other to assist in its capture, and also specify the manner of assistance.

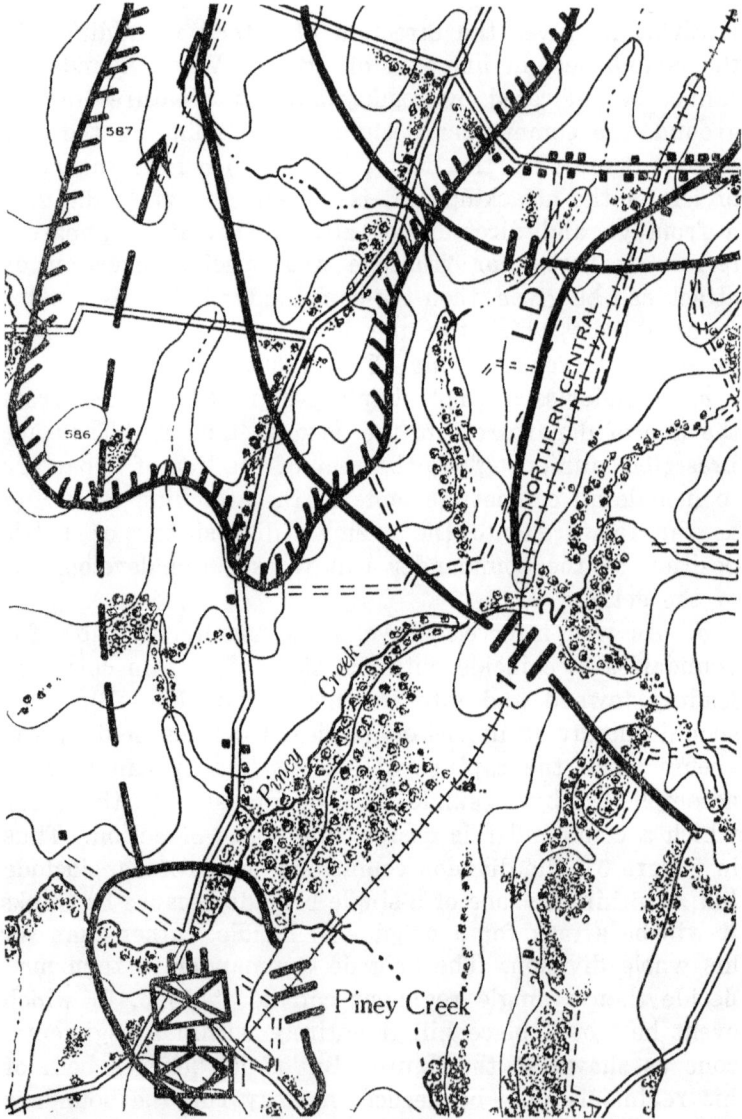

FIGURE 50.—An attack assignment without an announced line of departure. The 1st Infantry, from the vicinity of Piney Creek, will attack north, capture hills *586* and *587*, and drive the enemy to the north.

FIGURE 51.—Inclusion of an important tactical locality
entirely within the zone of action of a single unit.

8. *WIDTHS OF ZONES OF ACTION.—a.* Thus, it is generally true that the width of a zone of action assigned to a unit for an attack should be consistent with the mission of that unit. A unit assigned a decisive mission, requiring a deep advance, such as the main effort of a penetration, is given a relatively narrow zone because its elements must be disposed in depth in order to secure sufficient strength for successive impulses. But for a unit with a holding mission the reverse is true. The effective strength of a unit is also important in determining the width of its zone of action.

b. As we shall see in more detail in Chapters 9 and 11, an attacking unit does not necessarily cover its zone of action with a continuous line, extending from boundary to boundary, but operates with small combat groups— fighting teams—along parts of its zone, covering the remainder by fire, observation, or patrols. (This method is similar to that employed by advance guards when contact is imminent as described in Chapter 6.) The parts of its zone of action not actually occupied and passed over by elements of an attacking unit are usually one or another of the following: areas swept by hostile fire, gassed areas, gaps purposely left between elements of the command to permit the use of supporting fires, or gaps left in order to make possible a planned maneuver. Nevertheless, all hostile resistance in the zone must be definitely located, even if not attacked.

c. The formation to be used in any particular case depends upon the width of the assigned zone, the terrain, the resistance to be expected, the result to be accomplished, whether or not the flanks are protected, and on the organization and strength of the attacking unit. The various formations used in the attack are discussed in detail in Chapter 11.

9. *LINES OF DEPARTURE.—a.* A line of departure for the leading elements particpating in an attack is usually designated in the commander's plan. A line of

departure, in the usual case, extends across the zone of action of the unit. In his plan of attack the commander also states the hour at which the line is to be crossed by the attacking unit. The purpose of lines of departure is to coordinate the attack of assault units so that they strike the enemy in the order and at the time desired. In an envelopment, for example, the commander may desire to have the secondary attack begin two hours before the main enveloping attack. By designating lines of departure for both attacks, and the hour at which the leading elements of both attacks are to cross their respective lines of departure, he obtains the coordination he desires.

b. Lines of departure should be recognizable on the ground so that attacking units can find them with the least delay and be certain they have found the right line. Roads, small stream lines, railroads, edges of woods, and other well-defined terrain features often form lines of departure. (See Figures 49 and 51.) A commander should avoid designating lines of departure from untrustworthy or old maps, since woods, roads, and many other features may be different at the time of the attack. Much confusion resulted during the World War from the inability of attacking units to find nonexistent roads or no-longer existing woods edges that were indicated on maps.

c. Lines of departure should be approximately perpendicular to the direction of attack; for unless a unit attacks in general directly toward its objective, it is likely to lose direction. Sometimes, however, troops must begin their attack from a line that is not perpendicular to the attack direction. (See Figure 52.)

d. Where there is choice in the matter, a line defiladed from hostile observation and fire is far better than one exposed to either or both. (See Figure 53.) Sometimes, however, a higher unit designates a line of departure part of which is very exposed. In fact, a unit may have to fight to reach its departure line. When this is the case, the subordinate unit commander should select a better line farther back and direct his unit to cross that line at

such an hour as to cross the line of departure at the
time designated by higher authority. (See Figure 54.)

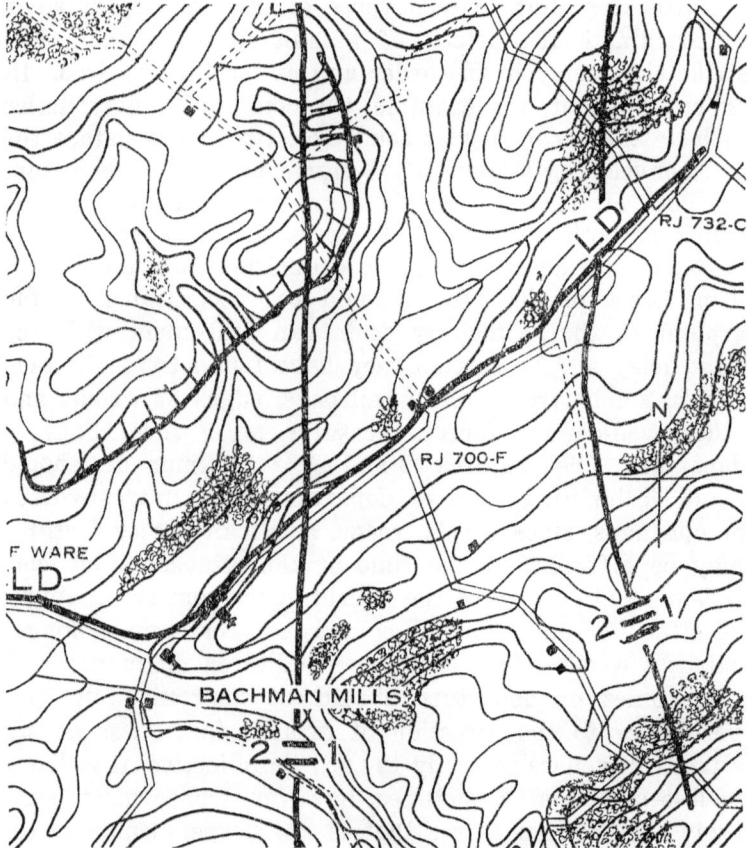

FIGURE 52.—Line of departure not perpendicular to
the direction of attack. *Line of departure*: the road:
RJ 732-C—RJ 700-F—Bachman Mills—F. Ware. *Direction of attack*: North.

e. When a readily recognizable line of departure cannot be designated because none exists, a commander may
designate an area from which the unit is to launch its
attack.

10. *THE HOUR OF ATTACK.*—*a.* In designating the hour for an attack, a commander must, of course, make allowance for the time required for numerous preliminaries to battle. The units that are to enter the fight

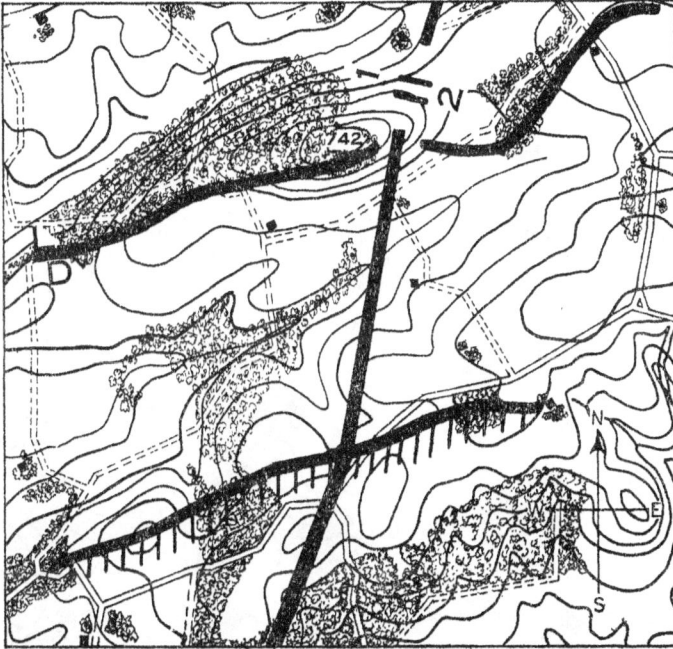

FIGURE 53.—Defiladed line of departure.
Line of departure, 1st Battalion: Southern edge of woods extending from hill 742 along spur to the northeast.

initially must reach their lines of departure. They each have a certain distance to go. One unit may be able to reach its line of departure largely on trails and roads and across cleared terrain, while another unit is making slow progress through mud or woods, across stream beds, or over rough broken ground. One unit may come under fire in gaining its line of departure while another may be able to follow well-defiladed approaches. Bad weather

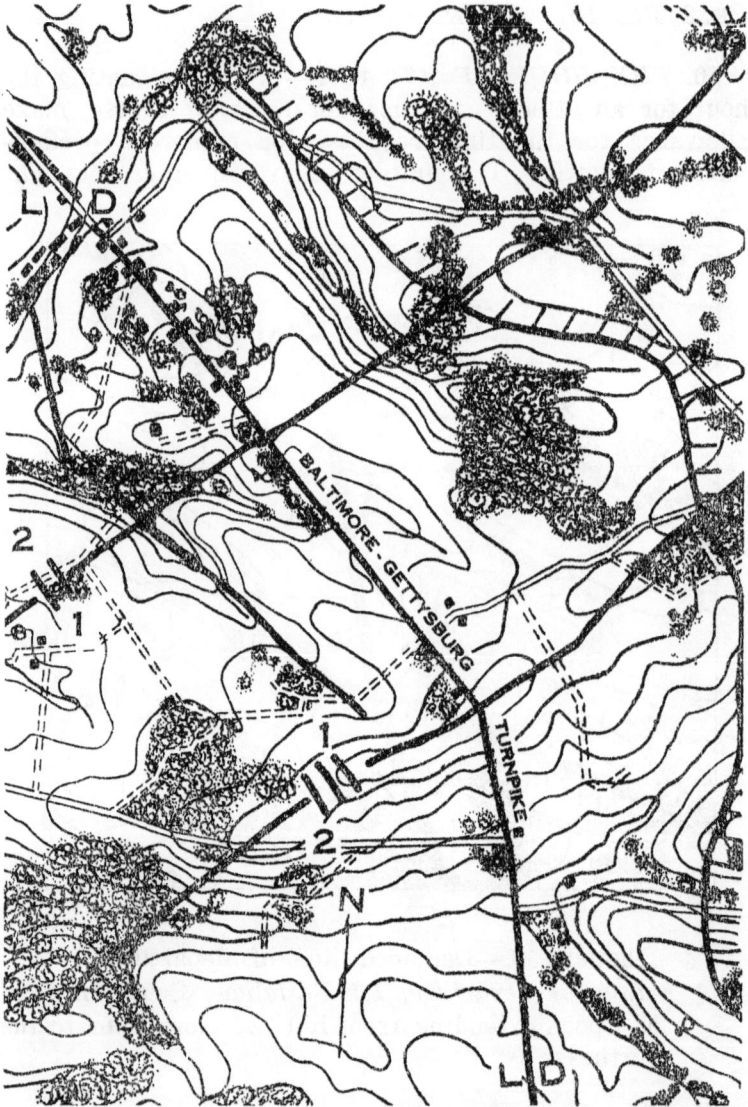

FIGURE 54.—An exposed line of departure. The regimental commander's order would state in the plan of attack: *Line of departure*: Baltimore-Gettysburg Turnpike. The plan of attack in the 1st Battalion commander's order would indicate also the preliminary line to the rear of the exposed line of departure.

may, in fact, slow up all movement. Moreover, all units not already completely deployed require some time for that purpose. Time is also necessary for each commander in turn down to platoon commanders to receive the orders from the next unit above, reconnoiter, and prepare and issue (or distribute) his own order. All these things must be allowed for. Their total usually runs into hours rather than minutes and for large units is actually a matter of one to several days.

b. But time, much more often than not, is a vital factor in the tactical situation. A commander needs to attack at the earliest possible moment to take advantage of an immediate opportunity which, in a few hours, will be gone. He may be able to strike at once in uncoordinated attack but realizes that a coordinated effort, if there is time to make it, will be far more effective. In such a situation, he must determine by calculation the earliest hour at which he can get his units in place to begin a coordinated attack.

c. To this calculation there are two approaches, both of which must be worked out. He must find—

(1) The hour by which all assault units (leading units of the attack) can reach and be ready to cross their lines of departures.

(2) The minimum time required by the commander and his subordinate commanders (to include platoon leaders) in turn to make their reconnaissances and issue their orders for the attack

11. *CALCULATING THE EARLIEST HOUR ALL ASSAULT UNITS (IN PARTLY DEPLOYED FORMATION) CAN BE READY TO ATTACK.*—*a.* As we have seen in the preceding paragraph, one unit, for various reasons, may take longer than another to reach its line of departure. It is evident, then, that if we find the time it will take the farthest away or slowest moving unit to reach its line, all others should be ready to attack before that unit is.

b. In calculating such movements the rates of movement shown in the table below are used:

RATES OF MARCH

UNIT	RATES OF MARCH (miles per hr)				REMARKS
	On Roads		Across Country		
	Day	Night	Day	Night	
Artillery					
Horse drawn	3½	3	3	2	
Truck drawn	20	15 (lights) 10 (no lights)	5	4	
Tractor drawn	3½	3	3	2	
Trains					
Animal-drawn	3½	3	1½	1	
Motor	14	12 (lights) 10 (no lights)	5	2	Under favorable conditions movements may be at considerably higher rates.
Infantry					
Foot troops	2½ (88)	2 (70)	1½ (53)	1 (35)	*Figures in parentheses are yards per minute and are to be used for calculating marches of less than a mile.*
Motorized elements	14	12 (lights) 10 (no lights)	5	2	Under favorable conditions movements may be made at considerably higher rates.

RATES OF MARCH (miles per hr)

UNIT	On Roads		Across Country		REMARKS
	Day	*Night*	*Day*	*Night*	
Tanks, fast	20	15 (lights) 10 (no lights)	20	2	Under favorable conditions movements may be made at considerably higher rates.
Miscellaneous Dismounted individuals or small dismounted groups	3	3	3	3	
Mounted individuals or small mounted groups	8	8	8	8	
Light passenger cars cross country	20	15 (lights) 10 (no lights)	10	2	Under favorable conditions movements may be much faster.
Motorcycle messenger	20	15 (lights) 10 (no lights)			

c. Factors affecting rates of march.—(1) The rates given in the table are generally lowered when such adverse conditions as the following exist: heavy dust, high temperature, heavy mud, snow or sleet, and icy roads.

(2) In cross-country movements of foot troops, swampy ground, heavy underbrush, or very broken terrain, lower the normal rates of movement. (Such terrain also, of course, affects movements of all types of units; but for the most part, it is to be assumed that units other than foot troops will avoid such ground.)

(3) In measuring the distance troops have to march on winding trails in hilly and wooded country, allowance must be made for a greater distance of march than that shown on the map. For scales of 1/21,120 or smaller the distance of march on such trails should be doubled to allow for the many small changes of direction not shown on the map.

(4) The rate of march of a unit composed of elements with different rates of movement is that of the slowest type of element.

(5) Allowance must be made for delay in changing from column of fours to column of twos or files at defiles where this is necessary.

(6) In motor movements of troops time is allowed, as shown below, for normal delays:

		Minutes
(a)	Loading personnel only	15
(b)	Loading materiel and personnel	30
(c)	Unloading personnel only	10
(d)	Unloading personnel and materiel	15
(e)	Placing empty trucks in condition to move after receipt of orders	15
(f)	Placing trucks loaded with supplies ready to move after receipt of orders	60

(7) Of items (a) to (f) above, it is apparent that one or more may apply to a given movement. Thus, an infantry rifle unit already in trucks would, in moving up to battle, only have to unload; and (c) would apply. If the same unit were marching, and trucks were sent to bring them up (e), (a), and (c) would apply (in addition, of

course, to travel time). For machine-gun units, and other units that have to load and unload their weapons and men both, (b) and (d) apply rather than (a) and (c).

ILLUSTRATIVE SITUATION

d. *Situation.*—(1) The Blue 1st Brigade, reinforced, detached from a larger force, is advancing on an aggressive mission. This morning Brigadier General 1st Brigade decided that contact was imminent; and as the brigade moved out of its bivouac area at 5:00 AM, it took up the partly deployed formation shown in Figure 55.

(2) General 1st Brigade knows that the hostile force he expects to encounter is apparently about two-thirds as large as his own brigade and, like the Blue brigade, contains no motorized or mechanized units.

(3) The Blue advance guards met the enemy's leading elements about 10:00 AM, shortly after leaving Phase Line B. By 10:30 the advance guard commanders had thrown in their reserves and had pushed forward to the line shown in Figure 55. The right advance guard has apparently found the hostile left flank and is making slight progress on the extreme right of the line of contact. At all other points the advance guards are making practically no progress.

(4) The enemy is apparently organizing as a defensive position the area shown. General 1st Brigade has just sent word to the advance-guard commanders that he is taking over the command of the advance guards as a coordinated whole. The elements of the main body are now in the areas shown in Figure 55.

(5) At 11:00 AM, General 1st Brigade, whose command post is being set up at X, near the 1st Infantry command post, receives further information regarding the hostile force, to the effect that the hostile left flank is actually where earlier indications have placed it. At 11:05, he makes a brief reconnaissance and decides (at

11:20) to attack using the 2d and 3d Battalions 1st Infantry, and the 3d Battalion 2d Infantry as the enveloping
force; the attack to be made from the line of departure
shown on the map. The holding attack is to be made by
the advance guards and the 2d Battalion, 2d Infantry.
Colonel 1st Infantry is to command the enveloping force,
and Colonel 2d Infantry, the holding attack.

(6) Colonel 1st Infantry is already at the brigade command post with part of his staff, and Lieutenant Colonel
2d Battalion 1st Infantry. He knows the whole brigade
situation and hears the decision of General 1st Brigade as
to the method of attack to be used. At the brigade command post are motorcycle messengers, the brigade commander's cross-country car and driver, and other command post personnel. The radio operator is in touch with
the 2d Infantry command post.

e. *Calculating the hour of attack.*—(1) It is now necessary for General 1st Brigade and his staff to find out the
earliest hour at which the attack can be made, so that this
hour can be included in the attack order. Let us see how
this is arrived at.

(2) It will take General 1st Brigade and his staff some
time to prepare the details of the order, which he will
issue orally to his regimental commanders. Colonel 1st
Infantry is already at hand and Colonel 2d Infantry must
be sent for. Colonel 2d Infantry is directed by radio to
report at once.

(3) In the meantime, there is no reason why Colonel
1st Infantry should delay getting his enveloping force
started toward its line of departure. Remaining himself
with General 1st Brigade, until he gets the detailed plan
of attack, he sends a staff officer with two runners, in
his cross-country car to pick up Lieutenant Colonel 3d
Battalion 1st Infantry and Lieutenant Colonel 3d Battalion 2d Infantry and bring them at once to the brigade
command post. He also sends instructions to these battalion
commanders to change the direction of march of their
battalions before coming to report. The 3d Battalion 1st
Infantry he directs to move toward the woods in rear of

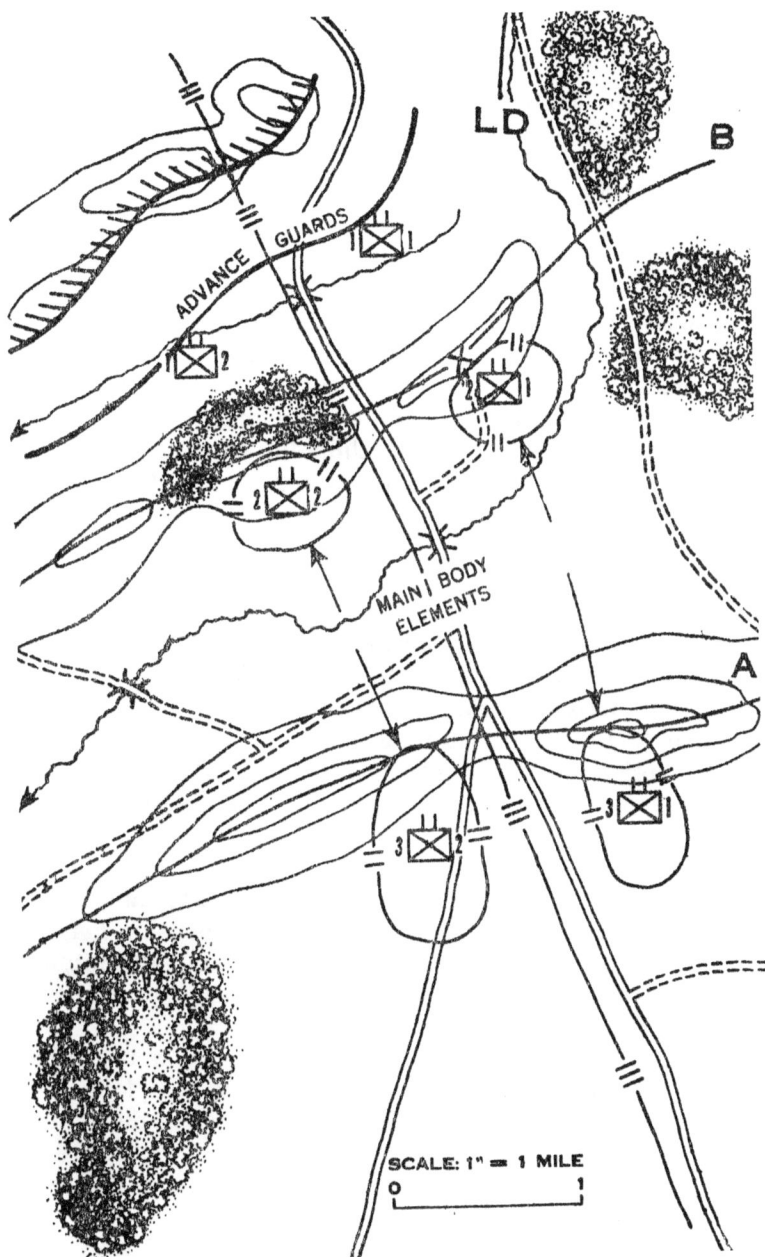

FIGURE 55.

the line of departure of the enveloping force. The 3d
Battalion 2d Infantry he directs to move to the woods
about 1000 yards east of the present position of the 2d
Battalion 1st Infantry, there to remain as the reserve of
the enveloping attack.

(4) It is the 3d Battalion 1st Infantry that has the
longest distance to march, however. General 1st Brigade
and his staff estimate the time it will take for this unit
to reach the line of departure, in the following manner:

Time required for Colonel 1st Infantry to
give instructions to a staff officer .... 5 minutes
Time required for staff car to go from the
1st Brigade CP to the 3d Battalion 1st Infan-
try (2.5 miles at 25 miles an hour) .... 6'   "
Time required to find Lieutenant Colonel 3d
Battalion 1st Infantry (estimated) 15   "
Time required for Lieutenant Colonel 3d
Battalion 1st Infantry to tell his executive officer to issue orders for change of direction 5   "
Time required for 3d Battalion 1st Infantry
to march across country to the line of de-
parture (2.4 miles at 1.5 miles an hour) 96   "
Completion of deployment at or near the
line of departure 15   "
                                                        _____
    Total ....142 minutes

(5) It was 11:20 AM when General 1st Brigade, in the
presence of Colonel 1st Infantry decided upon his general
scheme of maneuver. At 1:42 PM, then, 142 minutes
later, the unit that has the longest distance to go should
be on its line of departure.

(6) But while the troops are moving up, many other
things must be done. These, in fact, may take longer to
do, than the movement of units to their lines of departure.
Using the data in subparagraph (4) above, we can estimate the time required for battle preparation as follows:

Time required for Colonel 1st Infantry to
instruct staff officer .... 5 minutes
Time require for staff car to go from the
Brigade command post to the 3d Battalion 1st

Infantry .... 6 minutes
 Time required to find Lieutenant Colonel 3d
Battalion 1st Infantry ....15 "
 Time required for Lieutenant Colonel 3d
Battalion 1st Infantry to instruct his executive officer .... 5 "
 Time required for Lieutenant Colonel 3d
Battalion 1st Infantry to return in staff car
to the brigade command post .... 6 "

Total .... 37 minutes

(7) By 11:57 AM (37 minutes after 11:20), it is reasonable to assume, Colonel 2d Infantry will have reached the brigade command post and General 1st Brigade will have formulated his oral order. Beginning at this hour, then, we must make allowance for the actions and orders of each commander in turn, down to include the commanders of platoons. The time each will require varies greatly in different situations. Reasonable time assumptions, however, for determining the earliest hour of attack are as follows:

(a) For the brigade commander's reconnaissance, for the formation of his plan, and the preparation of his oral order .... 45 minutes

(b) For the issuance of the brigade oral order .... 15 minutes

(c) For the regimental commander's reconnaissance, for the formation of his plan, and for the preparation of his oral order .... 45 minutes

(d) For the issuance of the regimental oral order .... 15 minutes

(e) For the battalion commander's reconnaissance (after he will have arrived near his battalion's line of departure), for the formulation of his plan, and for the preparation of his oral order .... 20 minutes

(f) For the issuance of the battalion oral order .... 15 minutes

(g) For the company commander's reconnaissance .... 10 minutes

(h) For the issuance of the company oral order, for the reconnaissance of platoon leaders, and for the issuance of the platoon oral orders .... 15 minutes

(i) For the deployment, or completion of deployment, of the rifle companies .... 15 minutes

Total time for brigade ....3 hours 15 minutes.

(8) Of these nine steps, only the first will have been completed by 11:57 AM. Hence, if all of them must be completed, 3 hours and 15 minutes less 45 minutes, or 2 hours and 30 minutes, from 11:57, all preparations for the flank attack will have been accomplished. The hour for the flank attack will have been accomplished. The hour for the attack, then, is 2:27 PM.

(9) From this second calculation we see that it would probably take longer for all commanders to reconnoiter, form plans, draw up orders, and issue them, than for the most distant troops to reach their line of departure.

f. Additional factors.—(1) Sometimes, however, it is possible to make short cuts. The brigade commander and his staff might consider in detail whether all of the time allowances in subparagraphs (g) to (i) above would be necessary. The terrain sometimes favors rapid reconnaissance, for example. The use of accurate new maps, or aerial photographs, or a previous knowledge of the terrain on the part of commanders may do away with the necessity for a personal reconnaissance of the terrain. It is often possible for the commanders of more than one unit echelon to make their reconnaissances simultaneously and for brief oral orders to be given as the reconnaissance is made, especially commanders of small units; for example, a battalion commander and his company commanders. Again, reconnaissance in a cross-country car is much faster than on foot.

(2) On the other hand, many unforeseen minor delays may occur, and should ordinarily be allowed for. As an example, a point of observation from which a commander desires to examine the ground over which his unit is to

attack, may prove more inaccessible than it appeared. But when the point of observation is close at hand and readily reached, it is definitely known that the reconnaissance will take less than the time ordinarily allowed. For such unforeseen circumstances 30 minutes is a reasonable allowance for a commander to make, who deems it safest to do so in a given situation.

(3) In finding the earliest possible hour of attack to be 2:27 PM, as we did above, we assumed for the sake of discussion that the 3d Battalion 1st Infantry must reach its line of departure before the attack can begin. In all probability, however, in the situation assumed, Colonel 1st Infantry would carry out his attack with his assault battalions in column. This being the case, it may well be possible that time can be saved by having the 2d Battalion 1st Infantry attack as the leading assault battalion before the 3d battalion 1st Infantry arrives in rear. In other words, it seems possible for the 2d Battalion to begin the attack before the 3d Battalion comes up.

(4) To check this possibility, General 1st Brigade and his staff must calculate how soon the 2d Battalion 1st Infantry can reach the line of departure, and also how long it will take subordinate commanders to reconnoiter, form oral orders, and issue them as before. It is apparent from the short distance the 2d Battalion has to go, in order to reach the area just in rear of its line of departure—a much shorter distance than we saw the 3d Battalion had to march—, that the reconnaissances, etc. will take more time than the movement of troops. Hence, it is actually necessary to calculate only the time required by the commanders to make their combat preparations. Then, beginning at 11:25 AM, when Colonel 1st Infantry finishes giving instructions to the staff officer whom he sent to find Lieutenant Colonels 3d Battalion 1st Infantry and 3d Battalion 2d Infantry (subparagraph e(3) above), we can assume (with General 1st Brigade and his staff) that the following steps and periods of time will be necessary before the 2d Battalion 1st Infantry can attack:

Colonel 1st Infantry to instruct Lieutenant
Colonel 2d Battalion 1st Infantry to get his
battalion moving toward the woods in rear of
the line of departure .... 3 minutes
 Lieutenant Colonel 2d Battalion 1st Infan-
try: reconnaissance, formulation of order and
issuance of order (except for hour of attack) 35 minutes
 Company commanders, reconnaissances and
orders, platoon leaders, reconnaissances and
orders, and deployment of rifle companies ...... 40 minutes
 For a message to reach all commanders announcing the hour of attack (this last item
may be possible of accomplishment concur-
rently with the preceding items; but it is best
to make a reasonable allowance for it) .... 30 minutes

Total ....108 minutes
 (5) Thus, in 1 hour and 48 minutes after 11:25 AM,
or at 1:13 PM, the 2d Battalion 1st Infantry can attack.
 (6) But in his previous calculation (subparagraph *e*),
General 1st Brigade found that the 3d Battalion 1st Infantry could not attack until 2:27 PM, 1 hour and 14
minutes later. This is too long a delay for the succeeding
elements of a regiment attacking in a column of battalions to follow the leading. As we shall see later in this
text, 30 minutes is as much as a commander can safely
allow in such a situation. Hence, the earliest moment at
which the 2d Battalion 1st Infantry can attack is 30
minutes before 2:27 PM, or at 1:57 PM. General 1st
Brigade would therefore prescribe 1:57 PM as the hour
of attack.
 g. Artillery and other elements.—(1) In the foregoing
illustrative situation, there has been no mention made of
artillery, chiefly in order to keep the situation simple and
clear. Ordinarily, when routes to its positions are
available, artillery can be expected to reach its positions
before infantry, because its rate of march is faster. In
the situation discussed, General 1st Brigade could very
probably assume this to be the case, especially since the

infantry elements are clear of the roads. (When infantry is on roads that artillery must pass over to reach its firing positions, artillery is given the right of way; i.e., infantry must use the sides of the road to let the artillery by.)

(2) But sometimes artillery may have such distances to traverse or, if the infantry is motorized, may be the slowest moving of the combat elements. In that event, the commander of the whole force would calculate the hour of attack on the basis of the artillery's arrival in position.

(3) Other elements, too, may enter in. For example, a stream may have to be bridged; tanks may have to clear or pick a way through a wooded stretch; chemical elements may form the basis of the attack hour calculation; and sometimes the holding attack may take longer to arrange than the main effort. In fact, we have considered only the principal of the many factors that a commander may have to take into consideration himself, in estimating when he can begin his attack.

(4) In large units, naturally, the hour of attack must be worked out for days, rather than hours ahead. Moreover, a score of details must be allowed for where one must be considered in a small unit.

12. *CALCULATION OF THE EARLIEST POSSIBLE HOUR OF ATTACK FROM ROUTE FORMATION.*— When a force encounters the enemy unexpectedly, while it is still in route formation, the calculation of the earliest possible hour of attack ordinarily involves one step not necessary in a calculation for the same purpose involving a force already partly deployed. In route formation the elements of a force are in column one behind the other. If the commander decides to attack, he must, as we saw in detail in Chapter 4, wait for the units to come up in succession. At the hour he makes his decision, he is not likely to know just how far back every unit in the column is. But if he knows where the head of the main body is at that time he can measure back down the column, locate approximately the last unit that has to come up to its line

of departure before the attack can begin, and then calculate how long it will take for this unit to move up.
(A competent staff should have this data continuously
at hand, so that a commander can know without delay
how long it will take any unit in the column to reach the
point where the head of the main body is at any given
time.)

13. *SUPPORT FIRE PLANS.*—In most situations, it is
impossible for infantry to advance against even weak resistance unless that resistance is subjected to and reduced
by fire. Infantry units employing rifles and light automatic weapons should be advanced as close as possible to
the enemy without using their weapons, before having to
continue their advance under the protection of their own
fire. To this end, from the time they leave their lines
or areas of departure, infantry units must be closely supported by artillery, heavy machine guns, infantry mortars
and tanks. Attack plans must provide for continuous
support by prescribing initial tasks for these weapons.
Since information of the hostile dispositions is usually
vague at best, initial tasks are assigned as the situation
develops and the needs of the assault units are determined.
When, because of long range, fog, or other circumstances,
the infantry can no longer obtain the support of the artillery, it must rely chiefly on its own supporting weapons
—heavy and light machine guns, infantry mortars, and
tanks. These matters are discussed much more fully in
the next chapter.

14. *RESERVES.*—A commander also holds out a reserve, or reserves, in the initial stages of an attack in
order to insure the forward movement of the attack if it
be halted, to exploit the success of the attack, or to meet
unforeseen contingencies. The availability of a formed,
fresh body of troops that can be used where needed most
is of utmost importance and, in past history, has often
turned the tide of battle. The size and initial location of
the reserve are determined by the size of the force en-

gaged, the hostile situation, the mission, and the scheme of maneuver. The development of the situation determines where the commander subsequently moves and employs his reserve.

15. *FIRE PLANS FOR RESERVE UNITS.*—In his plan of attack a commander may give initial tasks of assisting the assault units to the supporting weapons of units in reserve. In assigning such tasks, the positions assigned to the weapons and the tasks themselves must be such that the weapons are always readily available to the reserve, at or before the time it is to be employed.

16. *FLANK PROTECTION.*—Each unit commander is charged with the protection of his flanks. This flank protection is usually performed by patrols. As a general rule, a unit making an enveloping attack is not charged with the flank protection of a whole force. If a commander deems additional security necessary for his whole force, over and above the security indirectly afforded by the enveloping unit itself, he uses part of his reserves for that purpose.

17. *ANTITANK PROTECTION.*—When the enemy is known to have tanks, a commander must include in his plan directions for the initial use of all antitank weapons that he has available. He must carefully study the enemy dispositions and the terrain in order to determine the enemy's most probable use of tanks. He then places his antitank weapons in position to cover possible tank approaches on his flanks and rear, or allots them to his assault units for use against possible counterattacks, or both.

18. *FLANK CONTACT.*—Each unit commander should establish and maintain contact with units on his flanks. This is usually accomplished by the use of patrols, which are specifically designated in a commanders plan. This matter is covered in detail in the next chapter.

19. *COOPERATION.*—*a.* Full cooperation between the various arms or weapons under one commander must be insured by the plan of attack. Subordinate commanders must do everything within their power to cooperate with the arms which are supporting them, or which they are supporting, as well as with the commanders of adjacent units.

b. The forward movement of infantry can only be insured by full cooperation between the units themselves, and between them and the supporting arms. Because of the uncertainty, however, as to the form and strength of the hostile resistance that must be overcome, initial arrangements for cooperation between the various arms can only be made on very broad lines.

20. *ADMINISTRATIVE MATTERS.*—*a.* Details of the plan of attack thus far covered in this chapter are usually contained in paragraphs 2 and 3 of a formal field order and in the equivalent parts of a dictated or oral order. But in addition to the tasks and directions given to the combatant elements of the command, certain administrative arrangements (usually placed in paragraph 4 of a formal field order) must be considered in the plan of attack. This is especially true in large units. Troops cannot fight without ammunition and they cannot make continued efforts for any great period of time without food and water. The collection, care and evacuation of the wounded must be provided for. The success of the plan of attack may be dependent upon engineering work—bridges, methods for crossing rivers, repair or building of roads—and this must be considered and included. The administrative portion of the plan is prepared with a view to assisting the main attack in every way, thus furthering the chances for success of the command as a whole.

b. The control of the operations and transmission of orders and information (usually given in paragraph 5 of a formal field order) are also an essential part of a plan of attack. The communications system is vital in obtaining coordination and cooperation. A plan of attack should, therefore, provide for as complete a system of signal communication as time, the means available, and the situation permit.

CHAPTER 9

THE CONDUCT OF THE ATTACK

SECTION I

GENERAL CONSIDERATIONS

1. *INTRODUCTION.*—In this chapter we take up the conduct of the attack. This section deals briefly with the conduct of specific forms of the attack, as we studied them in Chapter 7, and Section II covers numerous details regarding the conduct of a main attack, whether of a penetration or envelopment.

2. *THE CONDUCT OF A PENETRATION.*—*a.* In a penetration there is a main attack and a secondary or holding attack, which may be composed in turn of principal and secondary efforts, as in the envelopment. The holding attack in a penetration is often directed against a hostile exposed flank in order to make the enemy overextend his position and thus weaken the force he opposes to the main penetrating attack.

b. The main attack must be made in sufficient strength to cause a break-through or rupture of the hostile forces. It must also be made on a front wide enough to permit reserves to be maneuvered in the area penetrated. The penetrating force, in addition, must be disposed in great depth to insure continuity of effort.

c. Reserves should be located so that they are available to support either the main or secondary attacks. (See Figure 56.) A general reserve should be held ready to exploit the success of the attack after the penetrating force has broken through the hostile dispositions.

175

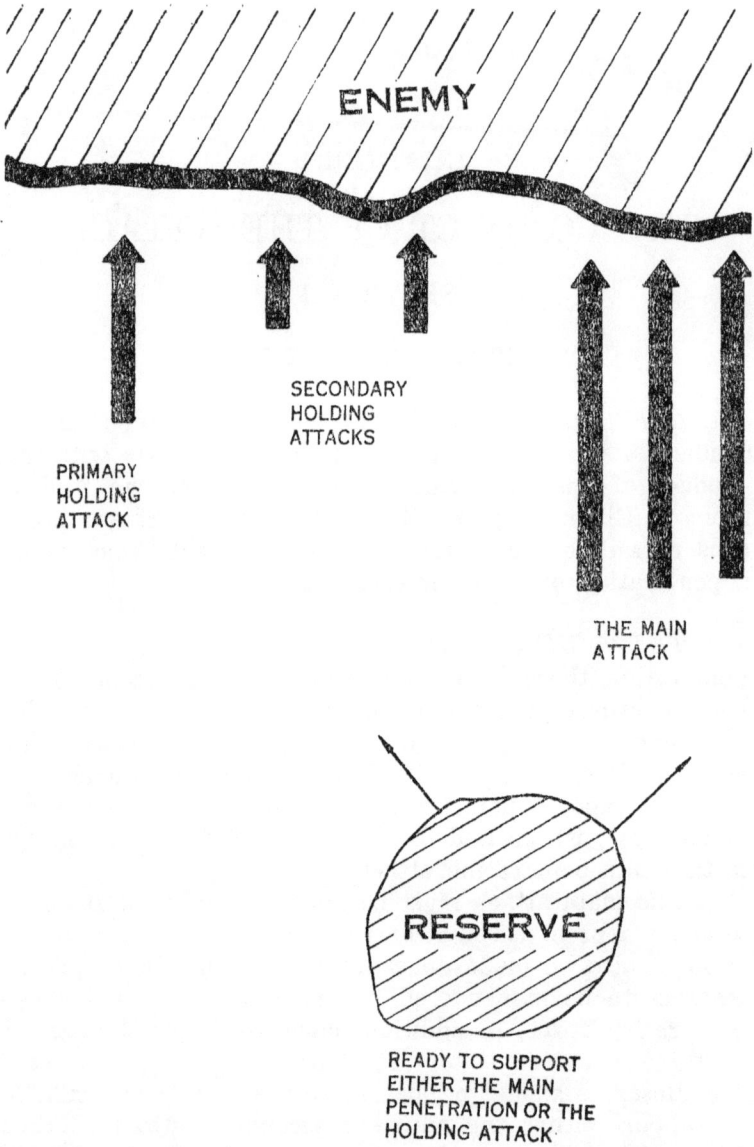

FIGURE 56.—Location of reserve units so that they can support either the main penetration or the holding attacks.

3. *THE CONDUCT OF AN ENVELOPMENT.*—*a. Formation of the enveloping force.*—In an envelopment, the force that makes the main effort—the enveloping attack —is usually disposed in depth. A division as part of a larger force, for example, making an enveloping attack would ordinarily be disposed with one brigade deployed behind the other, rather than with brigades abreast. A column attack is better adapted to fighting in any direction, since its rear elements can be maneuvered readily toward either flank, and is therefore more efficient and flexible when little is known of the hostile dispositions— the usual case in an envelopment. Moreover, when the enveloping force meets the enemy, its commander may be compelled to make both a main and a secondary attack of his own in order to carry out his mission. Furthermore, as in a penetration, successive groups of reserves behind the initial punch, afford the needed impetus to push ahead and continue the effort of the leading elements that have suffered depletion and disorganization during the first phases of the envelopment, and especially to exploit the success of those elements. This, too, can best be accomplished from a column formation. Therefore, it is obvious that the main effort—the enveloping force—in an envelopment should have a formation in depth in order best to exploit its initial gains.

b. Direction of attack.—Directions of attack are assigned both to the main effort—enveloping attack—and the secondary or holding attack. In a true envelopment as distinguished from a wide turning movement, the directions of these two attacks must converge. (See Figure 57.) The direction of one of these efforts—usually that of the secondary attack—eventually changes if the attack, as a whole, is successful. (See Figure 58.) The direction given to the secondary or holding attack is generally straight to the front toward the hostile position. The main attack is assigned a direction that aims at a decisive point or area within or behind the hostile position; such as a dominating terrain feature, a sensitive point in the hostile line of communications, or the hostile reserve. The direction assigned to the main attack should converge with the direction assigned to the secondary attack at a selected place—usually behind the hostile organized defensive area.

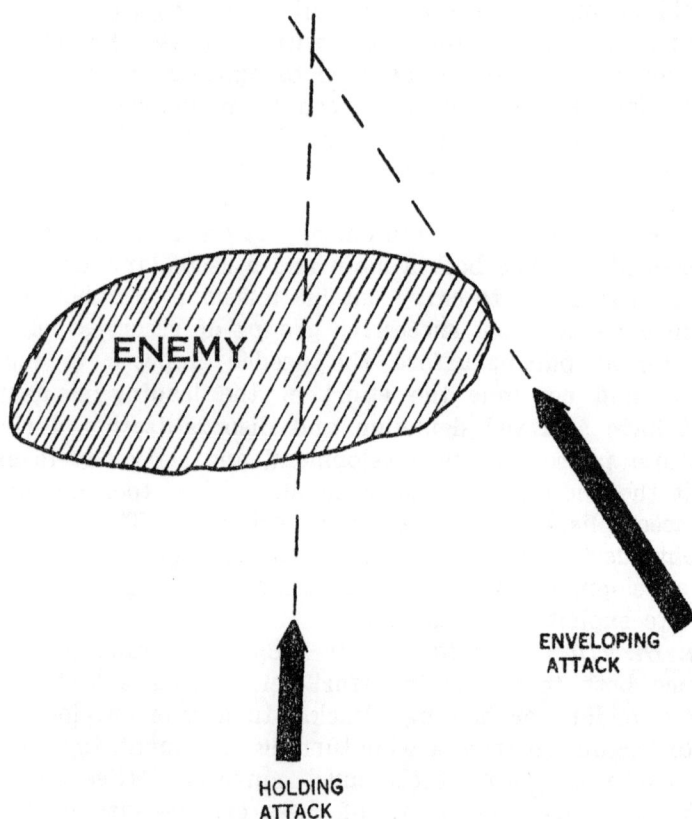

FIGURE 57.—Converging directions of the main envelopment and holding attacks.

Moreover, the enveloping force is usually assigned a direction of attack which leaves it free from participation in the attack against the front line of the hostile position—hence, a direction that permits it to pass by all, or most of, the main hostile position, preferably all of it. (See Figures 59 and 60.) If, however, the front of the hostile position is too wide for the troops available for

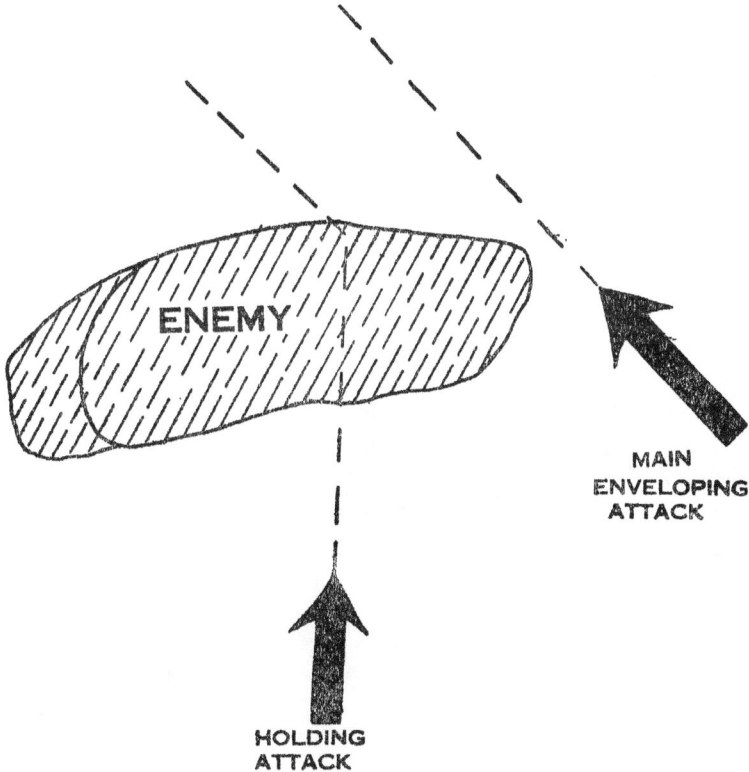

ENEMY

MAIN
ENVELOPING
ATTACK

HOLDING
ATTACK

FIGURE 58.—Change of direction by the holding attack.

the holding attack, the envelopment may be compelled to include part of the hostile front in its attack. (Here it should be noted that a commander, in many situations, may not know exactly where the hostile flank rests; and also, that the flank of hostile dispositions is not necessarily

ENEMY

MAIN
ENVELOPING
ATTACK

HOLDING
ATTACKS

WHEN THE
INTERVAL IS
TOO WIDE
THE ENEMY
MAY COME
THROUGH IT

ENEMY

MAIN
ENVELOPING
ATTACK

HOLDING
ATTACKS

WHEN THE
INTERVAL IS
NOT TOO WIDE
THE ENEMY MAY
BE STRUCK BY
EITHER THE
ENVELOPMENT OR
PART OF THE
HOLDING ATTACK
BEFORE HE CAN
GET THROUGH
THE GAP

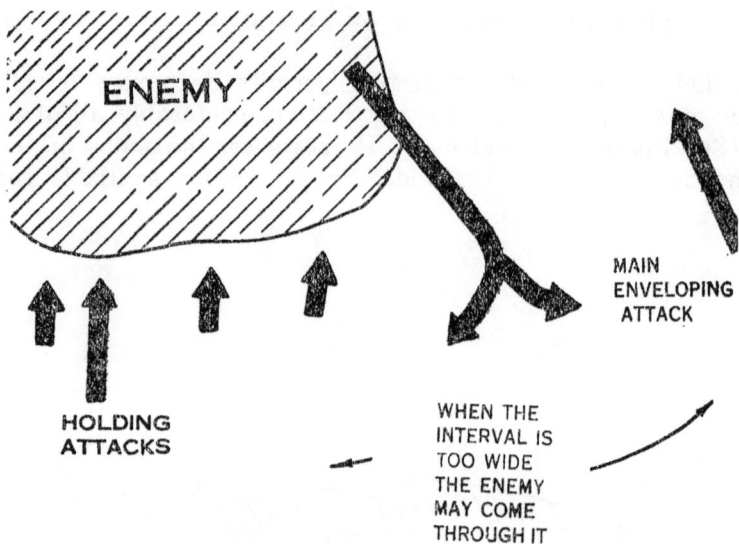

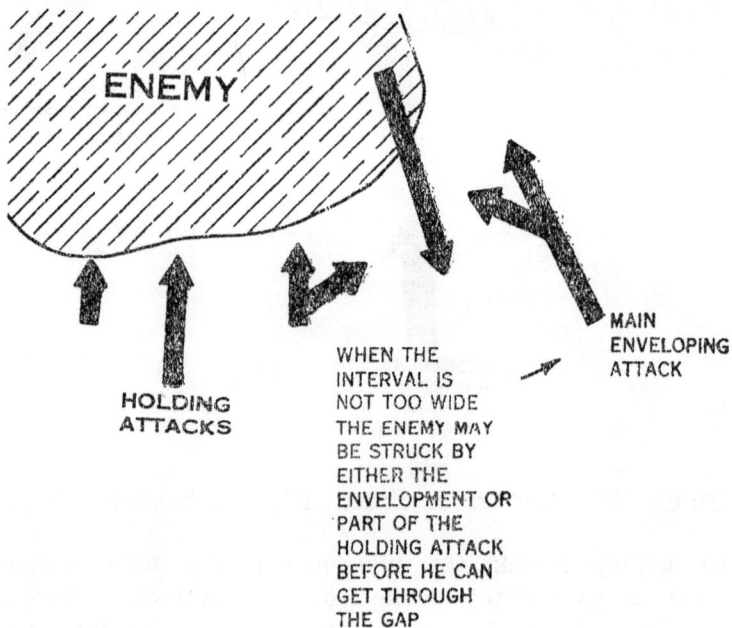

FIGURE 59.—Interval between the main envelopment and
the holding attack.

fixed. It may be changed by extension or refusal, even before a planned attack gets under way. For this reason, as well as for others, the interval between the main and

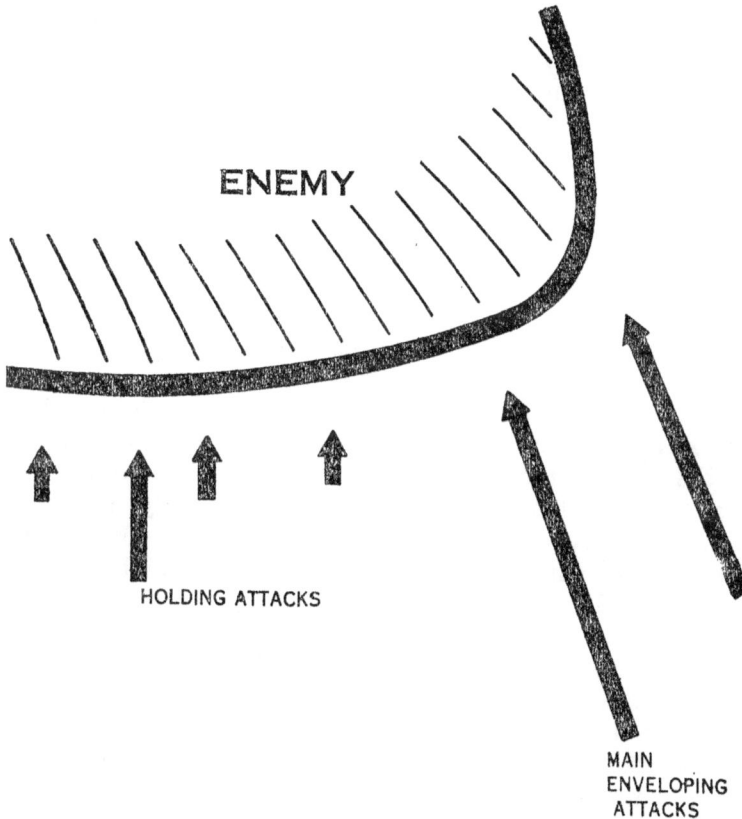

ENEMY

HOLDING ATTACKS

MAIN
ENVELOPING
ATTACKS

FIGURE 60.—An envelopment that includes part of the main hostile position.

secondary attacks is an important consideration.)

c. Interval between holding and enveloping forces.—The interval between the secondary, or holding, attack and the envelopment, or main attack, should never be so

great, however, as to offer the enemy an opportunity to
defeat each force in detail.   (See Figure 59.)   The two
must always be within supporting distance—that distance
in time and space within which one unit can come to the
assistance of the other before the second is defeated. Supporting distance is not alike in any two cases.   It is determined by the terrain, means of communication, visibility, mobility and condition of the troops, and the range
of weapons.   For small units, supporting distance is
generally limited to the effective range of the weapons
with which the units are armed as applied on the terrain
in question.   Where mechanized forces or motorized infantry are involved, their greater mobility effects a tremendous change in the time-distance factor.   For such
forces a distance of five or ten miles may be insignificant,
although the possible delays of loading, unloading, and
traffic congestion must always be considered.

d. *Reserve.*—In the initial deployment for the enveloping attack, a reserve, under the control of the commander
of the whole force is usually held out.   Here, as in any attack, the availability of a fresh, formed body of troops
may later prove a deciding factor.   The size of the reserve varies with the size of the whole force and with the
situation.   It should be located, initially, within supporting distance of both the holding attack and the envelopment.   Thus the best location, often, is in rear of the interval between the two.   However, when it is more
probable that the reserve will be used in one attack than
in the other, it is best to locate it nearer that attack, although for purposes of deception just the opposite is
sometimes done, especially when large forces are involved.
In general, the reserve is held for use in a decisive blow
against any part of the hostile position or dispositions.
(See Figure 60.)

4. *THE CONDUCT OF THE HOLDING ATTACK.—a.*
The holding attack is a succession of limited-objective attacks.   The effort of each assault battalion is generally
toward the capture of a specific objective, usually a ter-

ENEMY

ENVELOPING
ATTACK

PRINCIPAL AND
SECONDARY
HOLDING ATTACKS

RESERVES
READY
TO ASSIST
IN EITHER
DIRECTION

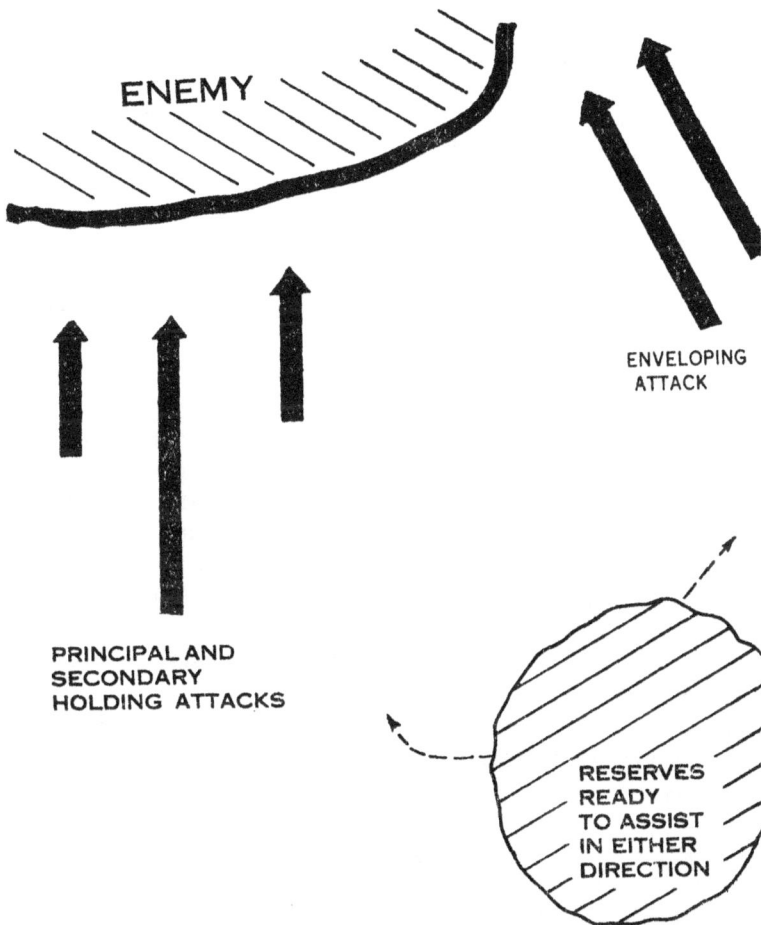

FIGURE 61.—Location of reserve units so that they can be
used to support either the main enveloping attack
or the holding attack.

rain feature from which more effective fire can be placed
on the enemy.   After a battalion has captured its initial
objective, control is regained, and then a coordinated effort is launched against another objective.   This process
of shouldering forward from one objective to another is
continued; advanced elements help their slower neighbors
by flanking fire and movement.   And thus the entire battalion, and its adjacent battalions move forward.   In this
way progress is equalized as far as terrain and hostile
resistance permit.

b. The unit that makes the principal effort of the holding attack, having more means of combat, can ordinarily
advance farther and more rapidly than those making the
secondary efforts.   Renewed impetus to its effort can be
given by the use of its reserves.   If the principal effort
of a holding attack succeeds in advancing and the adjacent secondary efforts do not, the artillery support may
be switched to the assistance of the delayed units.   If a
holding attack, at any point, meets with an unexpected
measure of success, it may, as we have seen before, be
changed into the main effort.

c. The force making a holding attack must be prepared
to defend.   The enemy may attack if he discovers its
weakness.

SECTION II

THE CONDUCT OF THE MAIN ATTACK

5. *SURPRISE AS A FACTOR IN THE CONDUCT
OF THE ATTACK.*—*a.* In the first chapter of this text
it was stated as a primary consideration that the main
attack should be a surprise as to strength, time, or direction—any or all of these.   So vital is surprise that it
is worth repeating here in brief what has been said before.

b. Surprise is a relative term.   The commander of a
large force may well feel that he has surprised the enemy
if he succeeds in concealing his intentions up to a day or

two before the attack. On the other hand, a small unit manifestly must gain surprise in an interval of minutes rather than days.

c. Just as secrecy, or disguise of intentions creates surprise, so does rapidity in delivering the blow reap the advantage of surprise, once the enemy is no longer ignorant of the impending attack. For as soon as he becomes aware of an intention to attack on a given front, he begins the movements of reserves to that front. Therefore, every minute of delay, even on the front of a small unit, aids the enemy. The better we conceal our intentions, the more we catch the enemy unprepared. The less time we give him to redispose his strength after awakening, the greater our chance of turning the surprise factor to profit. This is the foundation on which the conduct of the attack is based.

6. *ADVANCING THE ATTACK.*—*a.* When an attack leaves the line of departure, it does not ordinarily meet uniform resistance along its entire front. An enemy long in position, with every agency of defense in operation, might seem at first thought to be strong at every point. But this is seldom true, especially from the viewpoint of large units. And where the terrain furnishes the least variation of which the attacker can take advantage, small units, too, usually find that one place affords better going than another. (As we shall see later in this paragraph, it is where the hostile defenses are the strongest that infantry employs tanks, smoke, or gas, or uses the natural aid of darkness, to defeat the enemy.)

b. Thus, in the usual case, some weaknesses of the hostile dispositions, some favorable variations of terrain, some gaps in the defensive fires, or some influence of our own efforts in breaking down local resistance, ordinarily permit some elements of an attack to gain ground while others are checked. An advance is rarely uniform throughout; and is always more rapid through the avenues of easiest ingress.

c. When "soft spots" begin to develop, they must be

exploited without delay, for they are the channels leading to success. There is nothing to be gained by hammering desperately at the most difficult and stubborn localities of the hostile defense, when a way around them can be found. It is folly to strive against the enemy's strength if he has a weakness that can be struck. It is absurd to rain blows upon a stone wall when a gate can be pushed in. For the fire power of modern infantry weapons where favored by terrain is so great that it is suicidal to storm them. Consequently, commanders of all units, regardless of size, must seek weak points in the hostile defense and exploit them to the utmost. Only by so doing can they achieve economy of force and make the most of fire and movement. This is the big lesson taught by the World War; the one general rule, never to be forgotten, for advancing the attack.

d. The leading elements of units that find the enemy's weakness and succeed in penetrating his defense, continue to advance to their objectives. They do not stop their progress to engage enemy resistance on their flanks unless they are required to do so by the commander's scheme of maneuver, or unless it is necessary for their own further advance. This does not mean that the flank protection of these leading units is to be neglected, nor does it mean that defended localities which have thus been passed by are simply ignored. It does mean, however, that the primary concern of the successful leading elements is to press on to their objective, through the lanes of approach that they open, depending largely on support and reserve units to protect their rear and flanks. Therein lies the need for organization in depth; it enables the initial minor success to be expanded to eventual victory for the whole force.

e. It is not meant to imply that leading small units, (companies or battalion, for example) can make unlimited advances alone into a hostile position, ignoring their flanks. There is too much chance of the enemy's cutting off such a unit entirely. That is what happened to the famous "Lost Battalion" in 1918. Rather, when further advance is manifestly dangerous, they must hold their gains until adjacent units have come forward sufficiently to safeguard their flanks and rear or support units have forced back the sides of the salient created. As the gaps are enlarged and flank resistance reduced, the forward drive can be resumed. Thus along the entire front; where going is good small units forge ahead and hold, adjacent units and supports exploit the advantage and straighten the line. When greater mass can push through, the advance is deeper. In this way the operations of leading elements and their reserves are interdependent. Coordinated through the next higher echelon of command, they must be so controlled that the aggressiveness and success of the smaller elements are converted into a master-stroke of the larger unit.

f. While the more successful elements of an attack are forging ahead through the weak spots of the hostile dispositions, units whose leading elements have been held up by strong resistance do not merely halt and hold their ground. Instead they try to work around the strongly defended area, using if necessary the zone of action of an adjacent unit for this purpose. Part or all of a unit so maneuvering may leave its own zone temporarily; and in such an event, it should coordinate with the adjacent unit. Ordinarily, in a maneuver of this kind, it is necessary to leave a containing force opposing the strong hostile resistance, until the rest of the unit has had opportunity to act against the resistance from flank or rear, or until supporting troops, tanks, chemicals, or smoke have broken the hostile defense. Even when gas or tanks are not used, history shows that hostile resistance often withdraws when threatened by an outflanking movement. Sometimes it is not even necessary to complete the maneuver; the mere threat suffices.

g. As stated above, when such penetrations are effected, commanders must have their reserves promptly follow the successful leading units in order to protect their flanks and rear, deepen the penetration, or widen the gap. Whether the first reserve units to arrive are used to protect the flanks of leading elements, or expand the gap, must be determined by the commander from the tactical situation confronting him. But in any attack the expansion of a gap is eventually imperative. The deeper the drive, and the greater the resistance, the sooner this becomes necessary.

h. To repeat again for the sake of emphasis—units in the attack advance through the breaks or weak points in the hostile position; avoiding resistance; advancing where the going is easiest, and as far as prudence and the opposition allow. Units that succeed in thus penetrating are followed by reserves who bring lateral pressure to bear against hostile areas that continue to hold, and thereby enlarge the opening. Thus even a small penetration made by a squad results in advancing an entire platoon, that of a platoon a company, and eventually strong forces may be driven deep into the hostile position. Or where the enemy's dispositions are not yet those of an organized battle position, such tactics may eventually so threaten all or a part of his force that he is forced to give way. This, of course, may occur whether the enemy defenses are thoroughly organized upon the terrain they occupy or not. It is far more likely to occur, however, when an attack is pushed so vigorously that the enemy is kept too busy fighting to have opportunity to strengthen his position. (We should bear in mind, always, the various offensive situations that may be encountered in an attack, as outlined in Chapter 7, paragraph 3.)

i. As an attack progresses, the commanders of the more successful units are inevitably confronted with the problem of whether they shall render assistance to adjacent units stopped by hostile resistance or use all their combat means in exploiting their own success. In some situations it may be an advantage to assist a neighbor unit; in others, not. In still others the plan of attack may have already settled the matter by specifically directing assistance to an adjacent unit, to insure the capture of an important area. Again, it may be apparent that the failure of an adjacent unit to capture a critical area is jeopardizing the success of the whole command.   Here a commander must act on his own judgment in rendering assistance to the adjacent unit.

j. Throughout the discussion in this paragraph, the attack has been considered largely from the viewpoint of interior units.   There are only two flank units in any enveloping, penetrating or holding force; all the rest are interior.   (The functions and conduct of flank units will be covered in a later paragraph.)   All that has just been said, however, applies to any main attack; and for that matter, only in less degree, to secondary attacks.   The student should be clear at this point that interior units in any form of attack follow in general the method described above.

7. *DIAGRAMMATIC ILLUSTRATIONS.*—The method of conducting the attack as outlined above may be illustrated diagrammatically.   The diagrams that follow show a few of the many situations a commander may have to meet, and indicate methods he might employ in advancing his attack.

a. Situation No. 1.—(1) (See Figure 62.)   Units 3 and 4 are advancing.   Unit 2 is stopped by resistance at *B*.   Unit 1 is stopped by resistance at *A*.   Unit 5 is in reserve.

(2) In order to advance his attack the commander in zone 2 should push his reserve unit (5) forward to follow up the advance of unit (3).   As it advances abreast of the hostile resistance at *B*, he may use this reserve to take out resistance *B*, or push on beyond it in order to force the enemy out of *B* by the threat to his rear.

b. Situation No. 2.—(1) (See Figure 63.)   Units 1 and 2 are stopped by resistances at *A* and *B*.   Units 3 and 4 are stopped by resistances at *C* and *D*.   Unit 5 is in reserve.

(2) In order to advance his attack, the commander in zone 2 should use his reserve unit (5) to attack resistance at *B* in flank in order to relieve the pressure against unit (2), and thus permit the right of his com-

mand to advance.   An attack against *B* from both front and flanks would in all probability force *B* back at least abreast of *C*.   This in turn is likely to force *A* back and permit an advance along a considerable front.

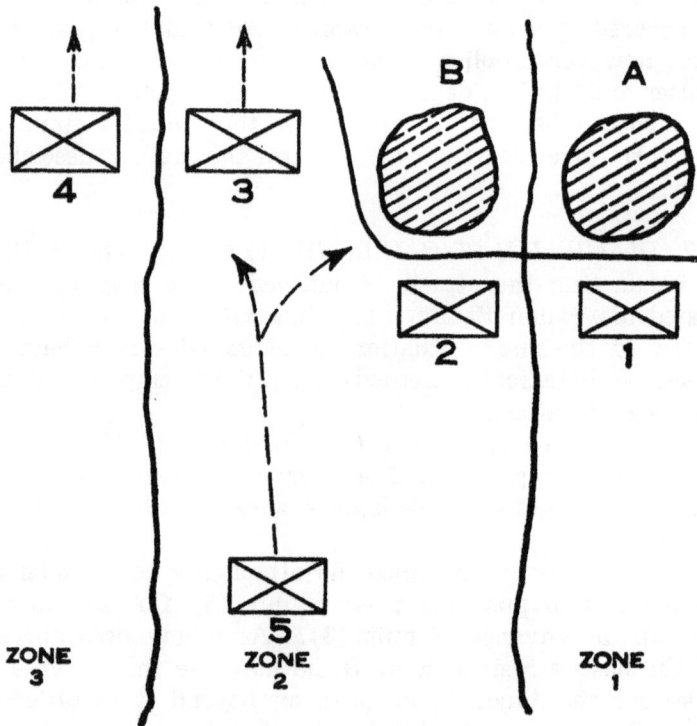

FIGURE 62.

c. *Situation No. 3.*—(1) (See Figure 64.)   Units 1 and 2 are stopped by resistances at *A* and *B*.   Unit 3 is advancing.   Unit 4 is stopped by resistance at *C*.   Unit 5 is in reserve.

(2) In this case the commander in zone 2 must decide

how far he can permit unit (3) to advance before widening the gap. The right of his line is held up; his left is

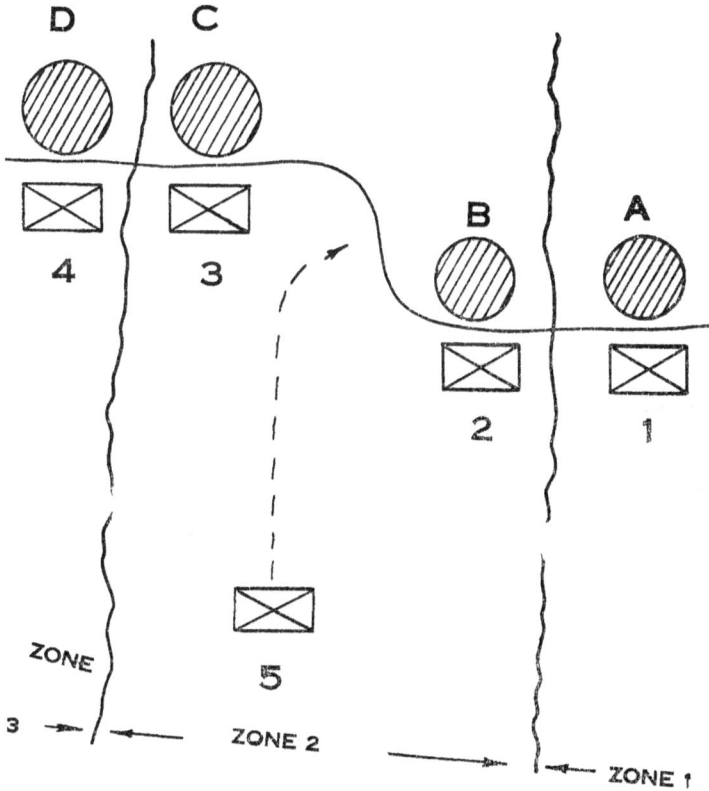

FIGURE 63.

advancing. But the unit on his immediate left is also held up. Here it is not necessary for him to take out *B* or assist the unit on his left in taking out *C*, because

neither of these areas is holding up the advance of his
unit (3).   However, he must push his reserves in rear
of his advancing unit, prepared to attack either *B* or *C*
if they threaten his advancing unit, or to widen the gap

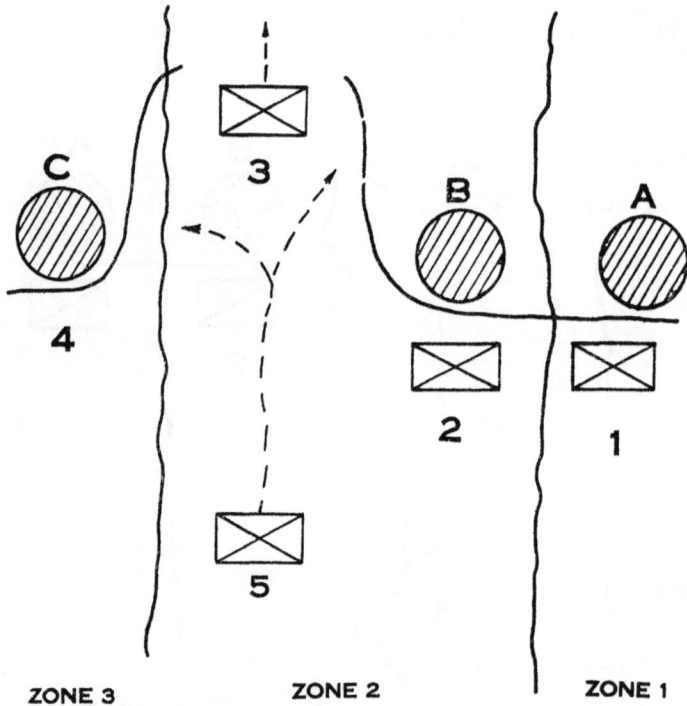

FIGURE 64.

in his own zone by pushing beyond the resistance *B* in the
right half of his zone, forcing this resistance out and
thus widening the gap to facilitate further advance.

  d. Situation No. 4.—(1) (See Figure 65.) Units 1 and

2 are stopped by resistances at *A* and *B*.   Unit 3 is

stopped by resistance at *C*.   Unit 4 is advancing.   Unit 5 is in reserve.

(2) In this situation the commander in zone 2 may use

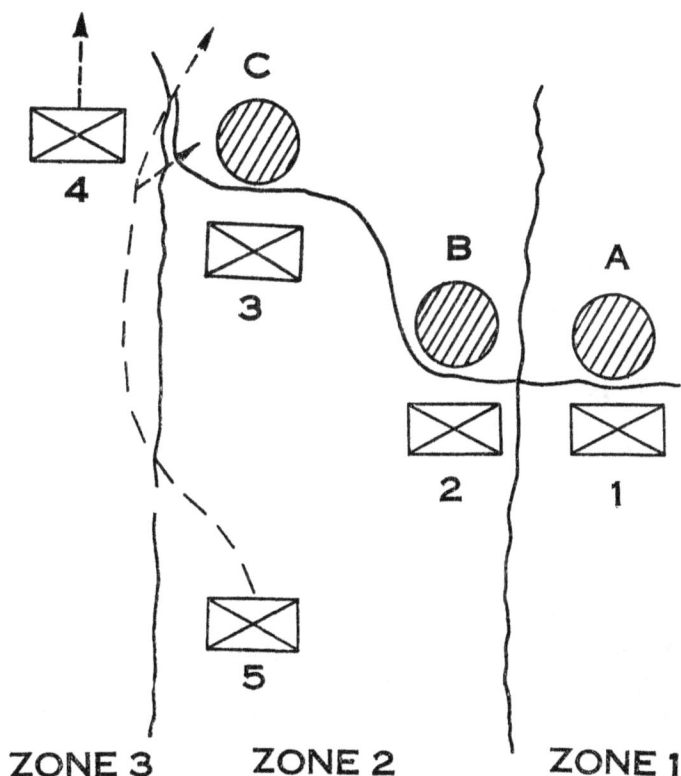

FIGURE 65.

his reserves to advantage in two ways.   First, he can at-tack the resistance at *B* from the flank.   This will permit the advance of unit (2) at least abreast of unit (3).   But here again he will probably be stopped.   The other method

—conducive, probably, to greater results—would be to take advantage of the advance of the unit on his left by following it up with his reserve, and engage resistance at *C* in the flank. This resistance attacked from two directions would probably give way permitting unit (3) to continue its advance abreast of unit (4). A continuation of this advance in zone 3 and on the left of zone 2 would soon force the resistance at *B* to fall back, thereby pushing the attack forward rapidly along a wide front.

8. *SUCCESSIVE OBJECTIVES.*—*a.* All this does not mean that troops push forward independently and without coordination. As we have seen in a previous chapter, there is a plan of attack embodying a scheme of maneuver and a plan of fires. This plan is prescribed only as far ahead as the action can be definitely foreseen. New and additional plans are announced from time to time to meet the development of the situation.

b. It is also true, moreover, that there are often key points within the hostile dispositions, the capture or neutralization of which is essential to the whole maneuver scheme. The capture of still other localities may be essential to the advance of an element of the whole force, and hence indirectly to that of the whole force itself. Whenever a commander can foresee the necessity of taking or neutralizing specific localities or key points, he prescribes their reduction in his plan. But when such a locality is unexpectedly encountered, it then becomes the immediate objective of the unit concerned, regardless of its previously designated objective.

c. As a unit advances towards its objective, it does not engage avoidable enemy resistance unless such action is necessary for the further advance of the unit. And even then, a commander only turns enough of his unit against the hostile resistance to permit the advance of the remainder towards its objective.

d. Only a first objective can ordinarily be prescribed in advance. Succeeding ones are prescribed from time to time during the course of the attack; occasionally prior to the attack. It is this that is meant by the seizure of "successive objectives." Upon the capture of one objective the advance is continued. The selection of the next objective is determined by a consideration of the battle task assigned to the unit as modified by the situation which then confronts the commander.

e. Large units, requiring more time for the dissemination of instructions, generally have fewer objectives assigned to them during an attack; and these are generally farther to the front than those prescribed for small units. For example, a platoon in advancing to a battalion objective, may have one or more objectives before reaching that of the larger unit.

9. *HALTS ON SUCCESSIVE OBJECTIVES.*—Except in a purely subsidiary operation directed against a definite and limited objective, an attack should be advanced, once it is launched, to the greatest possible initial depth by all units. Oftentimes, however, it may be necessary to halt temporarily on one or more of the successive objectives in order to reorganize the attacking forces; recoordinate the efforts of subordinate units; or give supporting-fire units time to displace forward. In addition to any or all of these requirements, resupply of ammunition or food, replacement of casualties, relief of assault units, and countless other contingencies may demand a temporary cessation of the advance. Such interruptions of the attack should be as brief as possible, and limited, to those actually necessary. Time thus lost may squander the whole advantage of surprise and inital successes.

10. *DIAGRAMMATIC ILLUSTRATION.* — Successive objectives in the attack are shown diagrammatically in the illustrative situation that follows:

Illustrative situation.—(1) In a regiment's zone of attack the ridge line: 1—2, is given as the regimental objective in the plan of attack of the brigade. (See Figure 66.) The regimental commander decides to attack with two battalions abreast, with the boundary between battalions as indicated. From his information of hostile dispositions and a study of the ground, he decides that he must capture the ridge line: 5—6—7—8, before he can take his objective, the ridge line: 1—2. He also decides that he must gain possession of hill 4 as soon as possible in order to have observation of the ridge line: 1—2, in his final assault.

(2) Accordingly, he would direct the 1st Battalion (on the right) to capture the ridge line: 7—8, and hill 4, and then continue its advance to the regimental objective. He would direct the 2d Battalion (on the left) to capture the ridge line: 5—6, and continue its attack to the regimental objective. Hill 3 being of no particular advantage to him, he does not order its capture, but instructs the 2d Battalion to pass by any resistance on hill 3, if it can reach the regimental objective without taking it.

(3) The commander of the 1st Battalion decides that in order to reach the battalion objective—ridge line: 7—8, he must capture hills 12 and 13. Company A is therefore ordered to capture hills 12 and 13 and continue its attack to the battalion objective—the ridge line: 7—8. Hill 11 is not considered essential; so Company B is instructed to pass by hill 11 and continue its attack to the battalion objective, the ridge line: 7—8.

(4) In planning for the attack beyond the ridge line: 7—8, the 1st Battalion must provide for the capture of hill 4, specified in the regimental order.

(5) The left battalion would plan its attack in a similar manner. The same scheme is followed in companies and platoons; each unit determines what it must do before it can reach the objective of the next higher unit and accordingly prescribes the capture or reduction of any intermediate positions blocking its advance to the larger objective.

11. *SUPPORTING FIRES.*—a. As stated before, a plan for the use of supporting fires is an integral part of a scheme of maneuver and must be coordinated therewith. The effectiveness of modern weapons in the hands of a defending enemy makes fire support absolutely essential

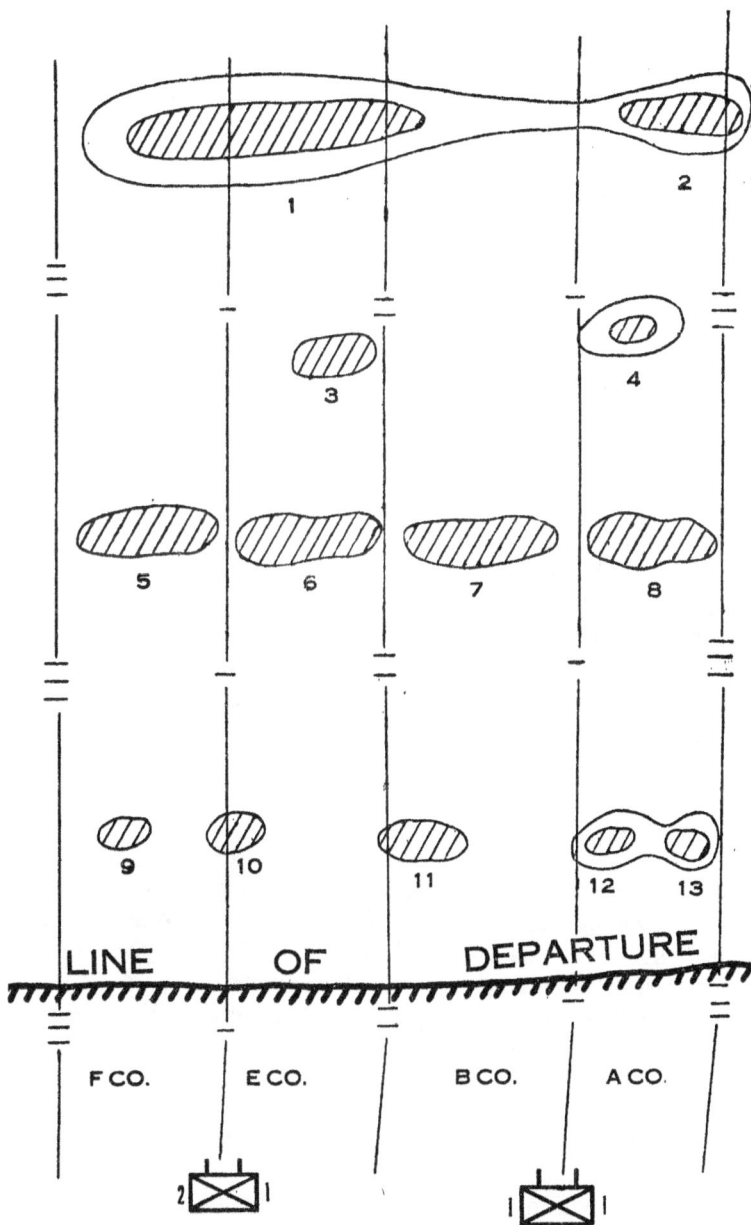

FIGURE 66.

in the attack. As we have seen, the foot elements of infantry, first use the natural cover of terrain to advance to their initial firing positions, and next combine the advantages of terrain with their own fires to push forward wherever advance is possible. But it is also true that these elements can rarely advance by themselves against modern defensive fires, especially those of machine guns, even when a hostile position is thinly held. These elements must have the aid of additional fire power. This fire power is furnished chiefly by artillery, mortars, and heavy machine guns, and, when available, by the use of chemicals. Under these supporting fires, riflemen and their automatic supports advance as close to the enemy as they can before opening fire themselves.

b. Nevertheless, except when this supporting fire is overwhelming, it may still be insufficient to enable the rifleman to close with the enemy. Additional fire power must then be provided by the organic weapons of rifle units themselves, or by tanks, or both. A lack of supporting artillery fires is not an excuse, on the other hand, for assault elements to discontinue their advance when the situation offers any chance of success. When artillery fires, either high explosive or chemical, are not available, infantry must then rely on its own supporting weapons, particularly the mortar and the machine gun.

c. A more detailed discussion of supporting fires and the supporting action of tanks is given elsewhere in this text. Two points, however, should be touched upon here. The first one is the frequent failure of fire support during the progress of an attack. Seldom sufficient at the "jump off" to cover all enemy targets, supporting fires become steadily less and less efficient as an attack pushes forward. (Only when chemicals are used against large parts of a hostile front, especially as delivered from attack aviation, is a complete "covering" of all hostile targets approached.) Hand in hand with this point is the second one, which is the need for every effort to advance the supporting weapons; for the closer they keep on the heels of the leading infantry elements, whether tanks or

foot troops, the more efficient—and hence the more help-
ful—are their fires in keeping the enemy under cover, or
in neutralizing the effectiveness of his fires.

**12. *THE POSITION OF A COMMANDER DURING
AN ATTACK.*—*a.*** The two main functions of a commander during an attack are—

(1) To keep in close touch with assault units in order
to influence their action.

(2) To control the employment of reserves and supporting weapons.

b. It is apparent that the first of these activities makes
it generally desirable for a commander to keep well forward, whereas the second indicates a post farther to the
rear. A commander, naturally, would like to be a dozen
places at the same time. But he can only be at one, and
the choice of that position is always something of a compromise. In small units, the tendency is to lean toward
the first of the two battle functions, and in large forces
toward the second.

**13. *THE INFLUENCE OF A COMMANDER ON THE
ACTION OF HIS FORCES.*—*a.*** A commander influences
the action of his forces in three principal ways:

(1) By the use of his reserve units and supporting
weapons.

(2) By coordinating the efforts of his subordinates.

(3) Lastly and most important, by personal leadership
—example, advice and help, encouragement, and energy.

b. Personal leadership on the battlefield is certainly of
utmost importance—an importance that needs no elabora-
tion. But in large forces, where a commander must remain in close touch with his reserves and supporting
weapons in order to best influence the action, there is all
too little opportunity for a leader's direct contact with the
fighting forces. We should not forget, however, that in
any operation, the personal appearance of a regimental,
brigade, or division commander on the battle front has a
profound influence on the morale of units, and thus on
the action. This is especially true in a withdrawal or re-

treat when encouragement is needed most. And in the attack, when the only means of attaining an objective becomes an advance involving heavy losses, visible personal leadership inspires troops with the will to win in spite of any fears. A battalion, regimental, or even brigade commander, who directs an assault entirely from a command post is courting failure. He must, when possible, move about the battlefield, if for no other reason than to let the troops see him.

14. *SUMMARY.*—*a*. Get through the hostile position rapidly.

b. Attack resistance only when:

(1) It is necessary for our own advance.

(2) Directed by higher authority.

c. Push reserves behind the units that get through.

d. Assist adjacent units:

(1) When it is necessary for our own advance.

(2) When directed by higher authority.

(3) When it will further the general scheme of maneuver of the higher command.

SECTION III

MAINTENANCE OF CONTACT AND FLANK PROTECTION DURING THE ATTACK

15. *DESCRIPTION OF "FLANK".*—*a*. A small body of men sent to a flank of a unit to maintain contact with the adjacent unit is called a "connecting group." Its primary mission is to keep touch between the two larger units. But if there is no adjacent unit, then the group is called a "flank protecting group" or "flank combat group" and has the primary duty of protecting the flank of the larger unit.

b. We see that the term "flank" appears in both these brief definitions, and it has been used in numerous places earlier in this text.

c. The infantry section fully deployed is the largest unit that has no depth to its flank. Thus deployed it consists of only a single, irregular line of men or small groups of men, as it works its way over the battlefield in an attack. A platoon generally deploys in two such irregular lines. Thus in Figures 67 and 68, which show diagrammatically a deployed platoon, and a deployed company, we see clear-

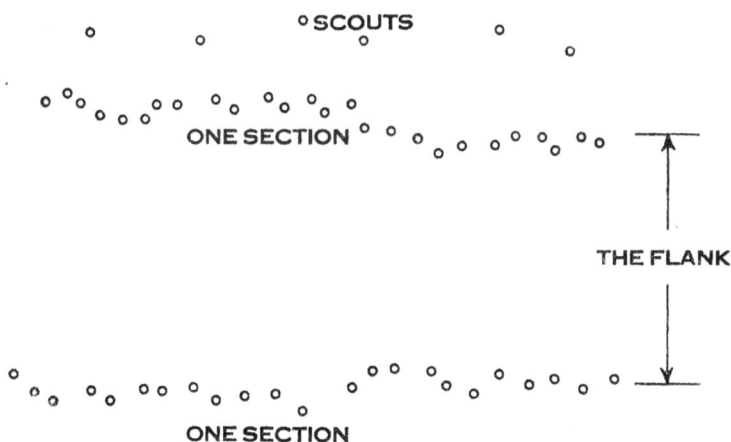

FIGURE 67.—The flank of a platoon.

ly what is meant by the "flank." And as we go on up through the larger units, there is an increase in the number of elements from front to rear, and hence a cumulative increase in the depth of the flank.

d. Thus flanks are not merely the points at the ends of the irregular line formed by the leading assault units (commonly called "the front line"), as they are often misconstrued to be, but are in themselves lines, roughly perpendicular to the front line. And wherever this line along the side of any force is exposed, it must be protected. Since any unit in battle is best disposed to fight in the

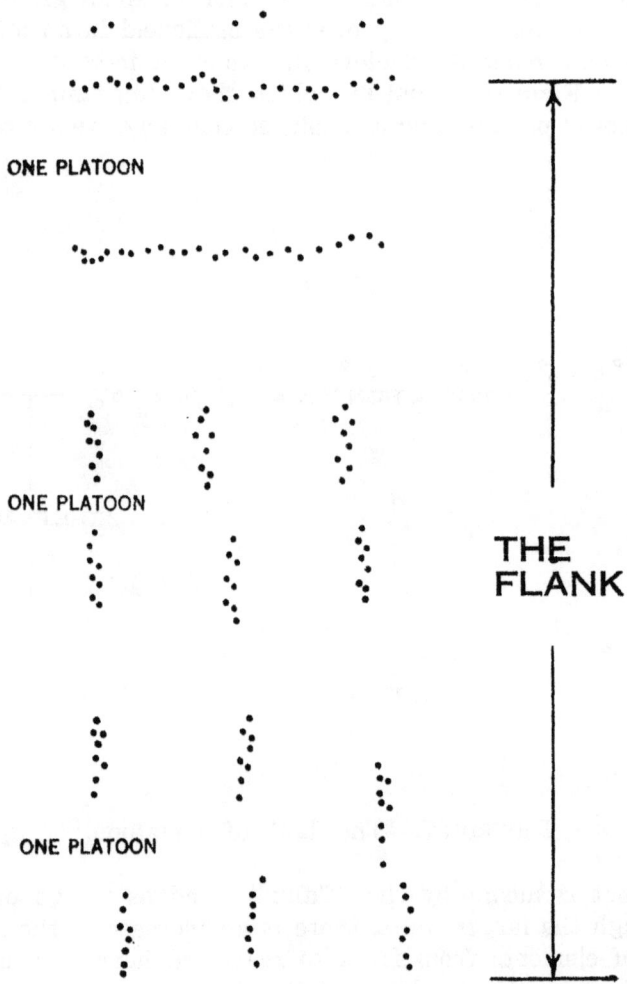

ONE PLATOON

ONE PLATOON

ONE PLATOON

THE
FLANK

FIGURE 68.—The flank of a rifle company in the attack.

direction it faces—especially units that are attacking—, a fraction of the force must, for complete protection, be so disposed that it can readily fight to a flank, if that becomes necessary, while the main force continues to fight toward its front.   (See Figure 69.)

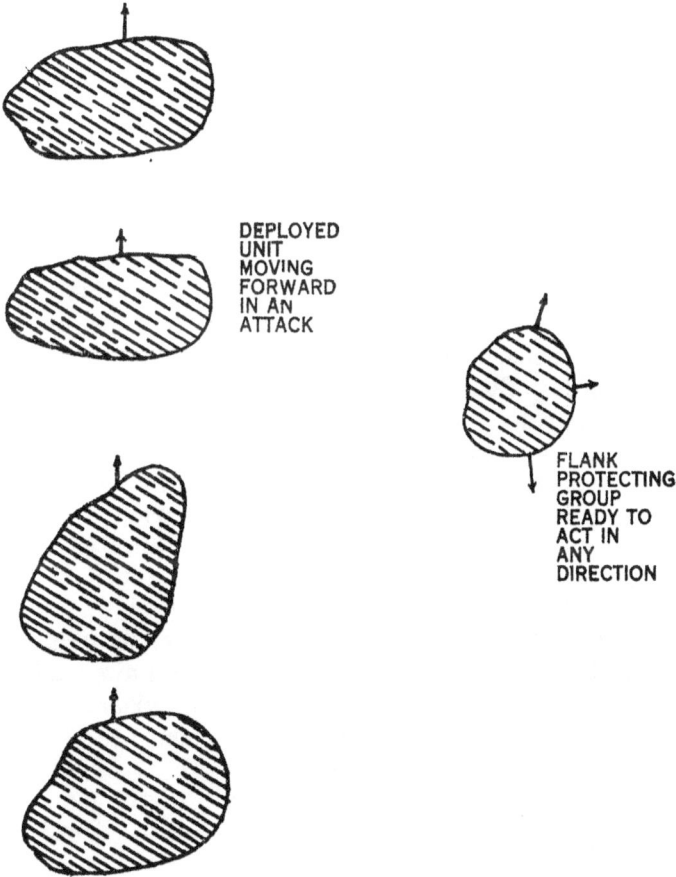

FIGURE 69.—Flank protecting group.

16. *FLANK PROTECTION.*—a. All units, of whatever size, are responsible for their own security.   In other words, there is no exculpation for the commander who

neglects to prepare against possible hostile efforts from
any direction. In an attack, of course, the location of the
main hostile force may be known. But rarely is it known
accurately. And almost never can the hostile dispositions
or intentions be so definitely determined that a commander can overlook protection for his flanks and rear
with impunity. This is especially true when mechanized
and motorized units are involved. Such units may strike
quickly from any direction. Well led cavalry has often,
in past wars, struck where least expected. Hence, if
there is a single principle of war that comes near being
an invariable rule, it is that a commander is always responsible for the security of his own force.

   b. So long as a given unit forms an integral, interior
part of a larger unit, its commander relies for protection
largely on the presence of other units on each side and in
rear of his own, and ordinarily needs to take no protective measures beyond keeping in continual touch with
his neighbors. But when a unit is on the flank of a larger
unit, or separated from that unit, even merely during the
process of deployment or partial deployment, its commander must at once provide for its safety. This he does
despite any security furnished by the larger unit. In
offensive combat this protection must be initiated during
the earliest stages of deployment and continued through
all stages of the attack, including pursuit.

   c. We have seen above that this protection is gained by
sending to the flank a group of men or units called a
"flank protecting group." The size of the group depends
on several factors, most important of which are: the size
of the parent unit, the type and strength of any protection provided by a higher unit, the depth of the flank, and
the probable character of any hostile threat. For a platoon or company, a half-squad or squad is usually enough;
for a battalion, a section or platoon; and for a regiment,
a company. An army often employs a whole division.

   d. In carrying out its mission, a flank group must conform to the movements of the unit it is protecting. If
that unit is halted, the protecting group stops. When
halted, the leader of the group disposes his elements for

possible action to the front, flank, and rear. In small groups each member selects two positions—one from which he can fire to the front and flank, and one from which he can fire to the flank and rear. One or two men are detailed to observe, and one or more are detailed to maintain constant communication with the unit protected. The remaining members of the group, after having selected their two firing positions, remain under cover, prepared to go immediately to either of these positions. Larger flank groups use the same general method with such modifications as are required by their size and distance from parent organizations.

e. If the troops to be protected are moving forward, the group moves rapidly from one position to another. Successive positions are selected sufficiently far forward, and occupied in adequate time to give the best protection to the advancing troops. The leader must decide when he should leave one position and go to the next. Keeping in mind that his mission is to protect the flank, he must base his decision on his mission, his best judgment, and what he knows of conditions.

f. A flank group does not join in the attack of the assault unit. Even when the nearest assault unit is held up and a favorable opportunity presents itself for the flank group to assist, a flank group keeps to its own mission.

g. A flank group should occupy the best available terrain for advantageously protecting its unit and should maneuver so that it can place effective fire upon any terrain feature from which the enemy could fire into the flanks of the force protected. No definite interval between a flank security group and the unit protected can be given. Where the latter is a platoon or company, the interval may be from 100 to 500 yards. The flank positions should be occupied with a view to stopping or delaying hostile action from a flank, and giving ample warning of a hostile flank attack. The enemy situation, the terrain, and the depth of the flank, all affect the selection of flank positions. Sometimes specific orders are given to flank units. Often, however, none are given beyond the general statement of the duty to be performed. When this is the case, the group leader must select the initial position and the route to it. In moving out, a formation suitable to the strength of the group is adopted. A squad or less takes up a patrol formation. A section should send forward a small point of three or four men and should protect its own exposed flank (or flanks) with a patrol of two, three, or four men.

h. Flank groups are usually taken from the reserve of the unit that sends it out. Thus the assaulting units are left with their full strength, and are relieved of all security responsibility other than their own immediate protection.

i. The commander of a large unit may delegate the duty of flank protection to a subordinate element. This is not often done, but when it is, the duty should be given to a unit not engaged in the main effort of an attack. All components of the main effort should be freed as far as possible from extra duties; and other units should furnish the flank protection. Nevertheless, the components thus freed are still responsible for their own protection, regardless of the provision made by higher units, and must therefore take reasonable precautions to that end.

j. In the company and larger units, flank protection is further assured by locating the reserve so that it can move rapidly to repel a hostile attack on an exposed flank or cover the flank with fire, especially that of machine guns.

k. Excessive caution denotes fear and weakness; but the commander who wilfully, or unwittingly, disregards his exposed flanks (or rear) invites disaster.

17. *MAINTENANCE OF CONTACT DURING AN ATTACK.*—*a*. Contact with units on the right and left—no matter what instructions or prearranged methods have been prescribed—is also a responsibility of all unit commanders. A regimental order, for example, may direct the right battalion to maintain contact with the left battalion; or the regimental commander may send forward

a group from the regimental reserve to maintain this contact. But in neither case is the commander of either assault battalion relieved from the responsibility.

b. Connecting groups between platoons, companies, and battalions usually consist of a squad or half squad. The size, of course, depends on the nature of the terrain and other factors effecting visibility. Connecting groups, like flank security groups, are usually taken from the reserve of the unit sending them out.

c. Owing to the fact that each attacking unit goes forward in its zone of action as rapidly as possible, almost unmindful of the progress of neighboring units, gaps between units are bound to result. Moreover, gaps open naturally where two units diverge to avoid fire-swept areas, obstacles, or strong resistance; and occasionally, merely from failure to maintain correct direction. Connecting groups watch for such gaps, and notify their commanders immediately one is observed, especially when a gap becomes so wide that either unit passes out of view. Connecting groups must constantly keep in contact with both units and, they must, at any time, be able to tell the commander of either unit where the other's flank is. A connecting group is ordinarily too small to fill a widening gap as a combat unit; hence the need to report the extension of a gap so that it can be filled by reserves, or other action taken as befits the particular situation. For example, where a gap had been caused by the more rapid advance of an adjacent unit, reserves of that unit might push forward in rear of their leading elements, and thus fill the gap, as has been illustrated earlier in this section.

d. If a connecting group loses contact, it tries to reestablish it. While doing this, and until the gap is filled, it remains out and acts as a flank combat group.

e. If a connecting group is sent forward from the reserve to maintain contact between assault units, regardless of this, as already stated, the responsibility for contact still rests on the assault unit commanders. It is preferable to employ groups from the reserve, in order to avoid duplication of effort.

f. A connecting group takes an initial position between the two units where it can best perform its duty. If both units move forward shoulder to shoulder, the connecting group merely conforms to the advance. If both are halted, the group places itself well forward in the lane between them so that it can observe the future actions of the leading elements of each unit. If either one forges ahead, the group also must move forward, but must extend itself in order to maintain contact both fore and aft—that is to right and left—especially with the unit from which it was sent. Smoke and darkness, and terrain features such as woods, standing crops, villages, and small hills, that obstruct observation of the two units, obviously influence the location, formation, and movement of connecting groups.

g. The last two paragraphs cannot be concluded better than by reiterating the responsibility for protection of flanks and maintenance of lateral contact that devolves on every commander, irrespective of protection or liaison prearranged, or to be provided, by a higher unit.

CHAPTER 10

RELIEFS TO CONTINUE THE ATTACK

1. *DEFINITION.*—A relief is the operation of moving forward a unit which is not in contact with the enemy, and effecting by means of the unit thus moved forward the replacement of a unit already in contact with the enemy. In a passage of lines—a form of relief used in the attack—the relieving unit passes through or by the other unit; and the relieved unit, instead of moving to the rear, generally reorganizes and follows as a support or reserve after the attack has progressed an appreciable distance.

2. *GENERAL.*—A commander may order a relief before an attack or during an attack. Reliefs may be ordered for both small and large units. In a relief of a unit larger than a battalion, the relief is coordinated and accomplished by battalions.

3. *PURPOSES.*—*a. General.*—In offensive combat, reliefs are ordered for one chief purpose; to maintain the forward drive of an attack.

b. Conditions necessitating reliefs.—During an attack a commander may order a relief for one or more of several specific reasons:

(1) To continue the momentum of the attack.

(2) To change the direction of the attack.

(3) To replace leading units that have become exhausted, depleted, or disorganized.

(4) To replace units in contact with the enemy by more experienced units capable of greater effort, especially when such troops have only become available after the launching of the attack.

(5) To withdraw temporarily from the attack, units that have been "pinched out" by converging boundaries between units.

4. *SECRECY.*—*a.* Every effort must be made to keep the time of a relief secret. While the relief is going on, both the relieving unit and the unit to be relieved are in a single area. Congestion is doubled and circulation is more than doubled. If one or both of the units lack experience, confusion may exist. Thus, unless the time of a relief is kept secret the enemy is liable to choose that time for a heavy bombardment or an attack, with serious results to both units taking part in the relief.

b. For these reasons, if the situation permits, reliefs are usually carried out at night, or when the operation is otherwise covered from hostile observation; for instance, by woods or fog. It may also be concealed from the enemy by placing smoke on the hostile observation posts or forward elements, or both.

c. Measures against hostile observation, both aerial and ground, are therefore very important during reliefs, especially in reliefs on a scale large enough to require more than a single night. During periods of visibility by day there must be no evidence to hostile observation that a relief is in progress. The troops, animals, vehicles, trains— in fact, every part and adjunct of the relieving unit— must be kept concealed.

d. The American preparations for the Meuse-Argonne thrust in the latter part of September, 1918, took three weeks. Every movement of troops was made at night, and the strictest orders against exposure obtained by day, and not even the cover of fog was trusted.

e. It is also important that the fact of an accomplished relief be kept secret from the enemy as long as possible, especially when the relief has been made as a preliminary to an attack from a stabilized position.

f. The necessity for maintaining continuity in an attack, however, often requires that a passage of lines be accomplished in the daytime. When this is the situation, secrecy becomes secondary to necessity. Nevertheless, all cover available should be used to hide the approach of the new unit so that its arrival in battle is as great a surprise to the enemy as possible.

5. *PRELIMINARY MEASURES.—a. Warning orders.* —Warning orders should be issued when possible. They should include: the approximate hour the movement is to begin, the sectors or zones in which units are to operate, and the restrictions imposed upon reconnaissance parties as to size, routes, and hours of operation.

b. Reconnaissance.—Reconnaissance, both by personal observation and conference with the commanders of troops to be relieved, is highly desirable in accomplishing a relief. But sometimes, in a relief to continue the attack, neither is possible before beginning the relief. The relieving units have to move forward to the attack without delay, reconnoitering as they go, and the commander of the relieving unit may not be able to find the other commander.

c. Guides.—Arrangements are made for the unit being relieved to furnish guides. Guides meet the relieving unit before it enters the area occupied by the unit to be relieved. They then conduct the relieving unit into the area and to its position for attack. Whenever possible, guides are provided for units down to and including the platoon.

d. Plan of troop movement.—(1) A plan of troop movement should be formulated and issued covering the movement of units to the area of relief. In forming this plan the commander should consider:

(a) Restrictions imposed by higher authority because of other traffic in the rear areas.

(b) The greater road spaces that may be required for the units because of increased distances between units.

(c) The road net and the practicability of cross-country movement.

(2) The plan should provide for secrecy and security.

It must also be flexible as to times and routes of movement.

(3) In a relief to continue the attack, however, there may be little time available for small units to form a complete plan. The emergency may require immediate movement to the forward area, with a minimum of delay for a hasty reconnaissance and a conference between the commanders. It is nevertheless desirable that the commanders and staffs of both units arrange necessary details such as guides, use of roads, fire support of the relieving unit by the unit being relieved, hour when command passes to the relieving unit, and other administrative details. In general, the size of the units concerned, and the haste with which the relief must be conducted because of urgency, govern the degree of thoroughness with which these details are considered.

e. Road spaces.—Road distances are usually increased during a relief. The distances between companies and platoons, whenever possible, should be greater than usual in order to make the enemy's observation more difficult, reduce casualties from artillery fire and hostile aircraft during the movement, and provide greater flexibility of movement. Relieving units approach the forward area either partly or completely deployed.

6. *CONDUCT OF A RELIEF TO CONTINUE THE ATTACK.*—*a.* The movement forward and the plan for the attack are the most important concerns of a unit effecting a relief to continue the attack. Passing through the unit to be relieved, although incidental to the whole operation, is an essential and sometimes critical phase of the forward movement. The commanders of both units must make every effort to avert confusion. No part of the relieving unit must be allowed to check its impetus. It must pass steadily through the areas occupied by the other unit, without allowing its probable state of disorganization to cause confusion.

b. Reconnaissance for this type of relief has to do chiefly with the terrain over which the deployment and attack are to be made. Whenever possible, information of the enemy's activities and probable intentions is secured from the commanders of the unit to be relieved. There may be much difficulty in locating the commanders and the dispositions of the troops to be relieved. Care must be taken not to fire on friendly troops during the initial stages of continuing the attack.

c. Depending on the plan for continuing the attack, the situation, and the terrain, the attack may have to be launched from:

(1) A line of departure coinciding with the forward elements of the unit being relieved.

(2) A line of departure in rear of the unit being relieved.

d. Relieving units reach their line of departure in the deployed formation to be used for the attack. Terrain permitting, a line of departure between a covered position just in rear of the unit to be relieved and the forward elements of that unit is usually a better line of departure than a line coinciding with the front-line elements.

e. When the relieving unit passes through the front line of the relieved unit, the latter does not leave the area as soon as the relieving unit is in position to attack. The relieved unit should remain in position continuing its fire support of the new unit until the attack has progressed far enough so that these supporting fires are masked and the old unit can assemble and reorganize without undue casualties.

f. Orders for the relieving units in a passage of lines are much the same as an ordinary attack order. They designate the specific units that are to be passed through.

CHAPTER 11

INFANTRY UNITS IN OFFENSIVE COMBAT

SECTION I

GENERAL

1. *GENERAL.*—In order to defeat the enemy, it is necessary to come to grips with him on the ground he is occupying and drive him from his position. This is accomplished by combining and coordinating fire and movement. The maneuvering element of an infantry battalion (which is the basic tactical unit of infantry) is made up of groups armed with the rifle, automatic rifle, and bayonet. Because of this fact, the rifleman with his bayonet is still to be considered the decisive element in combat. All the numerous agencies of fire power, both organic and supporting, that the battalion has available besides its riflemen, have the primary mission of furnishing fire support for the maneuvering riflemen—a support that facilitates their advance to positions from which they can come to grips with the hostile infantry.

2. *WEAPONS AND FIRE POWER.*—*a.* In offensive operations of the rifle platoon, we must realize that the rifle, considered specifically as a weapon, is no longer the principal source of fire power against the hostile elements, and is therefore not the weapon we depend upon basically to establish and maintain superiority of fire. However, riflemen with their bayonets, considered as small combat

215

teams, each having fire power, maneuverability and shock action, are still the decisive element in the infantry attack. The self-loading rifle (a weapon possessing many advantages over present equipment), the adoption of which we may expect in the near future, will have little, if any, effect on the application, in offensive operations, of the doctrine just outlined. The possible use of a high-velocity bullet, capable of penetrating an inch or more of armor, would likewise make no change, but rather increase the importance of the rifleman in battle by giving him effective fire against present types of mechanized elements such as tanks.

b. The Browning automatic rifle M1918 is not capable of delivering a sustained volume of automatic fire. (Regulations prescribe that this weapon be habitually used at semi-automatic fire.) For this reason, then, the automatic rifle is now used merely to increase the fire power of the rifle squad of which it forms an integral part. And the squad group as at present organized is thus the basic fighting group of the infantry, although the individual rifleman is the basic element.

c. In order for riflemen to deliver an effective volume of sustained fire, whether with the bolt-action, or self-loading rifle, they must deploy by groups in an irregular line, on ground from which they can see the hostile target to be engaged. (This we have studied in detail in the earlier chapters of this text.) This formation is most vulnerable and is difficult to conceal even when the utmost use of terrain is made; and therefore quickly subjects these riflemen to hostile fire, which in turn soon renders their fire ineffective, and causes casualties. Because of this, it is essential that we introduce into the rifle platoon some means of providing a base of fire of maximum effectiveness, delivering direct fire from a concealed position, and using as few men as possible. A light weapon capable of a sustained volume of automatic fire, equivalent in fire power to a considerable number of rifles, is the weapon needed.   Such a weapon can be easily concealed, both in movement and in position, and is most difficult for the enemy to discover when it opens fire.   Thus our vulnerability to hostile fire would be measurably decreased. Again, it delivers its fire from a concealed *point* on the battlefield as against the often greatly exposed *line* formation from which a group of riflemen have to deliver an equivalent fire power.   Herein also lies the effectiveness of this type of weapon: in all probability it can remain in action for a considerable time before discovery by the enemy forces it to move to a new firing position.   Thus, it releases the riflemen and enables them to maneuver. Under its cover they are free to work forward individually, utilizing the accidents of the ground, until they reach an area close enough to the enemy for the delivery of the final assault.   During this phase of the attack also, their base of fire may often be able to continue its support, and by continuing to neutralize hostile fire, insure success.

d. In the study of this question, two problems were presented to the infantry: first, to obtain a weapon with the desired characteristics; and second, to evolve a rifle platoon organization suitable to this method of advancing the attack. After considerable experimentation the first of these problems has been solved, at least temporarily.   It was found that by a slight modification of the present Browning automatic rifle we can obtain the requisite weapon.   This modification provides a bipod mount, fixed near the balance of the piece, and the possible introduction of a flexible, belt-fed mechanism to replace the present metal magazine-clip-feed type.   In considering the second problem—that of organization—, it was deemed essential, with this modification of the automatic rifle, to exclude it as an integral part of the rifle squad, and to organize all such weapons into a separate "light machine-gun" squad in each section, complete with ammunition carriers.   This leaves the riflemen unencumbered by extra weight and hence freer to maneuver.

e. Tests are now being conducted to determine the most suitable organization for the rifle platoon, no final conclusions having yet been reached.   Whatever the organization adopted, we can be practically certain that it will provide basically for separate "light machine-gun" squads within the platoon.   It is therefore timely and essential in discussing infantry units in the attack, to consider the place of this weapon in the tactics of the units that would employ it.

3. *ANTITANK GUNS.—a.* For purposes of instruction at the Infantry School, a machine-gun company consisting of three platoons, each equipped with four caliber 0.50 machine guns (See Special Text No. 14, Chapter 11), has been added to the regiment to provide antitank protection in all phases of operations.   This unit will no doubt soon be made an integral part of all infantry regiments.

b. In the attack this company should be disposed so as to best fulfill its antitank mission.   It may be retained under regimental control, or allotted by platoons to the assault battalions, depending upon the probability of hostile tank attacks.   In general, these guns should be so disposed that they can advance by bounds close in rear of the assault battalions most likely to be subjected to hostile tank attacks.   They should endeavor to advance over concealed routes to positions which have been reconnoitered in advance.

c. In operations where one or both flanks of the regiment is exposed, all or part of this company may be retained under regimental control, and placed in such a way that the flank and rear areas of the regiment receive maximum protection against attack by mechanized forces.

d. Situations will arise where the fire of these guns may be used with good effect to supplement the fire of the battalion supporting weapons against targets other than tanks.   When confronted with this situation, a commander should bear in mind the importance of the primary role assigned these guns, the necessity for their concealment and the fact that the ammunition supply is limited. Therefore, to direct the fire of these guns on targets other than tanks, must be considered exceptional, and justified only in extreme emergencies.

SECTION II

THE RIFLE SQUAD, SECTION, AND PLATOON IN THE ATTACK

4. *DEPLOYMENT.*—*a. General.*—As we saw in Chapters 1 to 6, a column on the march takes up a partly-deployed formation when contact becomes imminent; and if it strikes the enemy unexpectedly, without realizing the imminence of contact, the whole force, in this situation also, begins its deployment at once. Thus, a given rifle company will, in either case, soon find that partial deployment within the unit is a necessary step. Sometimes, when such a company is well toward the rear of the whole force, and roads forming covered and defiladed routes are available in the zone of advance, a company may continue forward for several hours in route formation. In the usual case, however, this is seldom possible, and companies, as well as larger units must partly deploy.

b. Platoon deployment.—The same things hold true of platoons. And when the partial deployment of a company becomes advisable, its platoons also adopt more open formations. The platoon must then take up a formation that can be easily controlled, that reduces the chance of casualties, and that permits a change in its direction of advance, or complete deployment, rapidly and easily, in order to meet any eventuality. The formation taken depends, largely upon the nature of the ground over which the platoon is advancing, the frontage assigned to it, and the mission of the platoon. In an attack, the frontage assigned to a platoon usually varies from 200 to 300 yards.

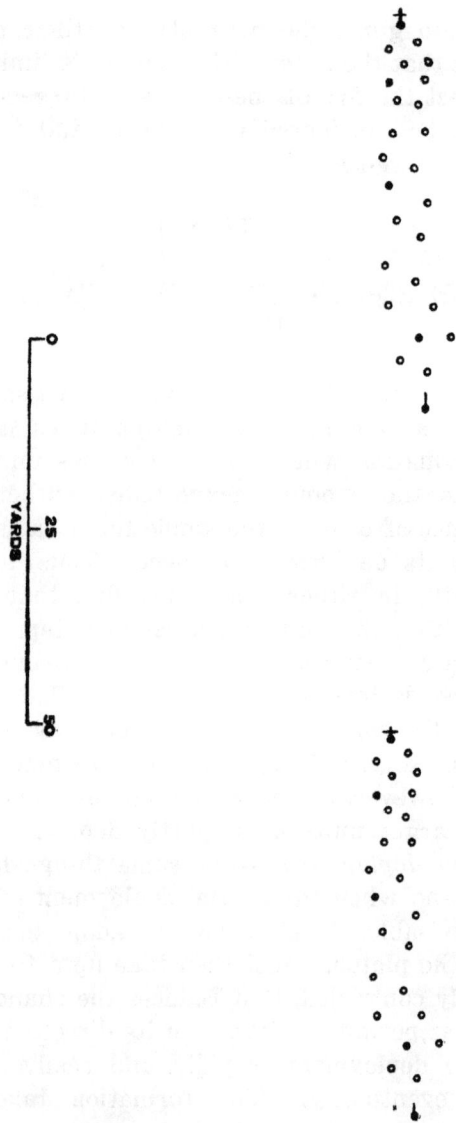

FIGURE 70.—Column of sections in section columns.

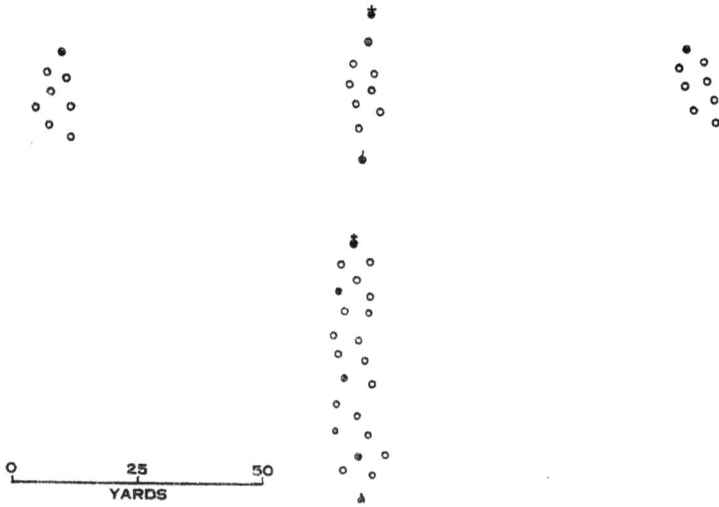

FIGURE 71.—Column of sections; leading section in squad columns, rear section in section column.

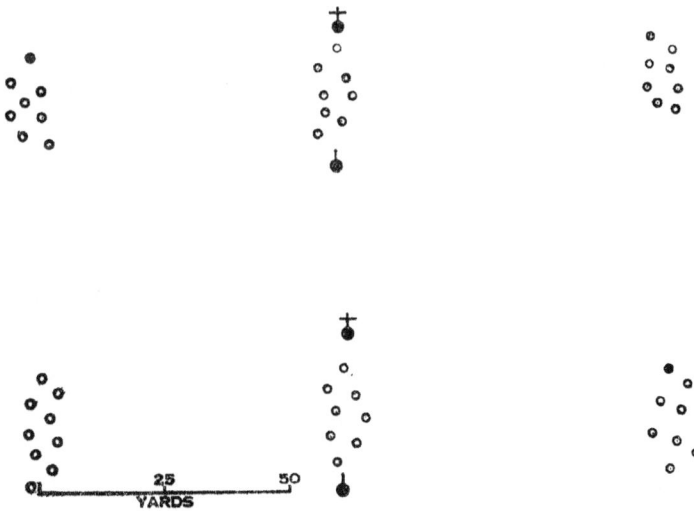

FIGURE 72.—Column of sections; both sections in squad columns.

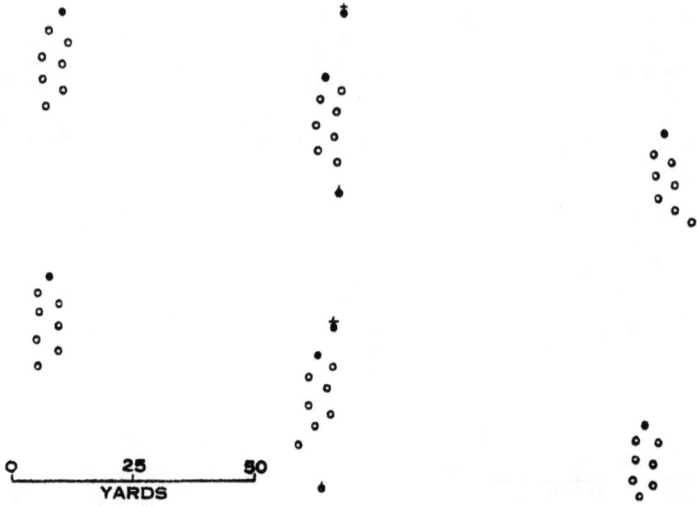

FIGURE 73.—Column of sections; both sections in squad columns echeloned.

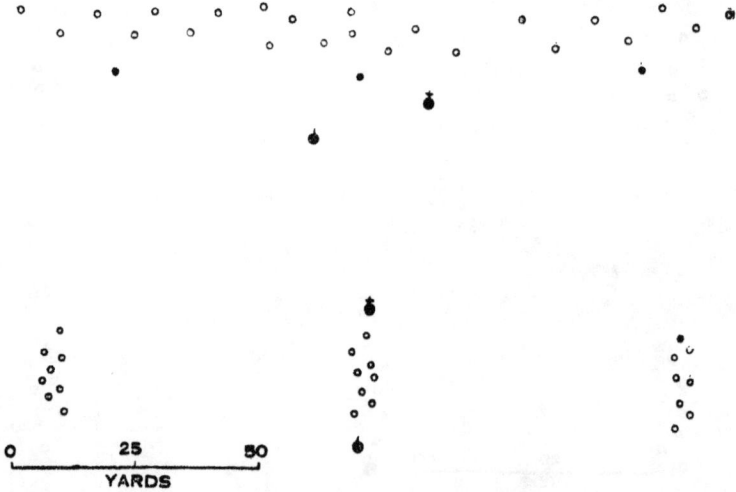

FIGURE 74.—Column of sections; leading section in irregular line; rear section in squad column.

c. Platoon formations in the attack.—Platoons attack in column of sections, sections abreast and with sections echeloned.

(1) *Column of sections.*—The formation of column of sections (see Figures 70 to 74) has the advantage of good security and the ability to maneuver. It has the disadvantage of hampering, initially, the employment of the maximum fire power of the platoon. This formation is used when the situation is obscure or the frontage assigned to a platoon is narrow.

(2) The formation of sections abreast (see Figures 75 to 77) permits an initial delivery of the whole fire power of a platoon. On the other hand, the maneuverability of the platoon is considerably reduced and its fire is somewhat fixed in direction. The platoon attacks with sections abreast when the enemy dispositions are known and it is desired to utilize the fire of the entire platoon frontally when contact with the enemy is first made.

(3) The formation of sections echeloned (Figure 76) is similar to that of platoons in column. It has almost the same advantages and disadvantages. It is used when there is a threat from a particular flank, in which case the rear section is on that flank.

d. Section formations.—Sections advance in section columns, or squad columns, or in an irregular line of individuals or small deployed groups. The formations used by both sections and squads are, in any given situation, those most readily adaptable to the best use of the ground for protection from the enemy's fires and the employment of our own. All elements of a unit, moreover, do not necessarily retain the same formation at the same time. One section or squad, for example, may have to pass over a dangerous, open area, while others are able to make use of cover. In this event, the former naturally use the full deployment of an irregular line, whereas the latter may best keep to a formation of small columns or groups. Section or squad columns are ordinarily used in avoiding

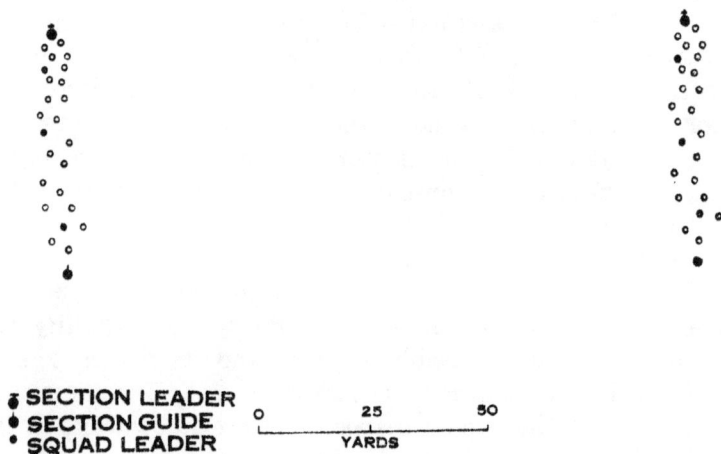

SECTION LEADER
SECTION GUIDE
SQUAD LEADER

0   25   50
        YARDS

FIGURE 75.—Sections abreast; line of section columns.

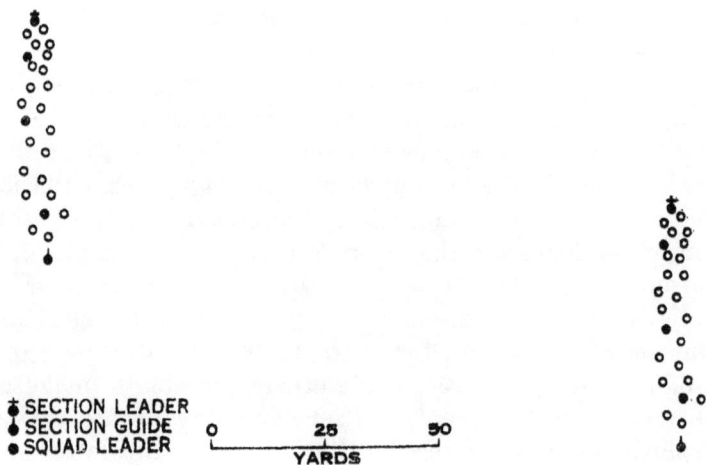

SECTION LEADER
SECTION GUIDE
SQUAD LEADER

0   25   50
        YARDS

FIGURE 76.—Sections abreast; section columns echeloned.

shelled areas, in passing through ordinary shelling, and in passing through woods, fog, or smoke. A section column is better for control; but as it is somewhat more compact, it is more vulnerable than squad columns.

   e. Sequence of deployment.—A platoon usually deploys into section columns first, and maintains this formation

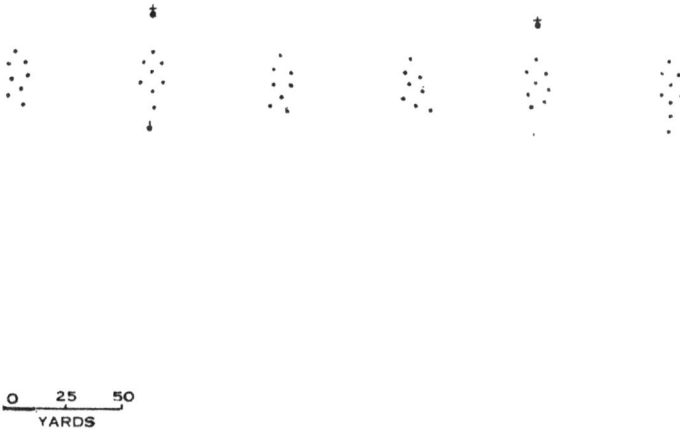

FIGURE 77.—Sections abreast; sections in line of squad columns.

until a more extended formation is necessary or desirable. It changes from section to squad columns ordinarily as soon as the advancing troops come under long-range rifle and machine-gun fire. This fire may at once necessitate, however, a complete deployment into lines of men at wide and irregular intervals, advancing successively, or by bounds from cover to cover. (See Figures 78 to 80.)

   f. (1) Figure 81 represents a platoon passing through thick woods in which long-range artillery fire is falling intermittently. The platoon leader has his unit pass through the woods in column of sections with each section in section column. The distance between the rear of the leading section and the head of the rear section is about 50 yards. This formation gives the greatest amount of control, and control is highly important under the conditions assumed. Sections in column, with each section in line of squad columns, might be used, but small groups easily become separated in woods. On the other hand, the disadvantage of using section columns is that a single artillery shell may cause great losses. If the zone of advance is under continual artillery fire, smaller columns must be used. In passing through very dense woods, a column of files is usually the only practicable formation.

(2) At the edge of the woods the platoon commander must change the formation. He knows that the enemy is on the hill across the stream to the front. The enemy has fair observation of the area to be crossed. Moreover, the edge of the woods is well within hostile rifle and machine-gun range. There is some cover, however, along the creek, and in two places this cover extends up toward the woods. He decides, therefore, to have his sections leave the woods in squad columns, the leading sections to go to the right and the rear section to the left as shown. Each section leader is to work his squads forward a squad at a time, and then move through the woods to the stream. In thus crossing the open ground squads should remain separated from each other by at least 40 yards distance and interval so that no two squads would be within the burst of a single shell. In crossing the gap the squads should crawl forward on the ground utilizing every irregularity as cover. This method of advance still permits control and at the same time is less vulnerable to artillery fire than section columns.

(3) Upon reaching the creek the platoon begins to receive small-arms fire from the hill to its front. The platoon leader then has one section complete its deployment with individuals roughly in line along the stream bed and about 5 yards apart. This is not a straight line, but on the con-

FIGURE 78.—A platoon not yet under hostile fire advancing in line of section columns.

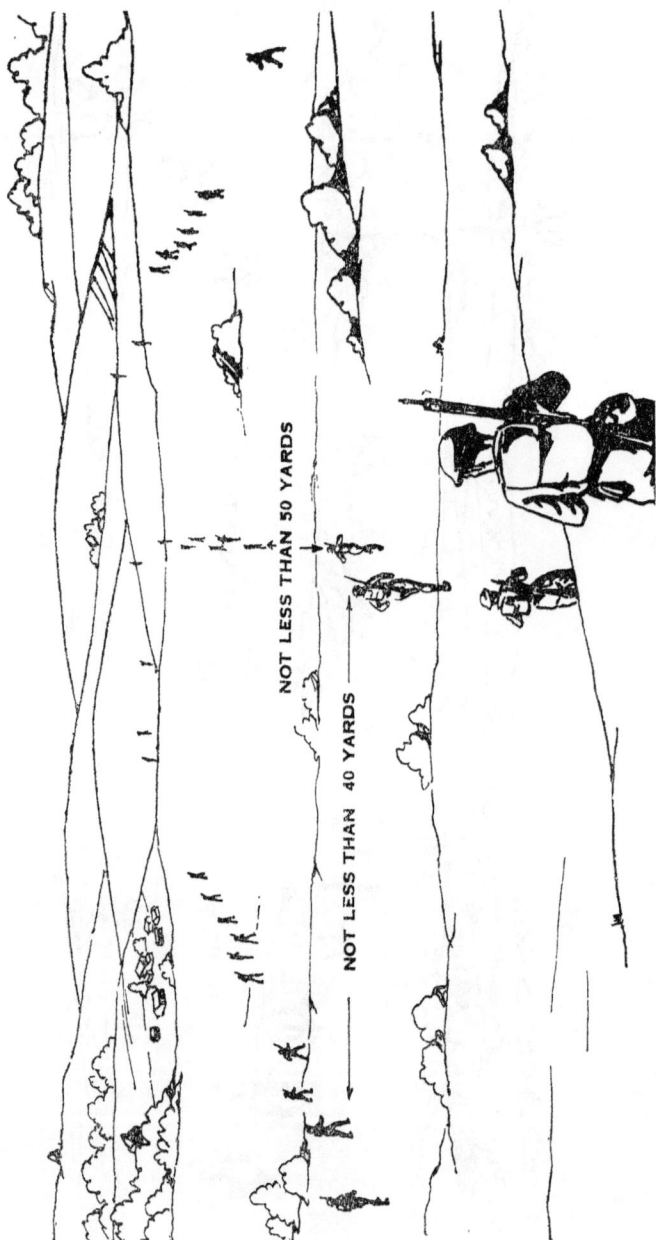

FIGURE 79.—Column of sections in line of squad columns. Unit not yet under small-arms fire.

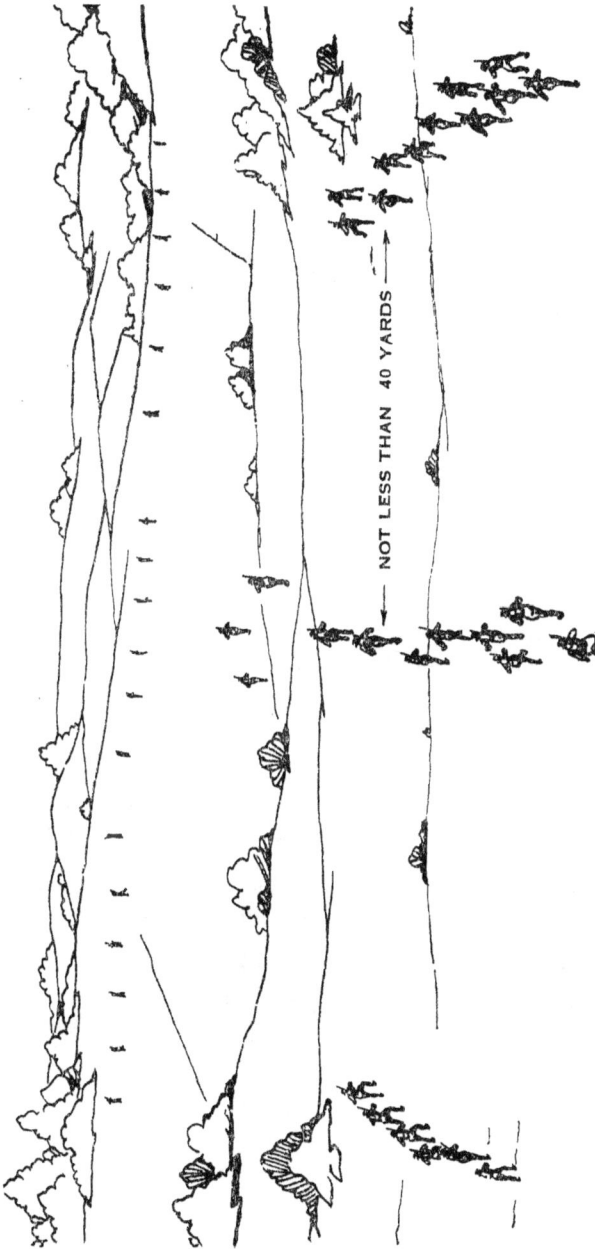

FIGURE 80.—Column of sections, leading sections in irregular line. Unit not yet under small-arms fire.

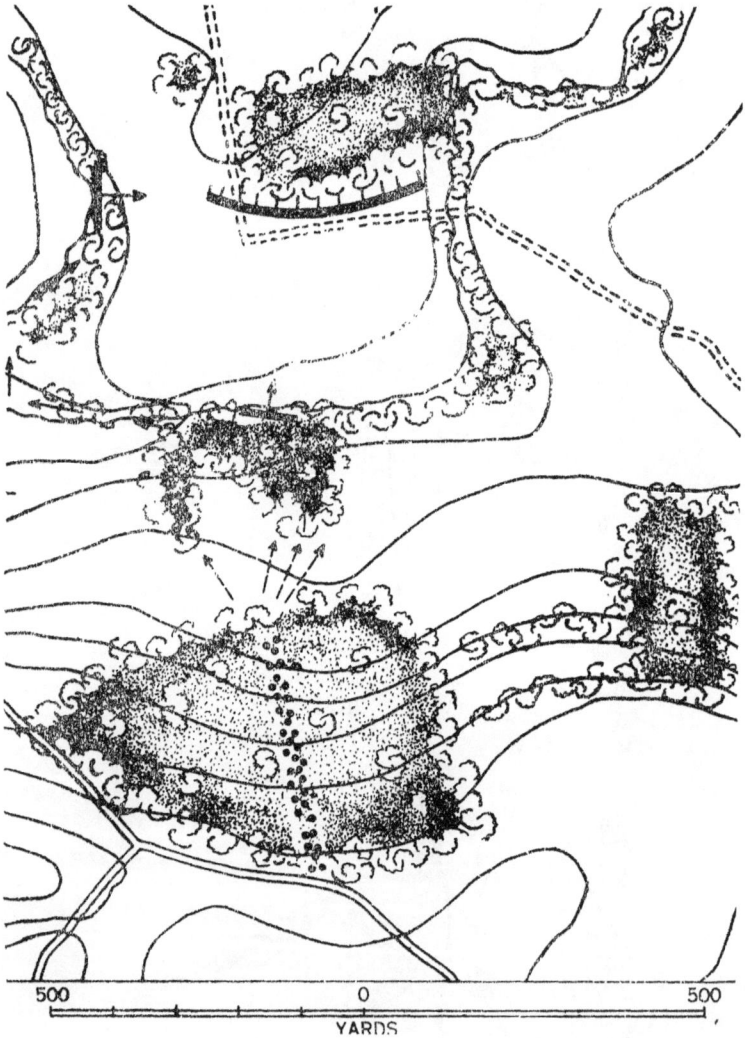

FIGURE 81.—A deployed platoon passing through thick
woods under intermittent artillery fire.

trary a very irregular line, because each man must take advantage of whatever cover is afforded by the terrain in his vicinity. This formation is not so easily controlled as squad and section columns, but is much less vulnerable to the fire of small arms and artillery.

(4) The platoon leader also directs the leader of the left section to work his squads along the stream bed around to the flank of the hostile position, giving him a definite length of time to accomplish this movement, and ordering him to attack at a given hour. He may take charge of the flanking section himself and give orders to the right section leader instead.

(5) Both sections carry out their attacks, of course, deployed in irregular lines working up the slope by creeping and crawling and by brief fast squad or individual rushes from one point of cover to the next.

g. (1) When a platoon first partially deploys, scouts from the leading section deployed at wide intervals (10 to 50 yards) are sent forward to precede the platoon. It is not an invariable rule for all six scouts of the leading section to be thus sent forward In advancing over open terrain, with unobstructed observation for long distances, as few as two scouts may be enough. But if the terrain is broken and observation to the front and flanks is limited, then the employment of all six scouts is advisable.

(2) The scouts locate enemy positions, overcome resistance by small hostile advance posts and patrols, and cause the enemy riflemen and machine guns to open fire and disclose their positions. Without such protection a platoon is liable to walk blindly into dangerous areas. The exact distance forward of the platoon at which scouts operate is not prescribed. They may be as much as 500 yards ahead of the leading section.

(3) In the advance, the platoon leader must maintain control of his command and keep it generally in place in the company formation. He maintains the direction of

advance by azimuth or by moving on distant direction points. His position is between the scouts and his leading section. Here he can observe the scouts or keep in contact with them by means of his runners, continually reconnoiter, and issue instructions promptly to his section leaders. The platoon sergeant is usually between the two sections. His principal duty is to prevent them from merging. Section leaders keep ahead of their units; and when the platoon is in column of sections, the commander of the leading section watches the platoon leader for signals. He looks back to exercise control and vary the formation of his unit in order to take advantage of cover. The leader of the rear section looks to the platoon sergeant for signals and orders. When the platoon sergeant is not between the sections, the leader of the rear section conforms the movements of his section to those of the leading section. Section guides follow their units. They see that the corporals keep their squads in hand and that orders and instructions are complied with Squad leaders lead their squads. In section and squad columns the second-in-command of each squad marches at the rear; when the squad is fully deployed he is in the firing line.

5. *ADVANCING THE ATTACK.—a.* The method of advancing the attack has been covered in considerable detail in the preceding chapter. Here we shall take up only special points regarding the platoon.

b. (1) It should be apparent from the illustrative situation in paragraph 4f above, that a platoon does not adhere to any particular "formation" in the parade-ground sense of the word, during any part of its deployment and attack. This is true from the first partial deployment until the enemy is assaulted. Numerous variations and degrees of deployment are necessary, but nothing of precision enters into their employment in battle. These various controlled dispositions of the squad, section, and platoon—and even those of the larger units to a great degree—have for their primary end, not precision but the utmost flexibility. And only through this flexibility can units take full advantage of the ground over which they advance and fight. The use of terrain in offensive combat has been emphasized time after time in this text. But here again it must be reiterated that there is no chance of a successful attack against an enemy on the defensive, unless fire and ground are effectively combined.

(2) Thus, in an attack a platoon leader does not in any sense parallel his drill-field activities. Instead he handles his two sections and six squads as moveable maneuver groups fitting their actions to the battlefield. The same is true of the section leader, and the squad leader as well. More often than not, at any given moment in the course of an attack, the dispositions of any small unit would be well nigh unrecognizable as those of a particular "formation".

(3) It is also true in battle, and especially in the attack, that actual personal control of more than a few men at any given moment, becomes sooner or later an utter impossibility. A platoon commander under ideal conditions would move about observing the progress of all his elements, continually giving new directions for their employment. This he does as much as possible. But in the usual case, once he has decided upon, and told his subordinates, what his scheme of action is to be, he can only exercise control at intervals thereafter. A section leader, under ideal conditions, would remain generally in rear of his squads, watch the effect of their fire upon the enemy and direct their advance and the routes they are to follow. The section guide would stay where he could watch the section and still be in touch with the platoon commander, receiving signals from him for the section leader. But once an attack is begun and the hostile defensive fires are fully active, it is far more likely that the squads of a section

will be so separated by the maneuver necessary in using and gaining ground, that neither the section leader nor guide can exercise these functions except intermittently.

(4) The squad leader, perhaps, can come closer to maintaining a continuous personal control of his unit than the leader of any larger unit. But even his seven men may, in taking advantage of the ground to work around a small area of hostile resistance, for example, be of necessity so separated that immediate control is out of the question.

(5) In the final analysis, then, squad, section, and platoon leaders use their units and their men in an attack, not in specific semi-rigid "formations", but as maneuvering elements, any one of which may at any time be freely despatched on a mission that will take it completely beyond control, at least for the time being. It is only in this manner that a leader can use his means of combat and the ground of the battlefield effectively in conjunction.

c. Upon receipt of the company attack order the platoon leader makes an estimate of the situation and issues his preliminary instructions. Unless it is necessary to open fire at once to continue the advance, the platoon moves forward from the line of departure in a deployed formation suitable to the terrain, covered by scouts. When the scouts are fired on, they take cover, locate the enemy, and, if within effective range, return the fire. If practicable, they outline the enemy position by the use of tracer bullets. When the scouts are halted by fire, the platoon commander goes immediately to a position from which he can see them, observe the ground to the front, and examine the enemy position. He then decides on a plan of action and assembles the platoon sergeant and the section leaders, if the terrain and the enemy fire permit, to give them instructions. Otherwise, he sends orders to them by runners.

d. The plan of attack should usually include pinning the enemy to the ground by frontal and flanking fire, under cover of which the elements of the platoon maneuver forward. The assault section, assuming the platoon to be in column of sections, may build up a base of fire generally on the front established by the scouts, provided the terrain where the scouts are is advantageous. But if the scouts are in a fire-swept area the base of fire is established on more suitable ground. Each squad may establish its own base of covering fire; or if the ground and the scheme of maneuver permit, the base of fire may be established to cover the whole section. Unless orders have been given not to advance, or to advance only when directed by higher authority, the leader of an assault section works his squads forward over the terrain assisted by the covering fires. Each leader, and even each individual, strives to carry out the original scheme of maneuver.

e. (1) Where it is necessary to advance under fire across open ground, the advance is effected by successive rushes of individuals or small groups. Here the base of fire is most important, since its purpose is to smother the hostile fire and give the manuevering elements a chance to gain ground. Enough weapons must continue in action to accomplish this end. The length of a rush depends upon the use that can be made of the ground—the existence of cover—, and upon the degree of hostile fire. It may be 5 yards; or it may be 40 or 50 yards to the next point it is desired to reach.

(2) An advance by rushes should never be made, however, when it is possible to make the same gain with less exposure. Every particle of cover or defilade that the ground affords—depressions, shell holes, fences, buildings, walls, trees, embankments, stream banks, ditches, furrows, and even the natural vegetation of the ground—should be used in maneuvering toward or around a hostile defensive locality.

f. Although both sections maneuver within themselves and both may be employed to maneuver in the platoon, the rear section is more usually the maneuver unit. This element is ordinarily more available than the leading section to work around the flank of hostile resistance and envelop it. In such a movement the section is protected by scouts and the same methods of maneuver described above are employed.

6. *THE ASSAULT.*—*a.* The assault is delivered at the earliest possible moment that promises success. A demoralized enemy may abandon his defenses in the face of a determined attack even before he is threatened with the bayonet. Or he may resist with every means of combat until we close with him. The moment of taking up the assault cannot be determined ahead of time. It may be started by an order of a platoon or section leader; but more often, perhaps, it follows the spontaneous action of several individuals or small groups who make their own decisions. For by the time the elements of a platoon have worked out their commander's plan, they are likely to be somewhat separated so that a command for a simultaneous assault by all units could not be heard if given.

b. It is in a maneuvering element of the platoon that the assault is more likely to develop than in an element engaged in fixing the enemy with fire. The latter, however, may, as a result of distraction caused by the original maneuvering element, have opportunity in turn to maneuver more freely; and may, in the end, make the assault.

c. Wherever the assault begins, and whoever begins it, it should receive the immediate cooperation of every individual and unit within sight. Elements farther away than the assaulting units should at once move forward more freely, or, if still under heavy hostile fires, redouble their own fires to assist the elements closing with the enemy.

d. This discussion of the assault should not be taken as a departure from the methods of the attack discussed in Chapter 9 under the Conduct of the Attack. A platoon, like any smaller or larger unit, seeks the easiest lanes of advance through the enemy's dispositions. But advance along these lanes often requires hard fighting, maneuvering, and a final assault on certain hostile localities, if any progress at all is to be made.

e. When tanks are cooperating in the attack, the necessity for assault is less frequent, since the tanks carry out much of the assault themselves. Their work in battle is treated fully in a later chapter.

f. The student should also be careful at this point not to confuse the assault with large scale frontal charges against organized defensive positions. The assault, in modern warfare, is simply the last brief stage in the attack, which clinches the skillful use of terrain, fire power, and maneuver, by a final application of shock action. The assault has lost none of its importance, except that it is now to be thought of as employed opportunely and independently by small units such as squads and sections, rather than by larger units acting under direct leadership in a coordinated charge against hostile defenses.

7. *REORGANIZATION.*—*a.* After a successful assault, a platoon may have to advance still farther to reach its objective. The maneuvering of fractions of the platoon preceding the assault, and the assault itself, are almost certain to cause some disorganization. But as soon as a platoon commander can determine that the elements of his unit are still sufficiently intact and maneuverable to continue the advance, he issues orders for the continuation of the attack to the objective, including if necessary a new scheme of maneuver.

b. If a platoon has had heavy losses, it may be necessary for its commander to reconstitute squads and sections before further advance. Reorganization, however, goes on continuously throughout combat within the smaller elements. When reorganization after an assault is necessary, however, the area occupied should be protected by scouts, or small patrols, including automatic riflemen.

c. In general, a platoon halts to reorganize only when it

is forced to do so. If possible, it is not done until the final objective is captured.

d. In regrouping units, at least two squads or groups should be organized as a new leading section. A rear group should also be formed, even if it consists of no more than a single squad.

e. The platoon leader gives orders for the reorganization and then reconnoiters to the front and flanks. The platoon sergeant supervises the reorganization. Section leaders first regain control of their comands, then appoint section guides and squad leaders, if any of these have become casualties. The section guide assists the section leader in reorganizing, personally supervising the squads and seeing that all orders are carried out. Squad leaders report casualties to the section leader and replace the automatic rifleman and his substitute if necessary. They also designate new men as scouts and new seconds-in-command, if the old ones are missing. Ammunition is collected from nearby dead and wounded, and reissued. Prisoners are disposed of in accordance with instructions from higher authority.

8. *PURSUIT.*—Pursuit at first is a mere continuation of the preceding action characterized by greater boldness and a more aggressive attitude on the part of the attackers. All units are thrown in. The platoon pushes rapidly forward; men are detailed to watch the flanks, and precautions are taken not to get so far ahead as to become isolated. Such effort cannot be maintained for long, however, without loss of control. It is usually soon necessary to reconstitute a reserve and, turning pursuit over to it, reform the remainder of the platoon as it continues forward.

SECTION III

THE RIFLE COMPANY IN THE ATTACK

9. *DEPLOYMENT.*—*a.* Rifle companies, like smaller units, adapt their formations to the terrain over which they advance and fight. If little is known of the hostile dispositions, the platoons advance in a formation suitable for meeting any emergency. Defilade and concealment are utilized so that the hostile artillery and aviation will have as little chance as possible to observe the advance and place fire upon it. Gassed and shelled areas are ordinarily avoided. Crossing exposed ground not under fire can best be accomplished by a rush of the entire company. If such an area is under fire, successive rushes of individuals or small groups are used. In passing through woods, distances and intervals are reduced, and connecting files put out, so that adjacent units can maintain contact.

b. As a company advances, its commander moves to points from which he can observe the ground to the front, continually reconnoitering in order to plan ahead. The company commander is responsible for the security of his unit and sees that it is covered by scouts and that exposed flanks are protected by patrols, or by the reserve. Contact is maintained with adjacent units by connecting groups of four to eight men. These groups are usually detailed from the reserve. They march approximately abreast of that element. A company commander maintains communication with his platoon leaders, with the battalion commander, and with the adjacent units during all stages of the advance. Whenever he is away from his company, as on reconnaissance, he keeps in touch with his platoon commanders through runners.

Upon receiving an attack order, a company commander forms his plan for the employment of his company and issues orders to his platoon leaders. He assembles them, if it is practicable, at a point which affords a view of the objective and the ground over which the attack is to be made, and there issues his order orally. His order includes all information of the enemy not already known to the platoon leaders, and all information regarding adjacent supporting artillery, tanks, or chemical units, and supporting mortar and machine-gun fires; the mission of the battalion; its initial formation; its objective; the mission of the company and its scheme of maneuver, formation, zone of action, direction of attack, line of departure, and hour of attack; the combat mission of each platoon, its objective, and its direction of advance; detailed instructions to attached machine-gun units; instructions relative to security and maintenance of contact; the location of the battalion ammunition distributing point and the battalion aid station; and the location of the battalion and company command posts. When it is not possible to issue a complete attack order, a series of partial orders may follow during the course of the advance.

d. Formations in the attack.—(1) *General.*—In an attack, a company commander employs the deployed formations that are best suited to the combined use of fire, maneuver, and the ground. The known dispositions of the enemy, the mission of the company, and the width of its zone of action, are also factors that he must consider. The frontage assigned to a company is usually about 400 yards. Either one, or two, platoons can be deployed in assault with two or one in reserve; or the company may advance in a column of deployed platoons. In exceptional cases all three platoons may be placed in the assault echelon.

(2) *All platoons in line.*—The formation with all three platoons deployed in line gives the greatest initial fire power and maximum distribution in width. It lacks driving power and maneuverability, however, and should not

be employed except when the position of the enemy is definitely known and the objective is close.

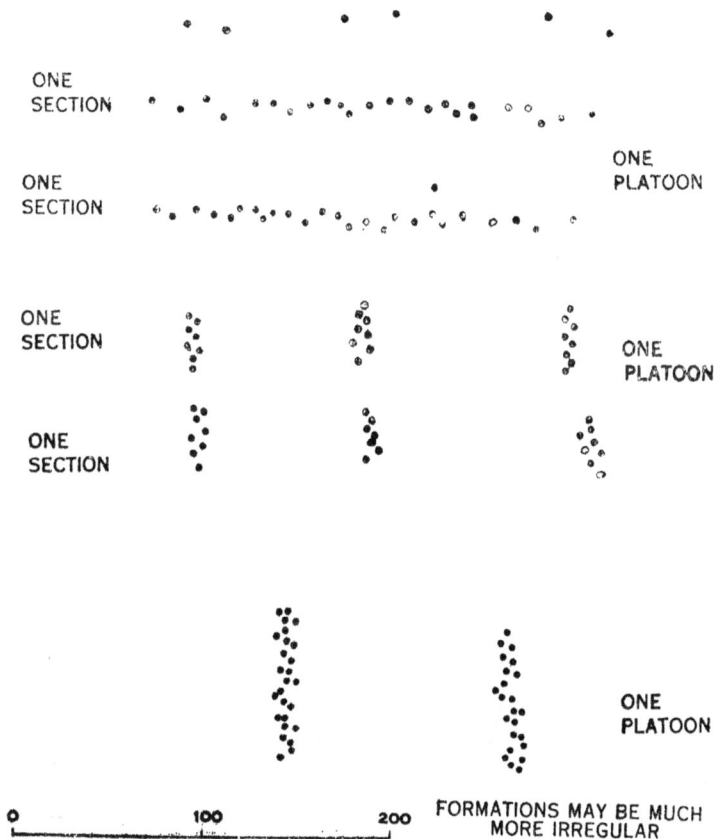

FIGURE 82.—A rifle company deployed in a column of platoons.

(3) *Platoons in column.*—The formation of platoons deployed in column allows the greatest latitude for maneuver and gives the greatest amount of security. It has the dis-

advantage that at first the company cannot employ its
maximum fire power. This formation should be used when
the zone of action is narrow, when the objective is at a

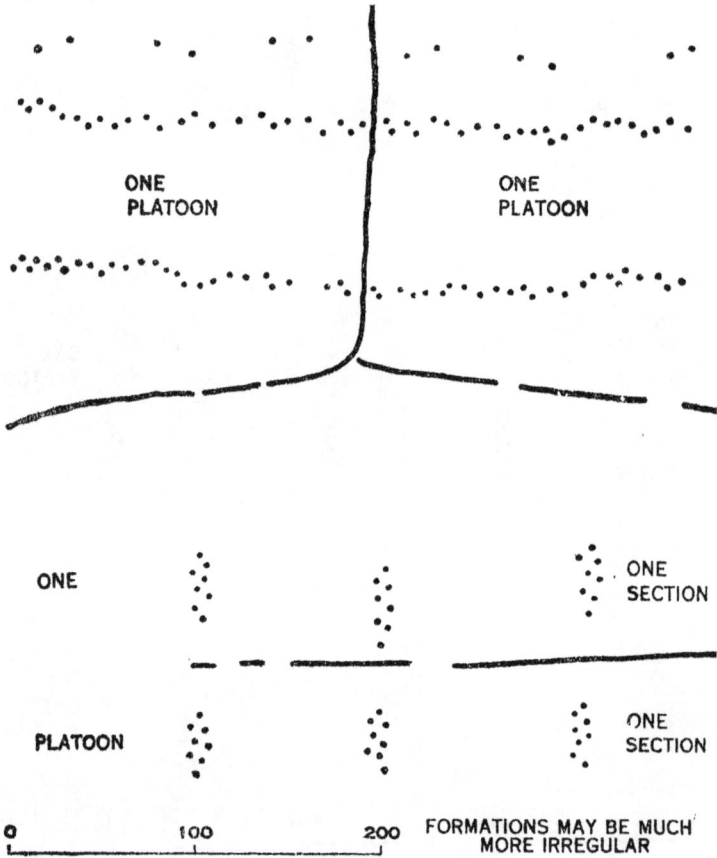

FIGURE 83.—A rifle company deployed with two platoons
in assault and one in reserve.

great distance, and when little is known of the enemy's
dispositions.   (See Figure 82.)

(4) *Two platoons in assault and one in reserve.*—At-

tacking with two platoons in the assault and one in reserve permits considerable initial fire power. This formation does not have the maneuverability of platoons in column or the initial fire power of platoons abreast, but, on the other hand, does not have the disadvantages of these. (See Figure 83.)

(5) *One platoon in assault and two in reserve.*—The formation of one platoon in the assault and two in reserve (diamond formation) has the advantage of furnishing excellent security and permits the company to face in any direction. (See Figure 84.)

e. (1) The above formations may be varied in many ways during the course of battle to meet any situation. Reserve platoons, for example, may be shifted from one flank to another to take advantage of defilade, cover a flank, or avoid gassed areas or areas covered by fire.

(2) It is not necessary for the whole zone of action to be physically covered, and this is seldom done. Gaps are often left purposely between the assault elements in order to permit machine guns to support the assault.

10. *ADVANCING THE ATTACK.*—*a.* (1) The assault platoons usually remain in small columns until they must reply to the enemy's fire in order to advance. This does not mean in any way, however, that small columns do not take every advantage of terrain in advancing when they first come under fire. From then until it is necessary to open fire, they work forward along the terrain, making no effort to stay in a regular formation. Once the leading elements of the company begin to fight, their further advance rests mainly in the hands of platoon, section, and squad leaders, who bend their whole efforts toward carrying out the scheme of maneuver originally ordered. The methods of advance of these elements were covered in the preceding section.

(2) The assault platoons make every effort to locate gaps and weak spots in the enemy's dispositions and work their way through them. They try to reach their objec-

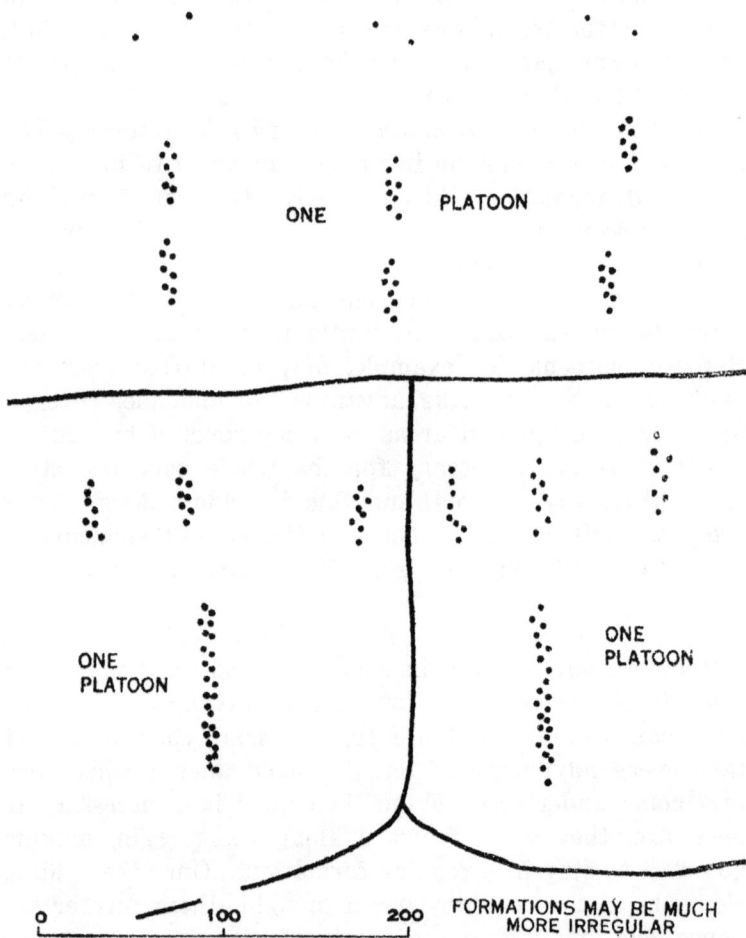

FIGURE 84.—A rifle company deployed with one platoon in assault and two in reserve.

tives with the minimum of losses and fighting. The company commander observes the progress of his assault platoons and helps their advance in any way he can. He reinforces the assault platoons whenever it becomes necessary, using the breaks or gaps made or located in the enemy's line by the assault platoons, as channels of advance for his reserve platoon.

(3) Elements that are held up are disengaged, unless the capture or neutralization of the hostile resistance is necessary for further advance.

b. (1) In Figure 85*a* scouts of the right assault platoon have located enemy resistance to their front and the scouts of the left assault platoon report no enemy in the woods to their front. The company commander directs the right assault platoon and the reserve platoon to follow the left platoon.

(2) In Figure 85*b* the right assault platoon is held up by the enemy while the left assault platoon is progressing satisfactorily. The company commander directs the reserve platoon to follow the route of the left assault platoon. When these two platoons have successfully passed the enemy resistance, the company commander has the right assault platoon disengage from the enemy resistance and follow the other two platoons. He would also in such a situation have the original reserve platoon take over the zone of the original right assault platoon.

(3) In Figure 85*c* both assault platoons have been held up by a resistance to their front and flanks. The company commander orders the reserve platoon to pass by the enemy resistance. He then disengages the other two platoons, designates one as reserve, and has the other continue on as an assault platoon.

(4) In Figure 85*d* the two assault platoons are held up by separate groups of the enemy at some distance from each other. The company commander has his reserve pass between the two points of enemy resistance and proceed toward the objective. The remaining platoons are disengaged, and follow.

(5) The situation in Figure 85*e* is somewhat similar to 85*d*. Here the company commander has the choice of sending the reserve platoon by three different routes.

(6) In Figure 85*f* the assault platoons have been definitely stopped, and it is not possible to pass the reserve platoon by the enemy resistance. The company commander uses his reserve platoon to attack the enemy resistance in the flank.

c. A company may have to move out of its own zone of action into that of an adjacent unit in order to avoid strong enemy resistance. When this happens, the company commander should get in touch with the company commander of the adjacent unit and coordinate the movements of the two units. The company returns to its own zone of action as soon as the situation permits.

d. The company commander intervenes in the conduct of the assault platoons only when it is apparent that his instructions are not being complied with. His principal duties are to supervise and coordinate the action of all three platoons, and assist the advance by calling for artillery, 37-mm gun, mortar, and machine-gun fire.

e. Several factors govern the distance of the reserve elements from those of the assault. The reserve must have space enough for possible movement or actions toward either flank. On the other hand, it should be close enough to be readily available for the reinforcement of the assault units in the exploitation of their successes, or for any other maneuver. The reserve, however, must not be permitted to merge with the assault elements. It advances by bounds from one covered position to another on orders of the company commander. It is held under cover when it is not advancing.

f. A company commander moves during the attack where he can best supervise the action of his platoons. His best position is usually with the reserve or between the reserve and the assault echelon, although he should at all times be able to observe the action.

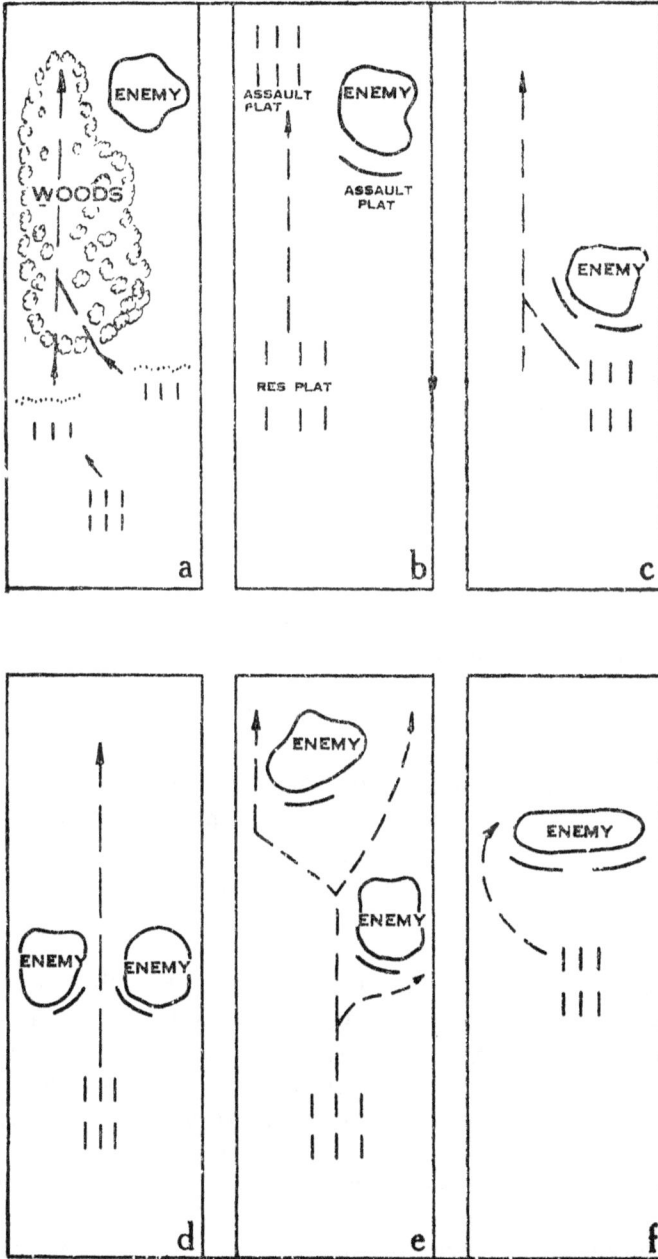

FIGURE 85.—A rifle company advancing the attack.

11. *THE ASSAULT.*—The assault is launched only from short distances as described in the preceding section. Troops cover their final advance by means of assault fire.

12. *PURSUIT.*—Pursuit is characterized by greater boldness and a more aggressive attitude on the part of the attacker. At first it is a mere continuation of the preceding action. All units are thrown in, and every effort is made to continue pressure on the enemy. He is given no pause that might give him a chance to reorganize or disengage. It is impossible to maintain such effort for a long period of time without loss of control. Therefore, it is usually necessary to reconstitute a reserve and, turning pursuit over to it, reform the remainder of the company while continuing forward.

13. *ORGANIZATION OF THE GROUND.*—When the advance is definitely stopped or the objective has been reached, the company commander organizes the area as directed by the battalion commander. In the absence of instructions he acts on his own initiative, reestablishes contact with adjacent units and the rear, assigns areas of defense and sectors of fire to his platoons, and arranges for mutual support by flanking fire. He makes a report of his dispositions to the battalion commander, preferably by sketch.

14. *A RESERVE COMPANY IN THE ATTACK.*—At least one rifle company is usually designated as battalion reserve. This unit operates under the direct orders of the battalion commander. However, the commander of a reserve company must be prepared to act without hesitation in emergencies, in repelling a sudden enemy counterattack or in protecting an exposed flank. A reserve company is moved from position to position or from cover to cover, as directed by the battalion commander, or in accordance with his general instructions. The commander of a reserve company reconnoiters to the front and flanks

whenever he has opportunity. He may precede his company and take station where observation is favorable; or he may accompany the battalion commander. When he does not accompany the battalion commander, he maintains contact with him by having his first sergeant and two company runners stay at the battalion command post. Before moving the company from one position to another, the company commander selects the route and decides upon the method of advance.

SECTION IV

MACHINE-GUN UNITS IN THE ATTACK

15. *GENERAL.*—*a.* Machine guns have been called "concentrated essence of infantry" because they combine tremendous fire power with considerable mobility. These qualities make machine guns one of the most effective means of supporting the attack. Moreover, they are so essential to the attack that the degree of effectiveness of their support may well mean the difference between failure and succeess.

b. A plan of attack should include full use of the fires of all machine guns in assault units, not only for initial support, but for continuous support thereafter. (Rifle company machine guns, which are primarily defense weapons, may be used in certain situations for additional initial support only.) Moreover, when it is practicable, the machine guns of reserve units should also be given supporting fire missions.

c. The machine-gun company is an integral part of the infantry battalion. Its employment is not confined to specific missions in support of the assault rifle units during the initial stages of an attack. Machine-gun units move forward with their battalions, supporting their advance whenever possible throughout the battle. They advance from one suitable position to another so as to furnish the maximum supporting fire, continually engaging new targets suitable to their fires, as they appear.

d. When the fire support provided by a part of a machine-gun company is adequate, a section or platoon may be directed to move forward close in rear of a rifle unit maneuvering to gain ground from which hostile positions can be attacked from a flank. The rapid action of a machine-gun element in such a manner, on ground well to the front and flank, provides a powerful close support, well located for breaking up hostile counterattacks.

e. The commander of an assault battalion is responsible for the proper tactical use of his machine-gun company. He employs it to further the scheme of maneuver of the whole battalion. He can use it in several ways.

(1) He may retain it under his control and employ the whole company as a single unit.

(2) He may retain it under his control and direct one or more platoons each to support an assault rifle company.

(3) He may attach one or more platoons each to a rifle company.

(4) He may use any combination of (2) and (3), such as placing two machine-gun platoons in support of a single assault rifle company and attaching the third platoon to another rifle company.

(5) In a single attack, he may use the company in several of the above ways in succession. For example, he may use the whole company as a single unit for its initial fire missions, and then place one or more platoons each in support of an assault rifle company as the attack progresses.

16. *MACHINE-GUN POSITIONS.—a. The ideal position.*—The ideal machine-gun position in an attack:

(1) Affords a long and wide area of fire over which fire can be safely delivered during much of the advance of the supported rifle units.

(2) Affords concealment from hostile observation. A gun seen by the enemy is a gun soon out of action, or at least forced to move.

(3) Affords protection from hostile fire.   Once the enemy places accurate fire upon a gun, that gun's supporting fire soon ends, unless the utmost use of defilade or accidents of the ground protect the gun and crew.

b. Limitations on positions.—Fires in support of advancing rifle units are limited to:

(1) Fires through gaps between supported elements.

(2) Fires over the heads of supported elements.

(3) Fires from positions on the flanks of the supported units.   The best firing positions are usually found in rear of the supported elements on high ground that commands the terrain over which those elements advance.   Suitable positions on the flanks of attacking units are rare.

c. Selecting positions.—(1) Positions are selected by machine-gun company or platoon commanders in conformity with the initial position of the supported infantry elements and their routes of advance.   The commander determines whether direct or indirect fire is best, and selects a general position for each platoon or section, or the exact position for each gun, if he has time.

(2) Guns in partial defilade have the greatest degree of protection from hostile fire, although the gunners cannot see the targets and therefore cannot fire as readily upon fleeting targets.   The simple indirect-laying methods used are very satisfactory, however, and cause little delay in laying and firing.   The observer controls and adjusts the fire from a position near the gun.

(3) Sight defilade can often be secured.   Here the guns are just behind a crest, bank, or ditch, or in a ground fold so that only the barrel and sight of the gun and the head of the gunner are exposed to hostile observation.

(4) When there is no position wholly or partially in defilade, concealment in shady areas, or behind brush, or grass, should be sought.

(5) Lacking a position from which fire can be delivered overhead, through a gap in the line, or from a flank, a machine-gun platoon leader informs the rifle unit commander and requests him to provide a gap.   As a last

resort, guns can be placed abreast of assault rifle units which are halted. This is objectionable because machine guns draw fire which endangers the supported troops and hinders their advance; and also because the fire of the guns may be wholly or partly masked soon after the supported elements move forward.

17. *MACHINE-GUN SQUADS AND SECTIONS IN ATTACK.*—*a. Deployment.*—(1) Machine-gun sections are under the control of the platoon leader until deployment is completed. When deployed, section leaders prescribe formations best suited to the conditions of the advance. An irregular column formation, with 20 to 100 yards between squads, is usually best because it requires reconnaissance and clearing for only a single route. Moreover, when the squads follow approximately in trace, they are not liable to become separated. The leading squad clears the route. If the advance is long and the route difficult, the work of trail breaking may be given to squads in turn. Under hostile observation and fire, machine-gun sections fully deploy like nearby rifle units.

(2) For crossing short stretches of open country under enemy observation and artillery and long-range machine-gun fire, a rough line of squad columns with intervals of from 20 to 100 yards is suitable. This formation is better than squads following in trace, because a leading squad may draw fire that will later strike succeeding squads. Moreover, a column formation is most vulnerable to enfilade fire. In the irregular line formation, the section leader usually marches midway between the squads, with both of them thus under his immediate control.

(3) A formation with echeloned squad columns combines distribution in width and depth. This is useful for crossing open country with scattered cover, and when marching on an exposed flank.

(4) All such formations, it must be well understood, can be greatly varied to conform to the conditions of battle.

In the strictest sense they are not "formations" but battle dispositions.

b. Supporting the attack.—(1) Platoon leaders usually precede their platoons and select the first firing positions. The platoon sergeant conducts the sections to an unloading point near the positions (a defiladed spot, if possible), and superintends the unloading of equipment. While the unloading takes place, the section leaders go forward to receive orders.

(2) The platoon leader tells the section leaders what the situation is, points out where the enemy and friendly troops are, and indicates general positions to be occupied by sections. He also assigns definite targets or sectors of fire to each section.

(3) Section leaders then go to their respective positions. If the squads have not already been sent forward by the platoon sergeant, section leaders first return to their units and conduct them forward. As his squads approach, a section leader signals for his squad leaders to join him. He points out the approximate positions for their guns and gives them orders for the occupation of the positions. Fire orders may be issued then, or later after both guns are in position. Each squad leader selects the exact position for his gun and directs its mounting. All men take the utmost care to prevent the enemy from observing the work of mounting the guns, for if the guns are seen they will immediately draw fire, and their installation and operation will be difficult, if not impossible. Whatever natural cover is at hand is used by the gun crews and other men of the squads as protection against hostile fire. The guns of a section are ordinarily located from 20 to 50 yards apart, but if good natural cover against enemy fire is available, this distance may be less. The gun carts, pack mules, or squad trucks remain under platoon control in rear under cover, in charge of the transport corporal.

(4) When a platoon leader assigns definite targets the sections engage these. When he assigns sectors of fire,

section leaders direct the fire of their guns on the most important targets that appear in their sectors. They must continually observe the enemy's position with glasses to determine organized areas holding up our troops. Lacking any definite targets, areas where a machine gun or automatic rifle may be concealed should be searched with fire. When the sector of fire is large, with fleeting targets appearing and disappearing rapidly, it may be necessary to assign half of the sector to each gun in order to engage such targets more rapidly. The section, however, is the usual machine-gun fire unit, and best results are obtained when the fire of both guns is concentrated on the same target. Surprise is obtained by secrecy in getting into firing positions and by causing both gun squads to open fire on signal from the section leader.

(5) Unless speed in getting into action is extremely important, it is best not to open fire until both guns are in position. The noise of firing may attract attention and the movement of the second gun into position may disclose the location of the section.

(6) Fire is opened at definite targets in bursts of from 10 to 30 rounds in order to destroy them instantly. When desirable, however, smaller bursts are used to conserve ammunition.

(7) Squads take care of their own supplies of ammunition and water, carrying them forward when the carts are unloaded. When the distance to the carts is great, or there is only one covered route of approach, section leaders combine the carriers of both squads into a single system of supply; in this manner ammunition is brought forward to a point convenient to both squads.

(8) Machine guns, because of their effectiveness in battle, are primary targets for hostile fire, and are sure to draw fire as soon as they are discovered. Whenever the enemy offers determined resistance to an attack, it is usually necessary to continue firing from one location for a considerable period of time. Consequently, section lead-

ers select alternate firing positions for use if the first positions begin to receive accurate hostile fire.   Section leaders may move their guns to the alternate positions at their own discretion.   They do not, however, have authority to withdraw their guns entirely because of enemy fire.   Alternate positions should have the approval of the platoon leader.

c. *Advancing the attack.*—(1) Platoon leaders decide when and how to move their sections forward for the close support of the advancing rifle units.   One section ordinarily moves at a time.   The section first to go forward goes out of action.   If the section transportation is to be used, the section leader sends a man to bring it up.   While waiting for the trucks (or carts), he has the ammunition collected and the equipment prepared for loading.   He announces the formation to be used in moving forward. As when contact is imminent, the section follows covered and defiladed routes where it can, and makes every effort to keep out of the enemy's sight.   When the distance of displacement is not far enough to warrant using transportation and the section is not under hostile rifle and machine-gun fire, the guns are moved by hand as shown in Figure 86.   The section as a whole takes up a deployed

FIGURE 86.—Moving machine guns by hand.

formation similar to that of near-by rifle units.   Under enemy observation and fire, the section works the guns forward to suitable positions, taking full advantage of the terrain for cover.   On arriving near the new position, the section is again halted in a defiladed place if possible. When practicable, the platoon leader has a messenger meet the section and conduct it to the position by the best route. The section leader goes forward, joins the platoon leader, and receives his instructions. The section takes up its new position as it did the old one.

(2) The section left in position continues to cover the advance of the rifle units. It remains in position, especially on the alert since it is performing the work of both sections, until the platoon commander directs it to move forward. This may be done at a prescribed time, on a signal, or when the leading section goes into position, or when it opens fire. The platoon commander prescribes the manner in which the rear section will move forward. In the usual case it can make greater use of trucks or carts than is possible for the leading section.

d. *Reorganization and consolidation.*—When for any reason an attack halts, machine-gun sections continue their supporting fires from the firing positions in which they are at the time. Changes in the location of guns at such times must be specifically ordered by the platoon leader.

18. *THE MACHINE-GUN PLATOON IN AN AT-TACK.*—*a. Deployment.*—(1) A machine-gun platoon operates either alone or as a part of the company. So long as the whole company operates as a unit, a platoon leader keeps his unit where it belongs in the general company formation. But in order to get his platoon forward with a minimum of losses, he suits its formations to the variations of cover and concealment available, and to the character of the hostile fire. When the front of the company is extended so that the platoons advance along separate routes, each platoon leader selects his own route in accord with general instructions from the company commander, and determines the distance at which his platoon follows the rifle companies.

(2) Cover, especially when animals and carts are used, is very desirable; and platoons follow the best concealed

routes available. It is also important when trucks are used, although trucks may often have to follow more open routes. Ordinarily, however, their speed will enable them to cross exposed areas rapidly. When the enemy has been forced back or is apparently withdrawing, a platoon does not usually advance from a concealed or defiladed position until the rifle units ahead have determined whether the next position in front is cleared of the enemy troops. Thus it follows the rifle units by a series of bounds, and is always ready to assist them in a resumption of the attack. When hostile fire and lack of cover make it inadvisable to continue the advance with trucks or carts, a platoon carries its equipment by hand, using formations and methods similar to those of a rifle platoon, where necessary working forward along the ground toward its new positions. When the platoon is advancing on an exposed flank, a portion of it must be continually ready to go into action quickly to protect that flank.

b. Supporting the attack.—(1) The principal mission of machine-gun units in an attack is to assist the rifle companies to advance. The primary targets of their fire are the hostile personnel operating the weapons that make it hardest for the rifle units to advance. This includes fire on low-flying attack airplanes that threaten the advance.

(2) For machine-gun support to be effective, good firing positions are essential. As stated in paragraph 16, positions should be selected, when they are available, that—

(a) Allow guns to fire with safety to friendly troops during the greater part of their advance (positions not soon masked by friendly troops).

(b) Have concealed and defiladed approaches.

(c) Have defilade and concealment from hostile fire and observation.

(3) When commanding positions are not to be found, fire must be delivered through gaps between rifle elements, or from a flank; or from positions abreast of rifle units.

(4) When cover is good, a machine-gun platoon may be able at first, during an attack, to advance without firing

until it reaches a position at short range from which it can support the rifle units during the final completion of their deployment and their subsequent attack against their first objective. When lack of cover requires early support to cover the completion of rifle-unit deployment, all or a part of the platoon may have to occupy a position farther back, later moving up to a better position for supporting the attack.

(5) (a) The platoon leader usually precedes his unit in order to reconnoiter routes and firing positions and de-- termine where changes of formation are necessary. When the company supports the initial attack as a unit, the platoon leader moves his command into the area assigned, and designates firing positions and targets or sectors in accordance with the instructions of the company com- mander. When a platoon supports an assault company, the first firing positions and targets may be prescribed by the machine-gun company commander or by the platoon leader. While the platoon leader is reconnoitering routes and positions, the platoon sergeant leads the sections to a covered place close in rear of the first firing positions. Ordinarily the guns are then taken from their carts or trucks and carried forward by hand.

(b) The platoon leader issues his instructions for the employment of the platoon at or near the first firing position. If he desires the section leaders to join him before their sections arrive, he sends a runner back to get them. In his instructions to section leaders, the pla- toon leader points out the location of the enemy and friendly troops and explains the plan of attack, making sure that they thoroughly understand the mission of the platoon. He points out positions for each section and their respective targets or sectors of fire. Meanwhile, the platoon sergeant superintends the unloading of equipment and sends the squads forward, or, if so directed, holds the platoon under cover, to await further orders.

c. Advancing the attack.—(1) When the company is supporting the attack as a unit, the company commander, often accompanied by one or more platoon commanders, usually goes forward to reconnoiter and select new firing positions and routes thereto. He may also prescribe, before thus going forward himself, when and how the platoons will move forward. When a platoon is operating apart from the company (supporting, or attached to, a rifle company), the platoon leader performs these duties. The platoon supports the attack from its initial position as long as its fire is of value to the advancing troops. When its fire becomes masked by friendly troops, or rendered ineffective by gas, smoke, or fog, the platoon leader moves his sections to new firing positions. In these circumstances, the platoon should, of course, move to more advanced positions where they are accessible, unless the situation does not favor such action.

(2) (a) The entire platoon may advance at one time, but when it is supporting an assault company, it is more common to jump one section at a time. The section that remains in position fires at targets in the sector of the other section as well as its own, as long as the safety of friendly troops permits. Gun crews must be alert, watch the flanks and gaps in the line of assaulting troops, and be prepared to engage hostile attack planes. They must also be ready at all times to meet any hostile counterattack. In most such situations, the rear section usually remains in position until the advancing section has occupied its new firing position. The platoon leader, before going forward, points out the probable location of the new firing position (if known), prescribes the method of advance to be used and the general route to be followed. He regulates the time of movement of his sections by prescribing the distance at which he is to be followed, and the hour of movement or the signal for movement. It may be necessary for him to send messengers for this purpose. When the platoon leader goes forward with his runners to reconnoiter the route of advance and select the new firing positions, he may, if necessary, mark the route of advance by messengers, slashings, broken twigs,

and rocks, or, at night, by white tape. As a rule both
sections follow the same route.

(b) When the section arrives near the new firing position, its leader is told where to put his guns, what targets
to engage, and what rate and kind of fire to use. He
also receives any new information of the enemy or of
friendly troops. Thereafter the platoon leader exercises
control as at the initial firing position. If one section
has remained in the former position to support the advance
until the other occupies its new position, its leader must
be instructed in the same way as soon as he brings his
unit forward.

d. Reorganization and consolidation.—When an attack
halts to reorganize, consolidate gains, or regain control,
or is stopped by hostile resistance, the machine guns must
be disposed to cover the front and flanks against counterattack. The attack may terminate at a predetermined objective; it may halt to enable commanders to reorganize
forces that have become scattered during the fighting, or
to allow fresh troops to pass through and continue the
attack; or it may be stopped by the strength of the hostile
resistance. In any case, great reliance is placed on the
machine guns to hold off the enemy while reorganization
is taking place, or while a position is being prepared for
defense. The positions of the guns during temporary halts
often require no changes except where they cannot cover
probable routes of counterattacks. The location of rear
platoons or sections may be altered or these units may
be ordered to forward or flank positions. Any changes
in the location of guns or units must be rapid and entail
as little movement as possible. All guns should be constantly ready to meet a hostile counterattack.

19. *THE MACHINE-GUN COMPANY IN THE AT-
TACK.*—*a. Deployment.*—(1) The disposition of the machine-gun company when the battalion is deployed or partly
deployed is prescribed by the battalion commander. It
should be such that the company can use the best concealed

approaches within the battalion zone of action; that is, far enough behind the rifle units to afford protection against surprise fire from advanced hostile detachments, yet close enough to the front to provide prompt support when resistance is encountered.   The formation of the subordinate units in the machine-gun company during the advance to contact and the various stages of deployment corresponds roughly to those adopted by the rifle companies, usually small columns.   The purpose of using similar formations is to dissemble the appearance of the machine-gun units and minimize casualties.

(2) Concealed routes for the animals and carts (or trucks) are especially important, because the transportation discloses the identity of the unit to hostile observers. Guns and equipment are moved forward on their carts or trucks as far as practicable in order to spare the crews and retain mobility.   However, a point is usually reached where direct observation by the enemy demands that the guns be advanced by hand.   During the advance by hand, the carts or trucks are kept as close to the guns as circumstances permit.   They may utilize covered routes not tactically suitable for the guns, or they may be held under cover until the enemy is driven from his position.   As we have already seen in an earlier paragraph, the company commander precedes his company to reconnoiter routes, decide when and where formations should be changed, select positions, and gain information of the situation.   If the platoons are assigned separate routes of advance, the company commander regulates their movements to insure readiness for the support of any battalion mission.

(3) The battalion commander habitually takes the machine-gun company commander with him on reconnaissance.   During or immediately following his reconnaissance, he formulates his plan of attack.   The machine-gun company commander assists by reconnoitering for suitable positions from which his platoons can most effectively support the rifle units initially; and, when called upon to

do so, he submits his recommendations for their employment. After the battalion attack order has been issued, the machine-gun company commander completes his plan, making such additional reconnaissance as time permits. He has his platoon leaders meet him at a point where observation is good, and issues his orders for the attack, pointing out on the ground all important terrain features, visible friendly and hostile positions in the battalion and adjacent zones, and the areas to be occupied by the machine-gun platoons. His orders should follow the sequence of the five-paragraph form of combat orders. They may contain such of the following items as are applicable:

(a) A brief summary of the enemy situation as known to him.

(b) The general plan of attack of the regiment and any special support to be given the battalion.

(c) The battalion plan for the employment of the rifle companies and any attached units.

(d) Any special signals authorized.

(e) Initial positions and missions for platoons not attached to rifle companies.

(f) The hour of opening fire; and, in some cases, the initial rate of fire.

(g) Administrative arrangements, particularly the method of ammunition supply.

(h) The location of the battalion and company command posts.

b. *Supporting the attack.*—(1) Machine guns can only promote an attack when they are firing. Their value in battle is in direct proportion to the amount of ammunition effectively expended. The first mission of machine guns in the attack is to smother the enemy with fire and force him to withdraw or take cover; or, failing that, to neutralize the effectiveness of his fire to such an extent that friendly rifle units can complete their deployment and maneuver to within assaulting distance of hostile defense areas without great losses. After a position has been taken or ground gained, machine guns have a second mission which is to assist by fire in holding the gains.

(2) The primary consideration in assigning tasks to the machine-gun units of an assault battalion is to insure that their fire support furnishes the maximum assistance to the battalion plan of attack. In most situations this requires the bulk of the fire for the support of the rifle element, or elements, known or believed to have the best chance of getting forward.

(3) To insure the greatest success from machine-gun fire the guns should be sited, whenever practicable, to secure either surprise or flanking fire, or both. Surprise may be obtained by careful selection and undisclosed occupation of positions, by frequent changes in positions, and by the time of opening fire. No effort should be spared to work guns up close to the enemy's position, utilizing natural cover as well as that afforded by the fire of other guns remaining in position. The effect of unexpected machine-gun fire at short range is demoralizing to the defenders. Decisive results demand prompt and aggressive action by subordinates and the utmost use of terrain. Flanking fire is far more effective than frontal fire, because more of the long, narrow beaten zone of the machine gun is placed on the target. Gaps caused by the irregular advance of different units appear frequently in an attack and present many opportunities for flanking or oblique fire. These gaps should be utilized when supporting fire can thus be made more effective.

(4) In order to obtain surprise and flanking fire, the machine guns of assault battalions must be used aggressively in close support of the attacking riflemen. There are four dispositions that facilitate close support.

(a) With the company operating as a unit in general support of the battalion under direct control of the company commander.

(b) With one platoon in support (not attached) of an assault rifle company, or one platoon in support of each assault rifle company, the remainder of the company operating under direct control of its commander.

(c) With one platoon attached to an assault rifle company, or one platoon attached to each assault rifle company, the remainder of the company operating under direct control of the machine-gun company commander.

(d) A combination of (b) and (c).

(5) (a) The machine-gun company of a reserve battalion frequently operates as a unit under the company commander; but this is not so often the case in assault battalions. The machine-gun company of an assault battalion may be so employed where there are suitable positions for all platoons close together from which hostile targets to the front or flanks of the battalion can be readily engaged. This method is most often applicable when the battalion is attacking in column of companies and therefore on a relatively narrow front.

(b) The allotment of one or two platoons for the direct support of assault rifle units, with the remainder of the company under the control of the company commander is the most common disposition. Closer and more effective support is thus provided. This furnishes better support: when the battalion is attacking on a broad front; when the flanks of assaulting units become exposed to hostile fire or maneuver; and on broken or wooded terrain where positions close together cannot be found for all platoons.

(c) The machine-gun company commander supervises, in so far as practicable, the operations of platoons in direct support of assault rifle units. In the absence of specific instructions platoon leaders perform all combat command functions, such as, determining targets to be engaged, positions to be occupied, times to open and cease firing, and the time and method of displacing forward. Although it is true that a supporting platoon is not controlled by the commander of the supported rifle unit, the platoon leader should nevertheless, at all times, cooperate with a view to furnishing the maximum support. To this end he must be in constant contact with the rifle unit commander in order to place the fire of his platoon where it is most needed when it is most needed.

(d) The attachment of one or two platoons to assault rifle companies while the remainder of the machine-gun company operates under the company commander is suited to situations wherein a rifle unit is sent on a semi-independent mission (as a wide envelopment or a long advance through woods), or when the advance is expected to be rapid. An attached platoon operates directly under the commander of the rifle unit to which it is attached, and the machine-gun company commander has no control over it. When attachment is not imperative, it is preferable to have a machine-gun platoon support without being attached, in order to relieve the rifle unit commander of the burden of controlling an additional unit.

(e) It is often best to attach one platoon to one assault company and place another platoon in support of another company. Various combinations of this kind may be used, in accordance with the needs of different situations.

(6) The support of an attack includes covering the completion of the deployment, the initial advance of the rifle units, and assisting them in the initial fire fight and maneuver, and close and continuous support during the successive stages of the whole attack. A single firing position for all phases is the ideal for machine-gun units. This is often possible for the first two phases. However, if hostile fire and lack of concealed approaches prevent the advance to a position well within effective range for covering the first phase, two or more positions may be necessary. The probable maneuver of rifle units frequently governs the selection of positions; it must always be considered. If available, commanding ground is occupied in order to secure overhead fire. Positions permitting flanking fire should not be overlooked.

(7) When the company is engaged as a unit, the company commander selects the firing positions for each platoon (in the general locality assigned by the battalion commander), and controls the fire in accord with the instructions of the battalion commander. As the attack

progresses, he diverts fire to new targets as required by the needs of the assault units.

(8) When one or more of his platoons are supporting assault companies, the company commander may select the first firing positions and designate the first targets. His object is to insure coordination of fire so that the platoons, in addition to supporting the designated assault companies, are also assisting adjacent units with oblique or flanking fire and, in general, are furthering the battalion scheme of maneuver. The company commander, by personal reconnaissance and by utilizing his liaison agents, keeps in close touch with the assault elements, and makes recommendations to the battalion commander as to the manner in which the company can best support the attack in the later stages of the action.

c. *Advancing the attack.*—(1) To make machine-gun support continuous, the company, when acting as a unit, ordinarily moves forward by platoons, using the leap-frog method. At least one platoon remains in position, firing or ready to fire, while the others are advancing. Where the platoons are widely separated, it may be advisable to advance a section from each, rather than to send forward a whole platoon. Platoons and sections ordinarily move from firing position to firing position. The rear platoons or sections may either remain in position until the advancing fraction is established in its new position; or they may follow at a specified time or distance, or upon a prearranged signal. Short moves are avoided, and no movement is begun until it is apparent that the assaulting troops have occupied, or shortly will occupy, the next firing position. Before the company or any part of it advances, a detailed reconnaissance is made for covered routes and for new positions that provide cover and sufficient elevation for overhead fire. If no such positions exist, machine guns may have to fire through gaps in the line of riflemen. The reconnaissance officer, accompanied by messengers, follows the assault echelon and keeps the company commander informed about the situation, routes, and firing positions. Agents from the platoons keep in touch with the assault companies and provide their platoon leaders with like information.

(2) The first units of the company to arrive at a new firing position are put into action without delay in the handiest suitable position that satisfies the requirements of the situation. As soon as arrangements are made to get the leading units in action the others are sent for, if not already on the way, and, on arriving, are installed in positions selected for them.

(3) When the battalion is halted because of resistance encountered, it is necessary to neutralize, drive out, or destroy the resisting elements before the advance can be resumed. In order to do this and regain fire superiority, all available fire power is employed. All guns of the company should be placed in positions from which they can fire on the hostile forces that have checked the attack. Quick decision and prompt action are necessary. Time must not be wasted in seeking ideal positions; guns should be placed in any available position, even in the firing line, and fire opened promptly. When a rifle unit is dispatched to outflank the hostile resistance, the terrain may permit a section or platoon of machine guns to accompany and support it. In that event, the machine-gun company commander should use his judgment, and, lacking battalion orders, detach the necessary guns if he can thereby further the plan.

d. Assault.—Under cover of the fire of artillery and infantry weapons, as well as their own, the rifle units advance to within assaulting distance of the enemy position. During the assault, the fire of the machine guns is directed, if practicable, against hostile elements in positions which enfilade the assault units; or else on probable assembly positions of hostile counterattacking units. When a foothold is secured in the enemy position, some guns should promptly be moved forward to resist counterattack.

e. Reorganization.—(1) When it is necessary, on account of casualties, disorganization, loss of direction or control, to halt the battalion and reorganize the assault companies, machine guns are particularly valuable for covering the reorganization. Guns are posted to cover the front and flanks of the battalion. Since counterattack may be expected, it is essential that positions selected have good fields of fire and afford maximum security to the disorganized battalion. This usually requires a distribution in depth of the elements of the machine-gun company.

(2) During the reorganization, fire superiority may be lost. It must be regained before the advance can be resumed. Machine guns are especially valuable in regaining fire superiority and assisting the resumption of the advance. In order to employ all available guns for this purpose, it may be necessary to change the position of some of the elements disposed to cover the reorganization. The new positions should be occupied in time to assist in the preparatory fire. In moving into new positions gun squads must be alert to avoid being caught en route by an enemy surprise action.

f. Organization of the ground.—When a battalion has reached its objective, machine guns are employed in a manner similar to that used for covering the reorganization of assault units. The first duty of the machine-gun unit is to cover the rifle troops it is supporting. The platoon leader keeps close touch with the commander of the company that he supports and assists him in organizing his position by providing protection for the front and flanks of his area. The machine-gun company commander, under the supervision of the battalion commander, coordinates all the elements of his company in order to provide the maximum protection for the front and flanks of the battalion area. He also arranges with the machine-gun company commanders of adjacent battalions for mutual flank protection. The means taken and methods followed are essentially the same as when a defensive position is organized. The organization of a defensive position is covered in detail in Special Text No. 9.

SECTION V

HOWITZER UNITS IN THE ATTACK

20. *HOWITZER COMPANY SQUADS IN THE ATTACK.—a. Deployment.*—As with machine-gun units, the primary reason why howitzer squads deploy is to permit the rapid advance of their weapons with the minimum of losses. The platoon leader prescribes the initial formation. Squad leaders, however, are responsible for maneuvering their squads to take full advantage of the terrain.

b. Supporting an attack.—(1) In supporting an attack, the platoon leader prescribes the first general firing position and squad leaders choose the exact locations for their weapons. Each weapon, if practicable, is placed so that it can cover the entire battalion zone of action as well as terrain features outside the zone from which the enemy can fire on the battalion.

(2) If for any weapon such a position cannot be found, a position is selected from which the weapon can cover that part of the battalion zone where hostile targets are most likely to be found. In such a case, supplementary positions are also chosen near by from which the remainder of the battalion sector can be covered by fire. All positions, whether for the 3-inch trench mortar, the 81-mm mortar (Stokes-Brandt), or the 37-mm gun, must be well concealed.

(3) The 3-inch trench mortar and 81-mm mortar fire at high angles. Accordingly, defiladed positions are used for these weapons whenever such positions are available.

(4) The 37-mm gun, unlike the mortars, has a flat trajectory. For most effective use, it requires firing positions on commanding ground, so that it can be fired over the heads of friendly troops without endangering them, and so that it can cover the entire zone of action of the

battalion. The firing of this weapon is difficult to conceal. Natural cover, such as a wood, thicket, tall grass, fold in the ground, or depression, forms its most suitable position. The reverse slope of a hill, near the crest, where the gun is hidden from the view of the enemy but can still be fired, is perhaps the most desirable position, especially when natural cover is also present.

(5) Both the gun and the mortar should, if possible, have positions near defiladed or covered approaches, in order to facilitate ammunition supply and communication to the rear.

(6) Alternate positions are always selected for the howitzer company weapons and prepared for use. This is particularly necessary for the 3-inch trench mortar because it takes several minutes to install the base plate of this weapon. Alternate positions, for use when hostile fire makes the initial position untenable, should have the same field of fire as the primary positions. They should, however, be far enough away not to receive fire directed at the original positions.

(7) Leaders of howitzer units must be where they can observe the effect of fire and control their units. The observer should be at the gun. If this is not practicable, a position on or near the line: gun-target, either in front or rear of the gun, is next best for preparing firing data and controlling the fire. The observation position should have concealment and cover.

(8) The ammunition supply for both weapons of a howitzer platoon is maintained by a single ammunition squad which brings the ammunition to points near the weapons, whence it is carried forward by members of the gun squads. Twelve rounds of 3-inch trench-mortar ammunition and two boxes (32 rounds) of 37-mm gun ammunition are held in reserve to be fired only on the orders of the platoon leader.

(9) It is possible to expend in a short time all the ammunition supplied for the howitzer-company weapons.

The selection of suitable targets is therefore of the utmost importance.

(10) The accuracy of its fire and the small area of burst of its projectiles make it desirable to employ the 37-mm gun against relatively small targets. Hostile machine guns that have been definitely located are its principal targets. Tanks are also appropriate targets and must be fired on whenever they appear within range. (For the present, the 37-mm gun must be considered purely as an antitank weapon in combat against an enemy equipped with tanks.) Other targets that may be engaged are troops in close formation, enemy small cannon, accompanying artillery, observation posts, and snipers. The tactical situation determines whether these targets are important enough to justify the expenditure of ammunition.

(11) The great value of the mortar lies in the large area covered by the burst of its shell. Its principal targets are machine guns, accompanying artillery, light mortars, and similar weapons, which have not been definitely located but whose location is known; hostile troops in counterattacks; and areas, particularly woods and ravines, known to be occupied by enemy troops. Mortars may also be used to provide local supporting fire for short periods and to put smoke on hostile observation posts and gun positions.

c. Advancing the attack.—After the attack is once initiated, howitzer-company weapons should be in position firing, ready to fire, or in movement forward to new positions. Since it is seldom practicable to locate the two weapons close together, squad leaders must be trained to recognize suitable targets for their own weapon. The movement forward is regulated by the platoon leader and is made by bounds from one firing position to another. The movement is made on carts or trucks when possible, taking advantage of defilade and concealment. Guns are kept within supporting distance of rifle units. The platoon

leader orders the movement, but squad leaders anticipate
the order and prepare to move without delay.

d. The assault.—Howitzer-company weapons may be employed in addition to artillery and machine guns, to cover
the advance and maneuver of rifle units to within assaulting distance of the enemy.   Their fire is stopped, or
changed to other targets, by signal or by a prearranged
schedule of fire before the advancing riflemen mask it.

e. Reorganization, and organization of the ground.—After a hostile locality has been occupied, both guns are
used to repel counterattack, firing on machine guns or
other weapons supporting it.   They also fire on counterattacking troops when they present suitable targets.   The
mortar is employed to fire on areas where the enemy is
likely to be concentrating to initiate a counterattack.

**21. *THE HOWITZER-COMPANY PLATOON IN THE
ATTACK.*—*a. Deployment.*—**A howitzer-company platoon
attached to a battalion is assigned a position in the battalion formation by the battalion commander.   Its position
should enable the platoon to use the best concealed route
of approach in the battalion zone of action.   It should
be far enough behind rifle units not to come under the
fire of advance hostile elements, but far enough forward
to provide prompt support.   Concealed routes for animals
or trucks are especially important.   The formation of
advance is adapted to the terrain.   When hostile fire and
lack of cover make it inadvisable to continue the advance
with carts or trucks, the platoon advances its weapons
and ammunition by hand, using formations and methods
similar to those of a machine-gun platoon.   The platoon
leader ordinarily precedes his platoon to reconnoiter routes
and firing positions and determine where changes of formation are necessary.

b. Supporting the attack.—(1) The main mission of the
platoon is to reduce the effectiveness of the enemy's fire
by combating his machine guns.   The 37-mm gun engages

hostile machine guns that have been definitely located. The 3-inch trench mortar, or 81-mm mortar (Stokes-Brandt), searches areas in which machine guns are believed to be located. Because of its great accuracy, the 81-mm mortar (Stokes-Brandt), can also be used against definitely located machine guns. If hostile tanks appear, they are engaged by the 37-mm gun. The 3-inch trench mortar, or the 81-mm mortar (Stokes-Brandt), as was stated earlier, shell probable assembly positions for hostile counterattacks, which cannot be reached by other weapons. When ammunition is ample, the 3-inch trench mortar, or the 81-mm mortar (Stokes-Brandt), can also be used to bombard important points in the hostile position and put smoke upon them if it is desirable to do so.

(2) Positions for both weapons should be near good observation points, from which hostile machine guns can be hunted, and fire on them observed, as well as the progress of the supported troops. Firing positions near the leading elements of the attack are only used when there are no suitable positions in rear.

(3) During the progress of the attack the howitzer-company platoon engages suitable targets as they appear, first ascertaining that its fire is not masked by friendly troops. Ammunition should be conserved for the principal mission of the platoon, and fire should be reserved for use against appropriate targets of primary importance.

c. *Advancing the attack.*—When assault elements mask their fire, the howitzer weapons are moved forward rapidly to other positions by carts or trucks if practicable. The platoon leader precedes his command to its new position. Before he goes forward he directs whether the platoon is to follow at once or whether it is to remain in its present position and advance on signal or order. If the platoon is to follow at once, he prescribes its route, the place where it is to halt or receive further orders, and also the place, if determined, where the animal transportation will halt. He reconnoiters the new position and

the route thereto, and endeavors to have firing data ready
and all arrangements made for putting the platoon in
action as soon as it arrives.

d. Reorganization and consolidation.—The howitzer company weapons are the chief defense of the battalion
against enemy machine guns.   When the advance of the
battalion is halted for reorganization or to consolidate a
position, the howitzer company weapons are disposed to
oppose possible counterattacks against the front and flanks
of the battalion.

22. *THE HOWITZER COMPANY IN THE ATTACK.*
—*a. General*—The howitzer company seldom operates as
a tactical unit.   The company organization is primarily
for administration and supply.   In the attack, one platoon may be attached to each battalion, one may be attached to each assault battalion, and the remainder of
the company held in reserve; or the whole company may
be held under the direct control of the regimental commander.   It is often best to attach one platoon to each
assault battalion.   The part of the company not so attached may operate under the regimental commander
in support of the initial phase of the attack, when it is
practicable for such elements to rejoin the reserve battalion or battalions before the latter are engaged.

b. Route column and deployment.—During the advance
in route column the regimental commander prescribes the
location of the howitzer company in the regimental column.   If platoons are to be attached to assault battalions,
they join their battalions at the first stage of deployment.
If the company is to act directly under the regimental
commander, he prescribes its position in the deployed
formation.   When hostile fire, or any other condition,
renders it inadvisable to march the company farther as
a unit, it is deployed by its commander.   As with the
machine-gun company, the presence of carts and mules

or trucks discloses the identity of the howitzer company
as an organization of supporting weapons. To conceal
its identity and minimize casualties the company takes
advantage of cover and adopts formations similar to those
of adjacent companies.

   c. *Supporting the attack.*—When the howitzer company
or a portion of it goes into action under the company commander, the regimental commander assigns missions and
approximate positions for the unit. Under these conditions, the company commander makes preliminary reconnaissance, decides on positions for the platoons, (when
they are not too far apart), and works out the details
for the most effective and economical accomplishment of
the assigned mission. When the platoons are separated
by a considerable distance, the company commander supervises the most important platoon. He directs the fire of
the company whenever practicable. Its fire cannot be
controlled by oral commands; hence he prescribes beforehand the time of opening fire, the rate and kinds of fire,
and the targets to be engaged.

   d. *Advancing the attack.*—It is exceptional for a howitzer
company or any portion of it not attached to assault battalions, to advance as a unit in general support of an
attack. At some time during an attack it is usually necessary to employ all battalions as assault units and at that
time they should each have the additional fire power of
at least one howitzer-company platoon. The regimental
commander, however, should never let an opportunity pass
to assist the assault battalions by fire from the howitzer-
company platoon with the reserve battalion. It is generally practicable for this platoon to assist one of the assault
battalions during the initial stages of the attack, reverting
to the control of the reserve battalion commander when
its fire is masked, or the attack has progressed beyond
range. It should not become so involved in the attack
that it cannot support the reserve battalion when that
battalion is committed to the assault.

SECTION VI

THE INFANTRY BATTALION IN ATTACK

23. *DEPLOYMENT FOR BATTLE.*—*a.* In Chapters 1
to 6 we saw the manner in which deployment for battle
is accomplished.   The formations that a battalion takes
up during the advance to contact vary with the terrain
just as in smaller units.   At the first partial deployment
of a force, a battalion may at once partly deploy or may
be able to continue farther in route column.   In the orders
for deployment, battalion commanders prescribe the formation, the zones of advance, the direction of advance,
the base company, security, reconnaissance, and flank
contact detachments.

b. The direction of march is designated by prominent
terrain features or magnetic azimuth, or both.   The advance is regulated as described in Chapter 6.

c. The formation of the battalion depends upon its
mission, the enemy situation, and the ground to be traversed.   It should permit the employment of the battalion
in any direction, and therefore should have depth.   Rifle
companies may be echeloned to one or both flanks.   The
machine-gun company may march as a unit; or its platoons may be attached to, and may march with, the rifle
companies.   In either event, rifle companies provide security for the movement of the machine guns.   When a
howitzer platoon is attached, it usually marches in the
rear or center of the battalion formation.   The headquarters company usually advances in the center of the battalion formation.   (See Figures 87 to 92.)

d. The combat train (less its kitchen section) marches
directly in rear of the battalion when the situation permits.   Otherwise it moves forward from cover to cover
under the supervision of the battalion S-4.

e. The battalion commander and his staff march with,
or in advance of, his leading units.   He continually reconnoiters to the front and flanks with a view to main-

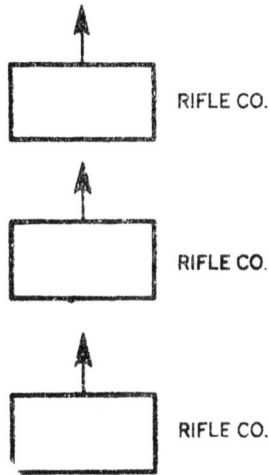

FIGURE 87.—Battalion in attack in column of companies (diagrammatic only); one company in assault; two in reserve in depth; supporting weapons are placed where they can best render support.

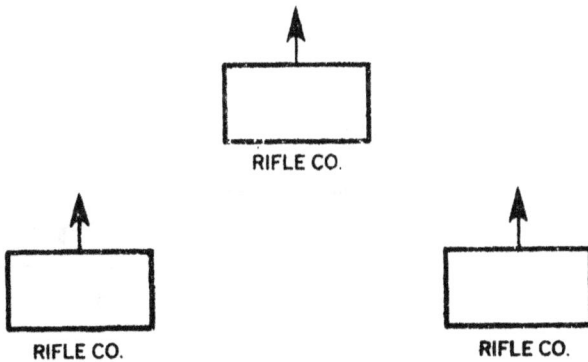

FIGURE 88.—Battalion in attack; one company in assault (diagrammatic only); two in reserve abreast and covering each flank; supporting weapons are placed where they can best render support.

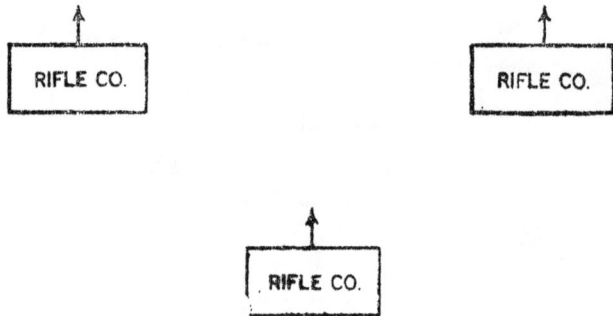

FIGURE 89.—Battalion in attack (diagrammatic only); two companies in assault; one in reserve centrally located for use on either flank. Supporting weapons are placed where they can best render support.

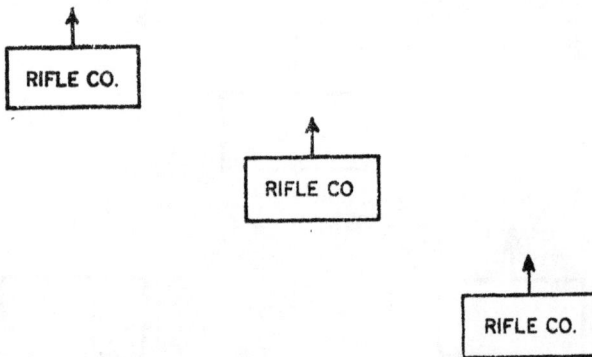

FIGURE 90.—Battalion in attack (diagrammatic only); companies echeloned to the right rear to cover a threat from the right flank. Supporting weapons are placed where they can best render support.

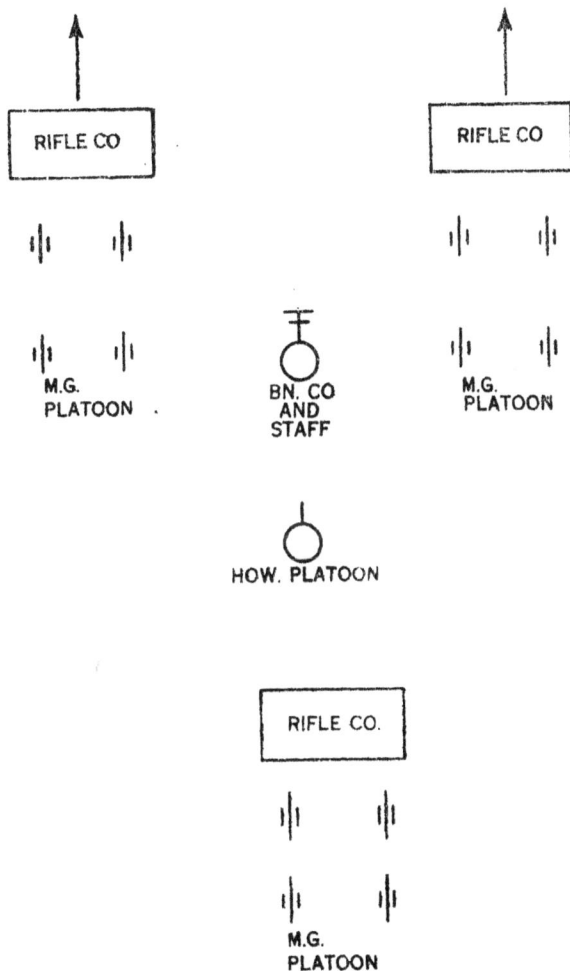

FIGURE 91.—An initial deployed formation for a battalion (diagrammatic only).

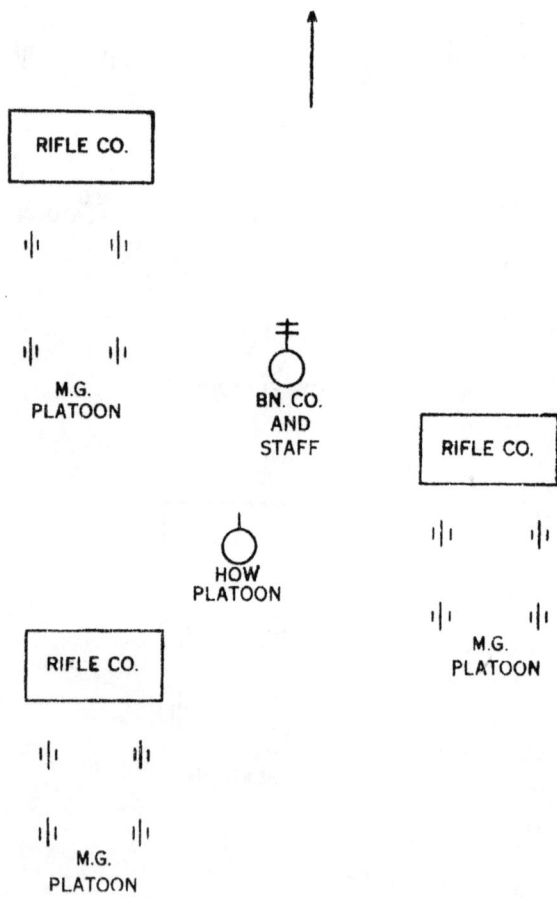

FIGURE 92.—An initial deployed formation for a battalion (diagrammatic only).

taining direction and determining favorable routes.

f. If the battalion is an advance guard, it moves forward as described in Chapter 6.

24. *ASSEMBLY AREAS.*—The advance thus deployed may continue directly to the attack or to an assembly area where final preparations for the attack are completed. (See paragraph 6, Chapter 5.)

25. *THE ATTACK ORDER.*—*a.* Attack orders are received and issued as described in Chapter 8. The battalion commander takes with him, when he reports for orders, such members of his staff as he deems necessary. In general, the more that hear the order, the better. But often considerations of congestion and confusion at the point of issuance of the regimental order, and the possible disclosure to the enemy of command group movements may keep the reporting party small. Also, enough assistance should be left with the battalion executive, who takes charge while the regimental commander is absent.

b. Upon receipt of the regimental order, the battalion commander makes his reconnaissance, and formulates and issues his order as described in Chapter 8. Before proceeding on reconnaissance, he may direct his subordinate commanders to meet him at a designated point and at a set time to receive his order. He may order the movement of the battalion to a position nearer the line of departure. In making his reconnaissance, he consults the members of his staff and his subordinate commanders that are present. For example, he may call upon the machine-gun company commander to recommend the use to be made of his company, this to include the gun positions, targets to be engaged and the general method of support. In a like manner, he may ask the commanders of the howitzer platoon and other attached units (as tanks) to make recommendations. If the time available for reconnaissance is limited, or the zone of action is large, he may require members of his staff to reconnoiter parts of the zone for specific purposes, such as finding locations for administrative establishments, and ascertaining what obstacles and cover exist.

c. The battalion commander usually issues oral orders to company and attached unit commanders and their staffs. assembling them when practicable. The order is issued if possible, from a point where the ground over which the attack is to be made can be seen. If time does not permit this, the order may be issued in fragmentary form and staff officers or messengers used for its dissemination. The battalion commander should release his subordinate commanders early enough so they can make their reconnaissances and issue their orders.

d. The battalion attack order should include as many of the following items as are necessary:

(1) Such enemy information as concerns the battalion.

(2) The mission of the regiment, to include its initial formation and its objective.

(3) The locations and missions of adjacent troops.

(4) The supporting fires to be furnished by artillery, tanks, and chemical units, and any other supporting fires.

(5) The mission of the battalion; its objective, formation, zone of action, line of departure, and hour of attack.

(6) Instructions for each rifle company in furtherance of the battalion plan of attack.

(7) Instructions to the machine-gun company, to include initial positions and initial fire missions; attachment of platoons to rifle companies, if any.

(8) Instructions to attached units.

(9) Instructions relative to security, antiaircraft and antitank defenses, and maintenance of contact both within the battalion and with adjacent units.

(10) Arrangements for lifting supporting fires.

(11) Instructions regarding trains; the location of ammunition distributing points and the aid station.

(12) The axis of signal communication, the battalion command post, and instructions for the establishment of company command posts.

26. *MOVEMENT TO LINE OF DEPARTURE.*—The attack order should be issued before the battalion leaves the last available cover in rear of the line of departure. It may include instructions for moving the companies from an assembly area to the attack. The formation varies with the terrain crossed, the distance from the hostile position, the cover available, and the battalion plan of attack. Platoons of the machine-gun company may be used to cover the advance, as described in Section IV.

27. *FORMATION FOR THE ATTACK.*—*a.* The formation of the attack depends upon the known or estimated enemy situation, the terrain, the mission, and the frontage assigned. The frontage assigned a battalion is ordinarily from 800-1200 yards. A battalion may attack with companies abreast, companies in column, with two companies in assault and one in reserve, or with one company in assault and two in reserve. The battalion commander does not ordinarily prescribe the formation to be adopted within the companies, but in order to insure that the front is adequately covered, he may do so.

b. The battalion usually attacks with companies abreast. when the position of the enemy is definitely known and when the battalion has to make a short advance on a broad front. The battalion usually attacks with companies in column when knowledge of the enemy is vague. The leading company, like the leading platoon in a rifle company, finds and fixes the enemy, leaving the remaining companies free to maneuver. This formation may also be used when the enemy situation is known, and when the battalion zone of action is narrow and a deep penetration is required. The formation with one company in assault and two in reserve is merely a variation of the column formation. The battalion may attack with two companies in assault and one in reserve when the zone of action cannot be properly covered by one company, and when considerable fire power and shock action are desired initially.

c. The position of the reserve companies in any of the above formations should be such that they can take full advantage of cover, furnish flank security for the battalion, and be available for prompt use in combat. It is not necessary for the companies to physically occupy the entire width of the battalion zone. Parts of it may be covered by the fires of supporting weapons such as machine guns, mortars, and artillery.

d. If tanks are attached to the battalion for the attack, the formation of the battalion may have to be modified. In his order, the battalion commander gives the tanks a definite task but does not specify the methods by which they are to carry it out. Unless there are vital terrain features that must be taken under any circumstances, tanks should not be employed until definite resistance is encountered; then they should be used for the reduction of this resistance. Since tanks cannot hold captured ground, the battalion must promptly occupy objectives reached by the tanks. To reach these objectives, the troops need not follow the path of the tanks if a better route is available. The battalion commander must not fritter away the power of this weapon on unsuitable missions like reconnaissance, but rather conserve it for the full exploitation of its protected fire power, mobility, and shock action, at the time when the foot troops can follow and profit most.

28. *ADVANCING THE ATTACK.*—*a.* After the battalion is committed to the attack, a battalion commander can only alter the formation of his units by the movement of the reserve. It is through coordination and control, and the employment of his reserve and his supporting fires, that he influences the outcome of the attack.

b. During the attack, the advance of the assault companies is mainly in the hands of subordinate leaders. The battalion commander assists the attack by the effective use of all supporting fires. He keeps in touch with the artillery liaison officer, continually informing him of his plans and of the progress made, and calls upon him for fires on specific targets, giving hours for fires to begin, and the hour or signal for lifting.   He directs the machine-gun company to place concentrated fires upon localities checking the advance of rifle companies, requiring the machine-gun units to change their positions if necessary.   He may use machine-gun fires to cover the flanks of the battalion, to cover gaps between assault units, or to protect the flanks of assault units in passing enemy resistance.   In a like manner, howitzer company weapons are employed against suitable targets to further the advance of the rifle companies.

c. During the attack, the battalion commander keeps in close touch with his assault units.   He moves the reserve to take advantage of favorable opportunities to push his advance deep into the hostile position.

d. He employs the reserve promptly when any of the following conditions exist—when the assault units have been held up or the attack is materially losing its forward momentum; when a gap exists in the enemy position which allows movement toward the battalion objective; and when a gap occurs through loss of direction by all or part of the assault units.   When it is ordered by higher authority, or when it is necessary to widen a gap before further advance can be made, the reserve is employed against the flanks or rear of the hostile resistance.   If no gaps exist in the battalion zone of action, the reserve may be employed in the zone of an adjacent unit to pass by the resistance and continue the advance toward its objective.   Movement in an adjacent zone must be coordinated with the commander of that zone.   This may be accomplished through the regimental commander, or directly with the adjacent battalion commander.   As soon as the reserve is thrown into combat, the battalion commander endeavors to reconstitute a new reserve.

e. When it becomes apparent that the attack will succeed in capturing the initial objective, the battalion commander designates a new objective and issues orders for the continued advance of the battalion.

f. The battalion commander is responsible at all times that proper security measures have been taken. This may be accomplished by making the assault units responsible for the protection of flanks, but preferably by detailing security detachments from the reserve. In addition, the formation of the battalion in depth provides security in that it facilitates effective action against hostile counterattacks or threats against flanks or rear. The battalion commander should provide for connection between units within the battalion and with adjacent units. This is done either by observation or by the detail of small groups from the reserve or assault companies.

g. The battalion may be ordered to assist the advance of an adjacent unit; but in the absence of such orders, the battalion commander should render such assistance to adjacent units as he can without jeopardizing or delaying the accomplishment of his own mission. For example, it may be possible to give fire support to an adjacent unit when part or all of his weapons cannot be employed to further his own attack.

29. *REORGANIZATION.—a.* When a battalion, on reaching an objective, is so disorganized that further advance is impracticable, it must hastily reorganize and continue the attack. Reorganization may involve regaining control, re-coordinating the efforts of subordinate units, constituting a new reserve, bringing up supporting weapons, or replenishing ammunition, any or all of these.

b. When the battalion is unable to advance farther, it must hold the ground gained. The battalion should be quickly reorganized and disposed for defense and provision made against hostile counterattacks. Whenever additional supporting fires or the advance of adjacent units reduce the hostile resistance, the battalion commander should resume his advance.

30. *THE BATTALION STAFF IN THE ATTACK.*—
The regimental order for the attack prescribes the initial
location of the battalion command post and its general
axis of movement. The post is established by the battalion
S-1 assisted by the communication officer. The battalion
S-2 establishes such observation posts as he considers
necessary, or as directed by the battalion commander.
S-3 is usually with the commander to continually assist
him. He aids in the preparation and issuance of the
battalion attack order, frequently drafting the written
order and coordinating details with other staff officers.
S-4 establishes the battalion ammunition distributing
point and coordinates the movement of battalion trains
when they are under battalion control. The executive
coordinates the activities of the staff and assumes command of the battalion when the commander is absent or
becomes a casualty. As he may have to make major
decisions during the absence of his superior, he generally
remains at the command post.

31. *COMMUNICATION AND OBSERVATION.*—*a.* As
the attack progresses, it usually becomes necessary to move
the command post in order to facilitate communication
and maintain control. Ample warning of a move should
be given by the battalion commander so that S-1 or the
communication officer can reconnoiter and make arrangements for the new command post. A portion of the
personnel operating the command post should be sent
forward to establish the new post, the remainder continuing at the old location until the new installation is ready
for operation. Normally the executive remains behind
until the new command post is opened.

b. An observation post is established on a commanding
spot that affords a view of the ground over which the
attack is to take place, and is connected with the command
post by telephone or other rapid means of communication.
In combat, there is no prescribed position for the battalion

commander; he usually stays at the observation post to watch the progress of his units and the fires of supporting weapons. In order to expedite artillery supporting fires, the artillery liaison detail should establish its observation post close to that of the battalion commander. The battalion observation post, like its command post, moves from time to time as the attack progresses so that the battalion commander can keep constantly in touch with the situation. The movement of both should be coordinated to insure rapid communication between them.

c. The battalion commander often keeps the commanders of reserve units with or near him in order that they may keep informed of the situation and thus be better prepared to carry out his plans and orders.

32. *THE RESERVE BATTALION IN THE ATTACK.*—*a.* A battalion as the reserve of a regiment, brigade, or division is disposed so that it can execute promptly any mission assigned to it. It may cover the flanks of assaulting battalions, repel counterattacks, relieve an assault battalion, or reduce isolated points of resistance that have been avoided by the assault elements. It is imperative that the reserve battalion commander keep informed of the situation and maintain contact with his immediate commander. He must anticipate the probable use of his battalion and reconnoiter frequently so that he can move his battalion promptly when he receives orders. The battalion commander obtains his information from the headquarters of his immediate commander, by personal reconnaissance, through observation posts established by his intelligence platoon, and by reconnaissance patrols.

b. The reserve battalion moves forward by bounds from cover to cover in conformity with the orders of the higher commander. During movement, and whenever it is halted, the battalion adopts deployed formations that will minimize casualties and facilitate prompt employment in any direction.

SECTION VII

THE INFANTRY REGIMENT IN THE ATTACK

33. *DEPLOYMENT FOR BATTLE.*—The regiment begins its deployment for battle as described in Chapters 1 to 6. The regimental commander, assisted by his staff, supervises and coordinates the partial deployment of his unit as soon as contact becomes imminent. In general. the regimental commander should be with or in advance of his leading battalion. The same general principles that apply to the disposition of rifle companies within the battalion, govern the formation and disposition of the battalions within the regiment. It should be remembered that the leading battalion often becomes an advance guard as shown in detail in Chapter 6.

34. *THE ASSEMBLY AREAS.*—If a regimental commander leaves his regiment to reconnoiter, to receive orders, or for any other reason, the regimental executive officer then supervises the advance. If the regiment moves into an assembly area and battalion assembly areas have not been previously designated, the executive, assisted by such staff officers as are available, conducts the necessary reconnaissance and subdivides the assembly area among battalions and special units. He also establishes a temporary command post in the assembly area.

35. *THE ATTACK ORDER.*—a. When a regimental commander receives an attack order from brigade, he reconnoiters, and formulates and issues his plan as described in Chapter 8. Lack of time may make it necessary for him to make his reconnaissance from a map only. During his reconnaissance and preparation of plans, he ordinarily keeps his battalions moving forward as described in Chapter 8. The battalions, however, should receive the attack order before they leave the last covered position in rear of the line of departure.

b. The regimental attack order is usually issued orally and preferably from a point on the ground where the terrain of the zones of action can be seen. It is preferable for battalion and special unit commanders to be assembled for the order; but if this is impracticable the order can be issued in fragmentary form, and staff officers used to assist in its dissemination. (The subject of orders for the attack and their issuance is covered in detail in Chapter 8.)

c. The regimental order includes as many of the following items as are necessary:

(1) Enemy information that concerns the regiment.

(2) The mission of the brigade to include its initial formation and its objectives.

(3) The locations and missions of adjacent troops.

(4) The supporting fires to be furnished by the artillery; and, in addition, any other supporting fires that may be furnished, such as chemicals; also the allotment of tanks, and the scheme for their employment.

(5) The mission of the regiment, its objectives, formation, zone of action, boundaries between battalions, line of departure, and hour of attack.

(6) Instructions for each battalion and designation of attachments to them.

(7) Instructions to special units not attached to battalions.

(8) Instructions relative to security, antiaircraft and antitank protection, and maintenance of contact both within the regiment and with adjacent units.

(9) Arrangements for lifting supporting fires.

(10) Instructions regarding trains; and the locations of the ammunition distributing points.

(11) Axis of signal communication and the location of regimental and battalion command posts, and aid stations.

36. *FORMATION FOR THE ATTACK.—a.* After considering the frontage assigned to his unit, the mission of his unit and that of the whole force, the depth to which

the attack is to be carried, and the terrain over which it is to be made, the regimental commander determines the number of battalions to be assigned to the assault. If all three of his battalions are available to him, he may place two battalions in assault and one in reserve, one battalion in assault and two in reserve, or three battalions abreast. When one battalion has been held out as brigade or division reserve, or is already engaged with the enemy as an advance guard, the regimental commander must carry out his task with two battalions. Sometimes, he may have only one battalion left. (This was the case in the 2d Infantry in the illustrative situation given in Chapter 8.)   The same general methods that apply to the disposition of rifle companies in a battalion also apply to the disposition of battalions. Whenever a regiment is acting alone, a reserve should be held out initially.

b. Platoons of the howitzer company, not attached to battalions, and the machine-gun company of the reserve battalion, can, in many situations, be disposed to support the initial action of the assault units. Weapons of reserve units should, however, be readily available to them when they enter combat.

c. When tanks are attached to the regiment, tank platoons may be attached to battalions or all or a part retained to operate as a supporting unit under regimental control as will be shown in Chapter 12 on the employment of tanks.

37. *ADVANCING THE ATTACK.*—a. After his regiment is committed to an attack, the only way the regimental commander can alter its formation is by the movement of the reserve. Through the use he makes of his reserve and of his supporting fires, particularly artillery, chemicals, and tanks, the regimental commander exercises his chief influence on the attack.

b. The regimental reserve is located so that it can be employed against hostile counterattacks, especially on the flanks of the regiment, and so that it is available to give fresh impulse to the attack. The reserve also reduces isolated points of resistance that assault echelons may pass by. The regimental commander regulates the advance of the reserve, usually moving it by bounds from cover to cover as required by the changing situation. He keeps the reserve commander informed of developments and continually advises him as to the possible uses of the reserve. The reserve is held sufficiently in rear of the assault elements so that it is free to maneuver in any direction. A regiment should have a reserve at all times; hence, if the original reserve is put into battle, the regimental commander should constitute a new reserve.

c. The regimental commander assists the attack by using his artillery and other supporting fires to the maximum. He coordinates the artillery fires through the artillery battalion commander and decides where these fires are to be placed when requests from his battalion commanders conflict. Supporting fires of all kinds are primarily employed to assist the elements that make the greatest progress.

d. The commander is responsible for the security of his regiment, and for maintaining contact within the regiment and with adjacent units.

38. *REORGANIZATION.*—When a regiment reaches its objective, its commander must continue to maintain contact with the enemy. He has the battalions institute vigorous patrolling or continue to new objectives. Or if a reserve is still in hand, he employs it to exploit the success. He must also hold the ground his unit has gained from recapture by hostile counterattacks. When the assault units lose contact with the enemy and the danger of counterattack seems remote, the regiment is reorganized and immediate preparations are made for pursuit; but the pursuit is not inaugurated without orders from higher authority.

39. *THE REGIMENTAL STAFF IN THE ATTACK.*—
During an attack, the regimental commander usually has
to stay at his command post in order to keep in touch
with the situation through his communications with the
battalions, and the intelligence agencies operated by S-2.
The executive coordinates and supervises the operations
of the command post, leaving the regimental commander
free to devote his energies to the tactical control of the
regiment. S-3 keeps the operations map posted and assists
the regimental commander in any other manner that he
can. As an attack progresses, it usually becomes necessary to move the command post. This is accomplished
by the communication officer under the supervision of S-1.

40. *THE RESERVE REGIMENT IN ATTACK.*—The
regiment or the regiment less a battalion may be employed as a brigade or division reserve. Such a reserve
operates in a manner similar to a battalion in regimental
reserve.

SECTION VIII

THE INFANTRY BRIGADE IN ATTACK

41. *THE EMPLOYMENT OF THE BRIGADE.*—*a.* In
a great majority of combat situations a brigade operates
as a component of its division, and the use of a brigade,
as a separate or independent organization is unusual. A
brigade may be employed as a security detachment for a
large force, or may become, in the course of rapidly de-
veloping action, so separated from other units that it is
temporarily isolated and, in effect, independent. But for a
given brigade even this rarely occurs, since entire divisions are more likely than brigades to act as flank or
rear guards, or other security groups of a large force in
the field. It is possible, nevertheless, that a war in the

near future might see important independent tasks given to single brigades that were largely composed of motorized or mechanized infantry elements. But for the most part, the brigade should be thought of, not as a separate force, but as an integral part of the next higher unit.

b. In battle a brigade invariably has supporting artillery; and it may also have the assistance of chemical and tank units. These supporting elements may or may not be attached. A brigade, acting as a security detachment may also have cavalry units attached. Thus a brigade is the smallest unit in which the combined arms customarily operate.

42. *DEPLOYMENT FOR BATTLE.*—Brigades begin deployment when contact is imminent and advance to contact as described in Chapter 6. When a brigade commander designates assembly areas for his regiments, they are occupied as described in Chapter 5. He reconnoiters, and forms and issues his orders as told in detail in Chapter 8.

43. *MOVEMENT TO THE LINE OF DEPARTURE.*—*a. Artillery.*—A brigade commander should always employ part of his artillery to cover the advance to the line of departure. Ordinarily, after the advance guards have made contact with what appears to be the hostile main force, the advance-guard artillery is already in action. The commander uses additional artillery to cover the movement of his main body elements into combat.

b. Reserves.—Part of one or both regiments is held in reserve. The reserve is employed only on the order of the brigade commander.

44. *FORMS OF ATTACK.*—Chapter 7 contains a detailed discussion of principals and factors governing the forms of the attack. We shall not discuss these matters further here, except to emphasize one point. When large

units are employed in an envelopment, the interval between the main (enveloping) attack, and the secondary
(holding) effort becomes so wide that the maneuver
approaches a turning movement.   There are three reasons
for this:

   a. Modern defenses are usually organized in great depth.
   b. Motor transport gives infantry greater mobility.
   c. Communications are more efficient than formerly.

   45. *FORMATION FOR THE ATTACK.*—*a.* In deciding
upon his plan of attack the brigade commander studies

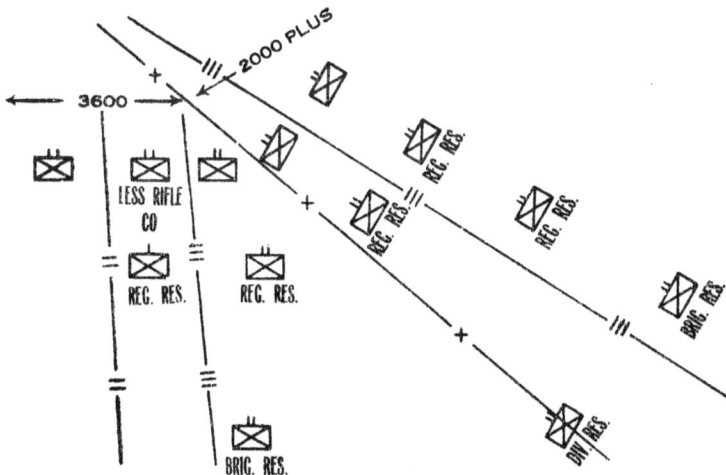

FIGURE 93.—A brigade as part of a division, making an
   envelopment.   Six battalions are available to the
   brigade.   No limit to the front of the envelopment.

first the orders received from the division commander
and then the situation as it stands, particularly the means
of combat in hand in the light of the terrain over which
the attack is to be made.   He decides what intermediate
objectives are essential, and then determines how many
assault battalions, and how many reserve, he will use,

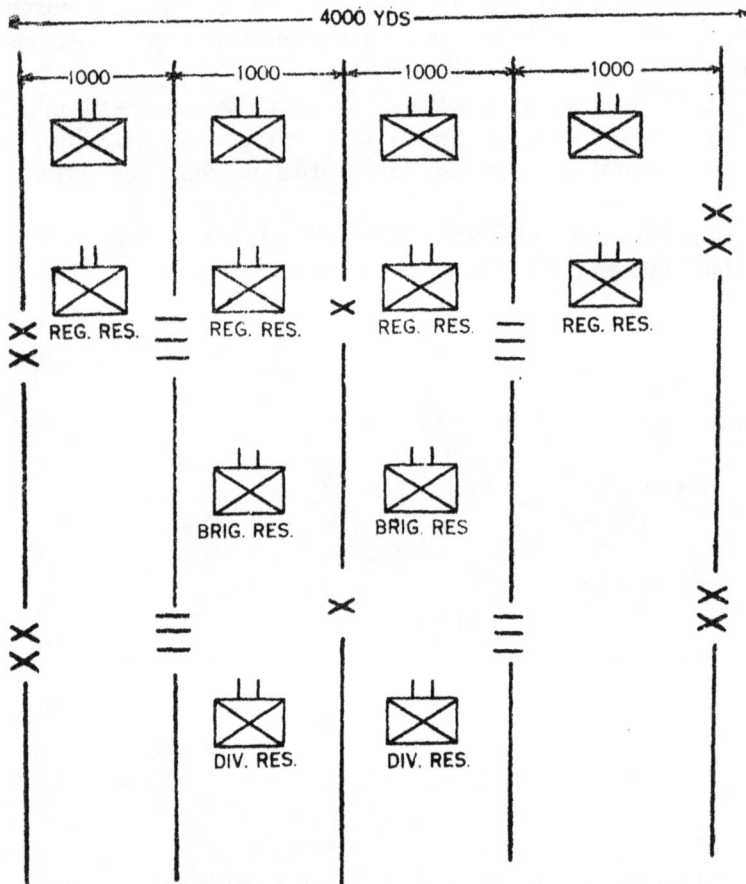

FIGURE 94.—A division making a deep penetration. Both brigades in line and regiments abreast; two battalions held in division reserve.

and where the main effort will be made. He invariably
holds out a reserve, the strength of which depends upon

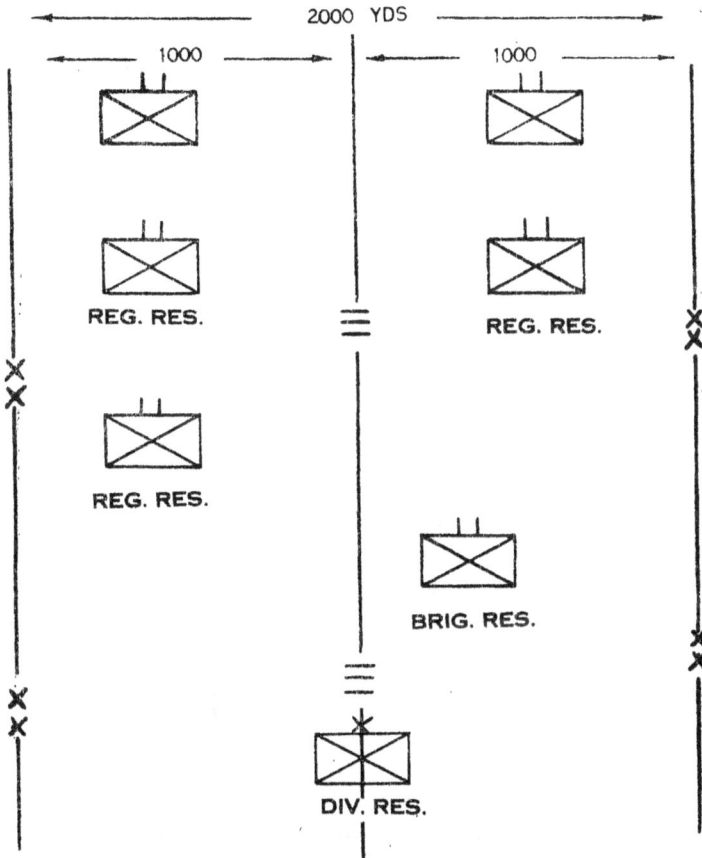

FIGURE 95.—A brigade, as part of a division, making a
deep penetration; brigades in column.

how much is known of the hostile dispositions. When
little is known the reserve is larger than when the information is more complete.

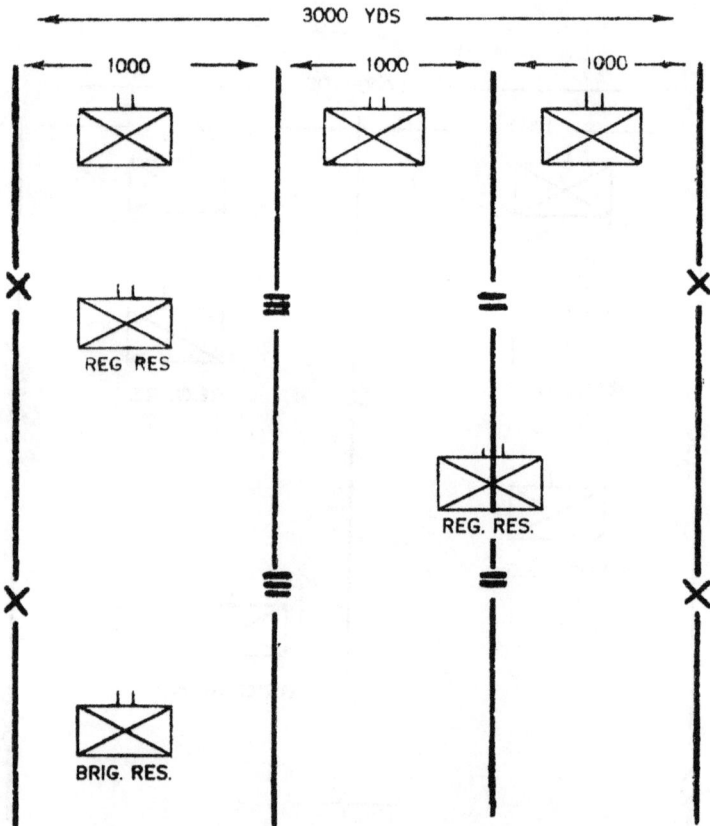

FIGURE 96.—A brigade as part of a division, attacking. Brigades abreast; interior unit. Six battalions available. To penetrate on a front of 3,000 yards. Main effort on the left along the boundary between brigades.

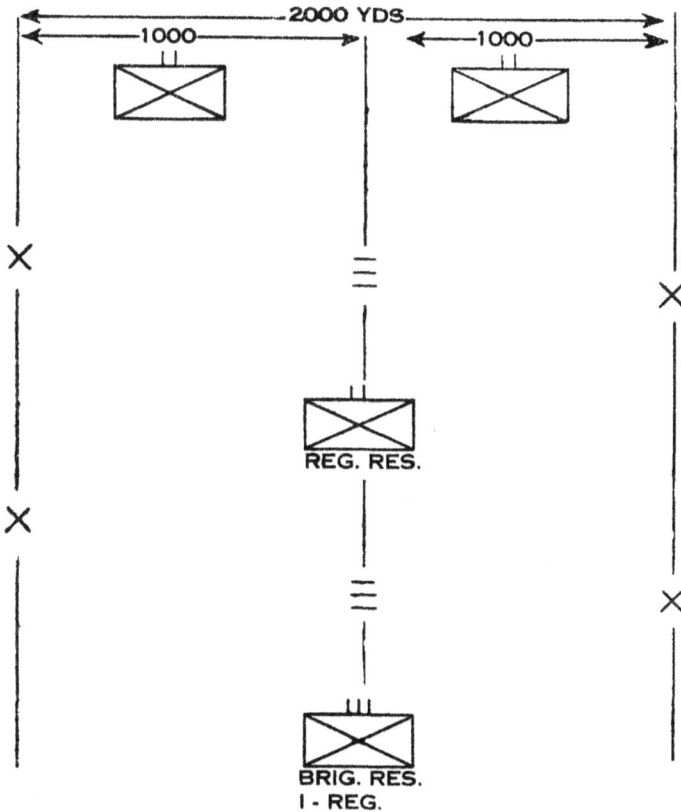

FIGURE 97.—A brigade, as part of a division, making a deep penetration into a zone defense, with brigades abreast; six battalions available; brigades in column of regiments.

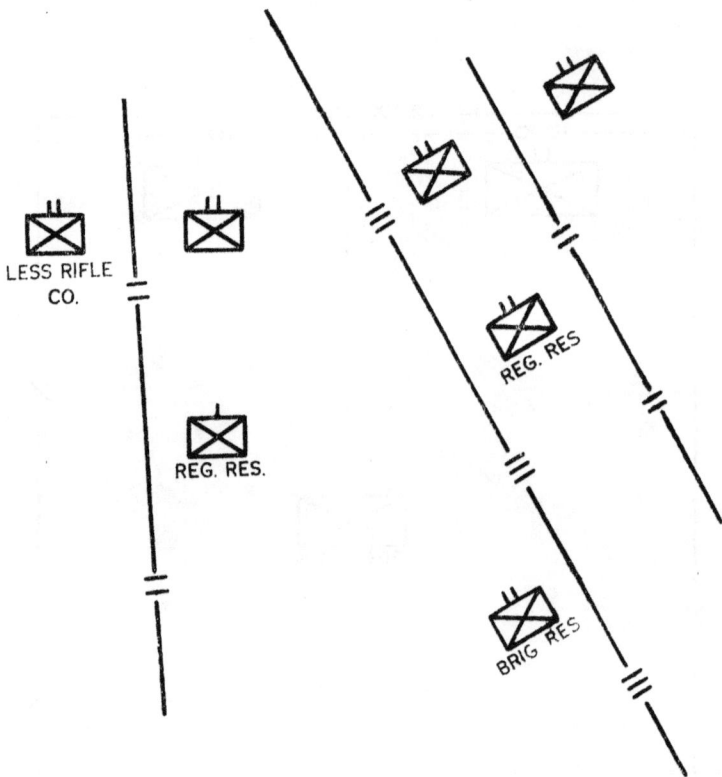

FIGURE 98.—A brigade operating alone, making an envelopment. Brigade reserve taken from the regiment making the holding attack. The brigade reserve may be taken from the enveloping regiment, which would reduce the strength of the enveloping attack to one battalion.

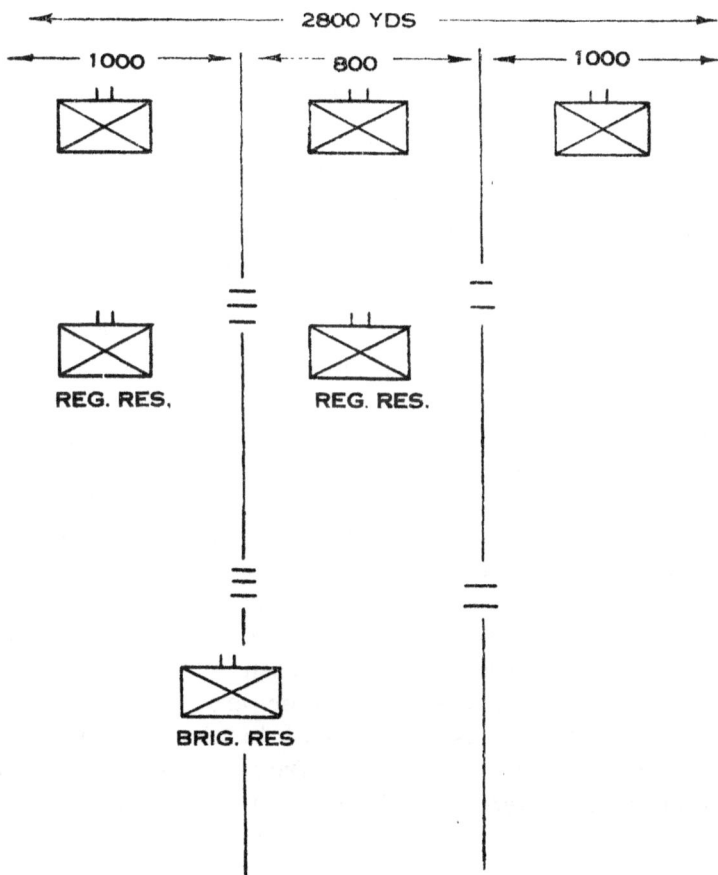

FIGURE 99.—A brigade acting alone, making a penetration along the boundary between the regiments, against an over-extended enemy position.

b. A brigade organization may be disposed for an attack in a number of ways. Figures 93 to 99 show several possible dispositions diagrammatically. These illustrations are in no sense rules for meeting specific tactical situations.

c. It was assumed in Figures 93 to 99 that the entire brigade was available to the commander. This would ordinarily be the case except in a brigade acting as an advance-guard element. After contact with the enemy has been established by leading advance-guard elements, the commander may at any time take over direct command of the advance guards as a coordinated whole. In this event division, brigade, and regimental commanders are likely to find their commands temporarily reduced in size. In this situation brigade and regimental commanders may also be placed in command of a group of battalions for the attack, rather than specific integral regiments. Although it is always desirable to maintain the tactical integrity of regiments and brigades, the considerations of flexibility of maneuver and availability of units for rapid action, sometimes make it essential for brigade and even division commanders to think in terms of battalions. When this is necessary, commanders should preserve tactical integrity as much as the scheme of maneuver will permit. Thus, when two battalions of one regiment and one of another are combined as a combat element, it is usually best to place in command the commander of the regiment from which the two battalions came. Here, too, the relative availability of commanders may outweigh the desirability of retaining integrity.

46. *ADVANCING THE ATTACK.*—*a.* Once a brigade leaves the line of departure its commander assists the advance by the use of the supporting weapons, primarily artillery fire—and if employed, tanks and chemicals—, and by the timely use of the brigade reserves.

b. Artillery is never held in reserve; that is, out of

action for future use. Artillery is often the most potent auxiliary a brigade commander has. Artillery fires are highly flexible; their entire mass can be rapidly concentrated upon any area within range. The brigade commander through his artillery commander coordinates these fires, which are employed primarily to assist the main effort; and, when requests for them from subordinate commanders conflict, he decides where they are to be placed.

c. In addition to this powerful and flexible support which can be shifted to vital areas, a brigade commander also has the weight of his reserves with which to influence the course of battle. These consist of one or more battalions and tank units. He throws in his reserve battalions to protect the flanks of assault units that have progressed beyond the units on their right and left, to widen the gap created by a deep penetration, to extend the envelopment, or to relieve assault units which have become exhausted. The reserve foot-troop elements, it should be remembered, are used primarily to exploit success rather than redeem failure. Reserve tank units are employed to attack and take out resistance that is holding up foot-troop elements, as well as to exploit the successes initially gained. The time of committing his reserves has a most decided bearing on the success of the attack. The brigade reserve may be released to a regiment, or it may operate directly under brigade control, whichever the situation (time and space factors, and terrain) dictates. When the entire brigade reserve has been committed the commander should reconstitute a new reserve as soon as he can, by withdrawing a unit committed initially. This may be done by taking out small units here and there where they can best be spared, and combining them.

47. *REORGANIZATION.*—When the attack reaches its final objective, or halts for any length of time, the brigade commander immediately takes steps to reorganize his unit

in order to hold the ground gained. He insures that contact is established with adjacent units, that proper coordination is established between his own regiments and battalions, and that the brigade is redisposed, if necessary, in sufficient depth to withstand hostile counterattacks.

CHAPTER 12

EMPLOYMENT OF TANKS IN THE ATTACK

SECTION I

GENERAL CONSIDERATIONS

1. *GENERAL.*—*a.* The tank is a supporting weapon of great power, which the supported unit commander can use to advantage. In the attack tanks help foot troops to overcome definite resistance that without tank assistance might be costly in time and casualties, or even insurmountable.

b. Thus tanks, like all other weapons, are a part of the offensive team. They do their share toward gaining victory, but while doing it often need help themselves. Hence, although tanks help other units advance; the other infantry weapons, and the artillery, air corps, and chemical warfare service usually in turn assist or protect the tanks. This, like all other phases of modern combat, is a matter of team work and mutual assistance.

c. Tanks are now such an important element of infantry and of armies as a whole that every infantry officer must understand their tactical uses. Tank organization, characteristics, and methods of employment are essential parts of this knowledge.

2. *TANKS OF OUR ARMY.*—*a. World War tanks.*— We have two World War types of tanks: the Mark VIII (heavy tank), weighing about forty tons and the 6-ton tank, Model 1917. The former carries a crew of eleven and is armed with two 6-pounder guns and five machine

guns. The latter carries a crew of two and has but one
gun, either a 37-mm gun, or a caliber 0.30 machine gun.
Both are slow moving (2-6 m.p.h.), subject to frequent
mechanical breakdown, have 10-20 mile radius of action,
and should be moved by rail (or motor carriers in the
case of the 6-ton tank) when the distance to be covered
requires more than three or four hours. The 6-ton tank
has been removed from the list of standard equipment.
The Mark VIII continues in service but is virtually obsolete also.

   b. *Modern tanks.*—A limited number of T2 (light) and
T4 (medium) tanks will be issued to the service in 1935
and 1936. These represent the latest development in
tank design. The former operates entirely on tracks, the
latter on wheels (on hard surfaced roads) or tracks. Both
can maintain speeds in excess of 30 m.p.h. on roads, but
when attacking cross-country are most efficient at 10-20
m.p.h. The T2 will carry two caliber 0.30 and one caliber 0.50 machine gun; the T-4, five caliber 0.30 and one
caliber 0.50 machine gun (or heavier weapon). The
light tank will have a crew of three or four; the medium,
four. Both are limited to approximately 100 miles of
operation on one fuel refill, but can make strategic moves
under their own power.

   c. *Future possibilities.*—Our army has no standard
tank; the latest tanks are considered experimental types.
What future developments may bring forth is a matter
of conjecture. It is reasonable to surmise that only a few
modern tanks will be available at the outbreak of any
future war. If tanks are needed, obsolete tanks may have
to be used until up-to-date combat vehicles can be turned
out in quantity. The employment of tanks discussed in
this chapter is based on our present experimental types,
but the probable necessity for using obsolete equipment
should be borne in mind in contemplating the future.

   3. *LIMITATIONS OF TANKS.*—All tanks require frequent mechanical attention, lubrication, and fuel refill.
All are vulnerable to artillery fire and, in less degree, to

37-mm and caliber 0.50 machine-gun fire.   Thick woods consisting of trees of large diameter, swamps, streams over 30 inches deep, wide trenches, and rock strewn terrain, all are obstacles to tanks.   Observation is limited owing to necessary protection for the crew and the unavoidable pitching and rolling of the tank.   Control is difficult after a tank attack is launched.   The effectiveness of their fire is limited by visibility and the instability of the gun platform.

4. *CAPABILITIES OF TANKS.*—In spite of their limitations, tanks are powerful weapons when skillfully employed.   Their chief asset is their ability to carry highly-mobile, protected fire power into the enemy position.   Their shock action is tremendous.   They can crush light guns, gun emplacements, and personnel; light frame buildings, trees under eight inches in diameter, and barbed wire entanglements.

5. *TANK ESTABLISHMENTS.*—*a. The tank park.*— The tank park is the base of operations of a tank regiment in the field.   Here is located the service company, and oftentimes an ordnance maintenance company. Regimental headquarters and the headquarters company are usually in the park or nearby.   It should be centrally located in a concealed position beyond the range of hostile medium artillery, and where an adequate road net facilitates movement both to the front and rear.   As a base for modern tanks, it will only move forward when it can advance from five to ten miles, or more.

b. The battalion bivouac.—The battalion bivouac is the base of operations of a tank battalion.   Its functions correspond to those of the tank park.   It should preferably be out of range of hostile light artillery, and have concealment, defilade, and access to a suitable road net.   It makes shorter moves than a tank park, endeavoring to keep as far forward as practicable to facilitate maintenance and supply.

c. Reserve positions.—A reserve position is occupied by

a tank platoon or company when it is not to be employed initially in the attack. Concealment and defilade, if available, are the primary considerations in selecting a reserve position. It may be from 1,000 yards to several miles in rear of the elements in contact with the enemy.

d. Assault positions.—An assault position is prescribed in the attack order—usually where recommended by the tank unit commander. It is the area occupied from 15 to 45 minutes before the hour to move forward, cross the line of departure, and attack. It should have concealment, or defilade, or both, and should be located as near as practicable to the point where the line of departure is to be crossed. To effect surprise, gain cover, or elude enemy fire, it is permissible with fast tanks to use an assault position as far as 4,000 yards in rear of the line of departure, although less confusion arises if it can be within 1,000 yards.

e. Assembly point.—An assembly point is also prescribed in the attack order. It is usually a defiladed or concealed area in rear of a local objective. At this point new instructions are given to the tank unit, and it reorganizes and prepares to continue the attack. The assembly point is an essential part of the means of coordination with, and control by, the supported unit.

f. Alternate assembly point.—An alternate assembly point is prescribed as a precaution in case the attack fails. It is usually the assault position or previously occupied assembly point unless otherwise specified.

g. Reservicing point.—A reservicing point, as its name implies, is a place in the forward areas where tanks assemble when necessary, for reorganization, reservicing, and maintenance. As a rule it is only established by companies or larger units. The battalion bivouac often functions as the reservicing point.

h. Headquarters.—Regimental headquarters is usually at the tank park; battalion headquarters, at its bivouac. Company headquarters, with its maintenance unit, should be close to its reserve platoons, if any, and near the command post of the supported unit. The location of platoon headquarters is always that of its tanks, for the platoon is solely a tactical unit.

6. *ORGANIZATION OF TANK UNITS.*—*a.* The organization of tank units having modern tanks has not been decided by the War Department (July, 1935). For the present the Tables of Organization for the 6-ton tank, Model 1917, with certain necessary modifications, are to be used.

b. Tanks are organized into regiments of three battalions containing three combat companies, each having three combat platoons of five tanks. Excepting the divisional tank companies, tanks belong integrally to the GHQ reserve.

c. In our present organization, there is one tank company as an integral part of each infantry division.

7. *COMMUNICATION.*—*a. General.*—Because commands by voice are out of the question and visual signalling is often very limited or difficult, radio communication between tanks is extremely valuable. Radio, however, does not solve all problems and can by no means be the sole dependence for communication, as it may fail from various causes. A combined use of all practicable types of communication must be employed to insure liaison.

b. The following types of communication between tank units, and tanks and foot troops, are now employed:

Flags
Messengers, foot and motorcycle
Radio
Pyrotechnics
Existing set-up of other units
Personal contact of commanders
Pigeons.

c. The radio installation that is now being experimentally tested, provides both sending and receiving sets for the command tanks of battalion, company, and platoon commanders. All other tanks are equipped with receiving sets only. An additional sending-and-receiving set is

furnished each platoon for use as a contact or liaison set.
It is carried in a light truck or replacement tank and
made available to the infantry commander with whom the
tanks are operating.

8. *CONTROL.*—The control of tanks by commanders of
units to which they are attached is accomplished by using
the various means of communication listed above, particularly personal contact, messengers and radio; and by
prescribing in orders: boundaries, assembly points where
new instructions can be given, and single missions terminating with occupation of assembly points.   Control of
tanks, once they have been committed, is difficult at best;
and unless definite measures are taken for its maintenance, becomes nonexistent.

SECTION II

EMPLOYMENT OF TANKS IN THE ATTACK

9. *PRINCIPLES OF EMPLOYMENT.*—*a. General.*—
As was stated earlier in this chapter, the principle mission of the tank is to carry protected fire power and
shock action into the hostile position and thus help the
foot troops forward.   Its high speed enables it to rapidly
maneuver around obstacles and antitank resistance, and
to cross fire swept areas in a few seconds.   However, we
have already seen that tanks are handicapped by heavy
woods, swamps, streams, etc., and therefore the need for
terrain favorable to tanks should be constantly kept in
view while considering their employment.   Moreover, in
this introduction to Tanks in the Attack, it should be impressed upon all infantry officers, that tanks must not
be thrown rashly or haphazardly into combat any more
than any other units.   Against weak or ill-defined resistance, tanks are not used economically nor to best
advantage.   They should not be committed, ordinarily, before the location of the enemy defenses has been deter-

mined. Of course, there may be occasions when a quick thrust by a few tanks is an economical and efficacious way of eradicating resistance before it can develop. Tanks are most profitably used when foot troops can immediately follow them and occupy the ground gained by the tank attack.

b. Effect of Surprise.—Surprise is highly desirable in any attack, whether tanks are present or not. An unexpected blow is more sure of success. Tanks should therefore be concentrated secretly, usually at night, and kept concealed; and every other measure possible should be taken to keep the fact secret that they are about to attack.

c. Use of Smoke.—Placed on known or suspected enemy observation points and on antitank gun locations, smoke is of great help to tanks in closing with the enemy. On the other hand, smoke used incorrectly may easily become a liability, blinding the tanks themselves or covering hostile targets so that tank fire becomes ineffective. Smoke laying by tanks is not at present favorably considered, but it is possible that some type of mortar or howitzer for firing smoke shells may become a necessary weapon in tank units. At present infantry mortars, chemical mortars, and artillery are the principal agencies depended on for smoke. In making up the scheme of fire support for the entire command, the needs of the tanks should be considered.

d. Fire support.—Tanks attached to attacking units seldom operate beyond the limit (roughly 1200 yards) of supporting fires furnished by other infantry weapons. Supporting tanks in exceptional cases may operate farther to the front. The fire support afforded tank units is ordinarily only what has been planned for the attack as a whole with due consideration to the needs of tanks. Arrangements may be made for infantry mortars and howitzers—or occasionally for artillery—to engage hostile antitank guns that have been located. The fires of all infantry weapons, smoke weapons, artillery, and aviation assist tanks just as they do foot troops. The rapid move-

ment of tanks, however, requires an even greater degree
of coordination of fires when they are present, than when
foot troops attack alone.

   e. *Assistance of fog.*—Fog not only conceals the enemy
from our attacking elements; it also hides our attack from
the enemy.   Viewed as protection, it favors tanks; on the
other hand, it hampers maintenance of direction and impairs the effect of tank fire power.   Therefore tanks are
mainly used in fog against definitely located resistance,
and are employed for short thrusts rather than deep
penetrations.   If the enemy is not definitely located, and
an attack is made in fog, tanks are usually held in reserve
close behind the assault elements, so that they can be
thrown into action with the least delay to destroy a hostile machine gun or combat group that is holding up an
attacking unit.   This use of tanks to perform a limited
task and then return, keeps them from getting lost or
beyond the control of the commander using them.

   f. Night attacks by tanks.—Poor observation and the
consequent difficulty of coordination with foot troops
limit tank night operations generally to bright moonlight
nights or situations where artificial light such as flares
can be used.   Moreover the terrain must present few obstacles to their advance, and the noise of their movement
up to the attack position should be covered by other dis-
tracting noise to avoid disclosing their presence to the
enemy.

   g. Unsuitability for reconnaissance.—Tanks are fighting machines, not searchers.   They are not suitable as
reconnaissance agencies, since they cannot hear because
of their noise of operation, and cannot see well because
of their restricted apertures for observation.   (They can
move, of course, with doors open.)   Moreover their noise
forewarns the enemy of their presence and enables him to
hide.   Even worse, he may have antitank weapons which
he can emplace and destroy the tanks.   Thus the tanks
may be merely thrown away without accomplishing anything.   But at times when power is desired in strong reconnaissance detachments, and other means are not avail-

able, tanks may have to be used for this purpose. If employed, they should use their radios constantly to transmit what they see to the commander who has sent them out. Ordinarily, tanks should not be given reconnaissance missions.

h. Mass (*Concentration of effort.*)—Tanks should be used in quantity, if possible, wherever they are employed. Given a certain number to use, the commander of a large force should carefully weigh the advantages of massing them for a concerted blow or distributing them for local assistance in several parts of his zone. The advantages of tanks used in mass were often demonstrated during the World War. Likewise many failures of tank attacks during the war are attributed to disregard of this principle. Mass is imperative for the successful penetration of a strongly organized defensive position on a wide front.

i. Width, depth and reserve.—The number of tanks available may preclude their employment in mass, as the term is generally understood. However, the commander who has a company of tanks, or more, at his disposal, should understand the comparative advantages of width and depth, and the need of a reserve. Width, of course, is desirable to assist the progress of the attack as a whole and to prevent the enemy from concentrating his anti-tank defenses. Depth, however, is more essential, even at the expense of width. Power, continuity of effort, and replacement of losses, all demand depth in tank formations. A deep penetration, obviously, needs more depth than a shallow limited objective attack. A reserve should be kept out to influence the later stages of the attack. Reserve tank units should not be attached to reserve foot-troop units but held under the orders of the higher commander.

j. Limited missions.—It has been mentioned earlier that control over tanks is retained by prescribing single objectives terminating with the occupation of an assembly point where new instructions can be given. It is well to reiterate this point here. Whatever the mission given to tanks it should be definite, and moreover simple, and seldom should more than one mission be given at a time. Individual platoons of a tank company may be given separate missions, but a platoon should not be divided— it should have a single mission for the whole platoon.

k. Resume.—We have seen that tanks are to be employed when, and where, and in the manner, they can best exercise their mobility, fire power, and shock action. This involves employment in mass and attacks in depth. It means withholding them from reconnaissance missions and attacks against indefinite resistance. It requires that the advantage of surprise and the assistance of smoke and supporting fires be turned to the best account. It necessitates definite, simple, and limited missions.

10. *ALLOTMENT OF TANKS.—a. General.*—Tank units are used in the attack as supporting tanks and as accompanying (attached) tanks.

(1) *Supporting tanks.*—When used in the supporting role, tanks are not attached to the subordinate foot troop units but operate under the orders of the higher commander and the control of their own commanders. Thus the higher commander retains a powerful and mobile unit which he can commit at the decisive time and at the place where it will be of most benefit to his entire force. Supporting tanks attack without regard to boundaries between subordinate units; consequently all units must know the plan for the employment of the tanks. Supporting missions are ordinarily assigned only to tank companies or larger units.

(2) *Accompanying tanks.*—The mission of accompanying tanks is to render close cooperative assistance to the advance of the assault echelons of attacking troops by destroying or reducing the effectiveness of the hostile fire, particularly that of machine guns. Tanks that are to be used as accompanying tanks are ordinarily attached—one or more platoons—to infantry battalions.

b. Factors governing allotment.—When commanders of units larger than a battalion have tanks allotted to them, they must either reallot them to subordinate units or re-

tain them under their own control to support the attack
when and where they can be used most advantageously.
Their decisions and the distribution they make, if any,
must be based on the scheme of maneuver, the terrain,
the enemy's defenses, and known defensive methods, and
the type and armament of the tanks.   The bulk of the
tanks is usually allotted to the main effort, or held by the
commander to support the main effort at the time and
place where the tanks can be most effective.   However
if the terrain in the zone of the main effort is unsuitable
for tanks, other arrangements may be necessary.   If the
terrain is unsuitable owing to an obstacle of no great
depth and width, a commander may use no tanks at all
in such a zone; or can require the tanks to move in an
adjacent zone where the terrain is passable, and then
shift into the zone containing the obstacle, once the obstacle has been flanked or encircled.   (See Figure 100.)
When it is known that certain areas are well prepared
for defense against tanks, they should be avoided if it is
possible to find weaker points in the defense.

c. *Battalion allotment.*—When tanks are reallotted to
subordinate units, the number attached to a battalion depends on what part the battalion has in the scheme of
maneuver.   In general, a platoon of five T-2 tanks is a
sufficient reallotment to a battalion.   But when a battalion is in the main effort of a deep penetration, more
than one platoon per battalion may be necessary.

d. *Advantages of tanks in general support.* — Mass,
width, depth, and reserve were stressed in the paragraph
on Principles of Employment as essential features in tank
operations.   This, together with the advantage of having
a highly mobile and powerful reserve to inject into the
attack where and when it will contribute most to the success of his entire command, should, in most situations of
open warfare, induce the commander that has a company
or even a battalion of tanks at his disposal, to retain them
under his control as supporting tanks.   This may also be
preferable in many situations of stabilized warfare.

e. *Combination of accompanying and supporting roles.*

—To insure a strong initial blow against a definitely located position, and still be able to employ his tanks as a single unit under his own control at a later stage of the attack, a commander may attach all or a part of the tanks to subordinate units for the initial phase of the attack, prescribing that they be used initially and then revert to the higher commander after the first objective has been reached. Or he may assign them by platoons to support the assault battalions with instructions to assemble and await further orders at a designated assembly point after they have accomplished this initial close support mission.

11. *TANK COMBAT ORDERS.—a. General.*—Orders regarding the employment of tanks are issued by the corps, division, brigade, regiment, or battalion to which they have been allotted. Such orders are contained in appropriate parts of the field orders of the unit employing them. When a unit has tanks attached to it and they are to be reallotted to subordinate units, the paragraph of the order covering tanks merely indicates that they are attached to such and such units. If the higher commander desires to insure their use initially, he so directs in the order, for example: "Company A 69th Infantry (tanks) is attached to the 12th Infantry; Company B 69th Infantry (tanks) is attached to the 13th Infantry. Tanks will be employed in each assault battalion zone initially." In case the higher unit commander decides to retain his tanks in general support of his force, his orders to it form an appropriate subparagraph of his field order.

b. Infantry battalion commander's order.—When a tank platoon is attached to an infantry battalion, it becomes subject to orders from the battalion commander. The battalion commander ordinarily outlines his plan to the tank leader, gives him opportunity to make a detailed study of the ground, and asks for his recommendations. The instructions for the tanks are then included in the battalion attack order. The part of the order referring in particular to tanks (if tanks are to attack initially) specifies the tanks' assault position, their objective, special mis-

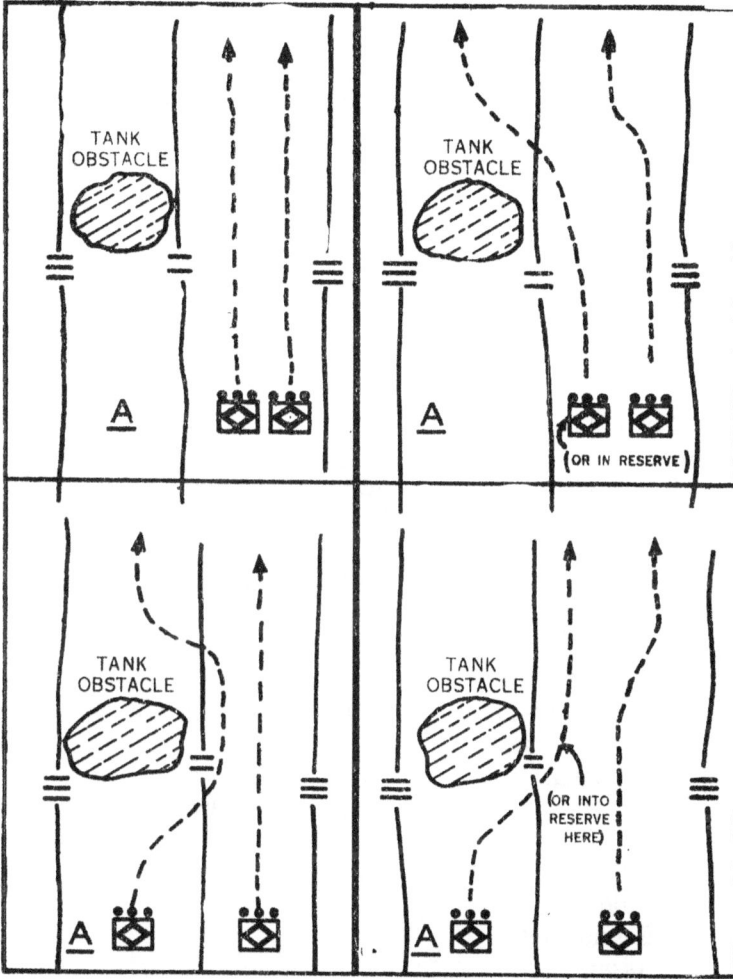

FIGURE 100.—Diagram of tank employment when an obstacle lies across the zone of attack.

Upper left: No tanks used in Zone A.

Upper right: Tanks begin their advance in the adjacent zone, entering Zone A when they have passed the obstacle.

Lower left: Tanks begin their advance in Zone A, and circle around the obstacle by entering the adjacent zone.

Lower right: Tanks begin their advance in Zone A; and, on nearing obstacle, move to the adjacent zone continuing their attack in that zone.

sion, assembly point, and alternate assembly point. The part of the zone to be traversed by the tanks may also be given.

c. Attack order of the tank unit commander.—Combat orders issued by commanders of tank companies or battalions that are to attack as units (instead of by attachment of platoons to battalions) follow the general five-paragraph form, and contain, in general, such of the points listed below in the check list for a platoon commander's order as are pertinent. Company and platoon orders are ordinarily oral.

d. Tank platoon order.—After receiving the battalion commander's order, the tank platoon commander assembles his tank crews and platoon sergeant—or the tank commanders only, if it is impracticable to assemble the crews—and issues his order. As a general rule, he should issue his order at a point where the terrain of the initial attack can be seen. A form for a platoon leader's order follows. This form should be used as a guide or check list, keeping in mind the several factors: available time, training and intelligence of the crews, and the necessity for brevity and simplicity.

(1) Information of the enemy.

Information of own troops to include formation, scheme of maneuver, and fire support.

(2) Mission of tank platoon.

Objective.

Boundaries.

Assault position.

Line of departure.

Time of attack.

Time to leave assault position.

Direction of attack.

(3) Special tank missions.

Missions for individual tanks.

Routes.

Special mention of attack of indicated points of resistance.

Time at which, or before which, certain things
must be accomplished.
Assembly point.
Alternate assembly point.
Route and formation to line of departure.
(4) Reservicing point (if known).
Aid station.
(5) Signals.
Radio, flag, pyrotechnics, etc.
Location of commander (this platoon).
Location of Company CP.
Battalion (infantry) command post.
Battalion observation post.
Second-in-command.

e. Administrative orders.—When a tank battalion or
company is broken up temporarily and attached to various
infantry units, only administrative orders are issued by
the tank battalion or company commanders. Such orders
ordinarily contain the designation of locations of tank
administrative, supply, maintenance, and repair establishments.

12. *FORMATIONS.—a. Battalion.*—The combat formations of a tank battalion in attack are essentially the
same as those of an infantry battalion, namely; column
of companies; two companies in assault and one in reserve; or three companies in assault. These may be
varied by echeloning the reserve company (or companies)
to the right or left.

b. Company.—The tank company attacks in like fashion,
substituting platoons for companies.

c. Platoon.—(1) The basic formations of the platoon
are column and line, and others are variants of these.
(See Figure 101.)

(2) A platoon ordinarily takes the column formation
when not heavily engaged, when passing through a defile,
and when under cover. This formation gives the best control but the least amount of frontal firepower. Distances
between tanks in any formation should ordinarily not be
less than 50 yards.

(3) A line formation permits maximum frontal fire-power but gives the least control. This formation is suit-able, however, in certain situations when the objective is close, where individual tanks can be given short definite missions, and where immediate control is not essential.

(4) A variation of these is the echelon formation which is suitable at certain times.

(5) A wedge, or inverted V, formation lends itself both to fire power and control, and has been found to be an excellent combat formation. The angle of the wedge may vary from a very small one, approximating a column, for movement through woods or defiles, to a large one ap-proaching 180 degrees (a straight line) when the platoon is attacking over flat and open terrain, or crossing a crest.

(6) The line of section columns is a variant of the wedge, the usefulness of which lies in quick control of both sections and the rapid dispatch of either section on special missions.

(7) The column-of-sections formation is designed to provide a means of rendering closer support to foot troops when the platoon is operating as attached tanks; it is also a convenient formation when the platoon must attack over a narrow zone, or when the situation indicates the desir-ability of having a portion of the platoon follow the lead-ing echelon as a support. (See Figure 101.)

13. *PREPARATION BEFORE ATTACK.—a. General.* —Because of the high road speed of modern tanks and their possible speed of commitment to action, officers of tank units should usually precede their units by several hours into the area of the unit to which they are attached, in order to make the necessary preparations for the ar-rival and future employment of their tanks. These prep-arations consist of obtaining instructions, followed by re-connoitering routes, attack terrain, and positions, and

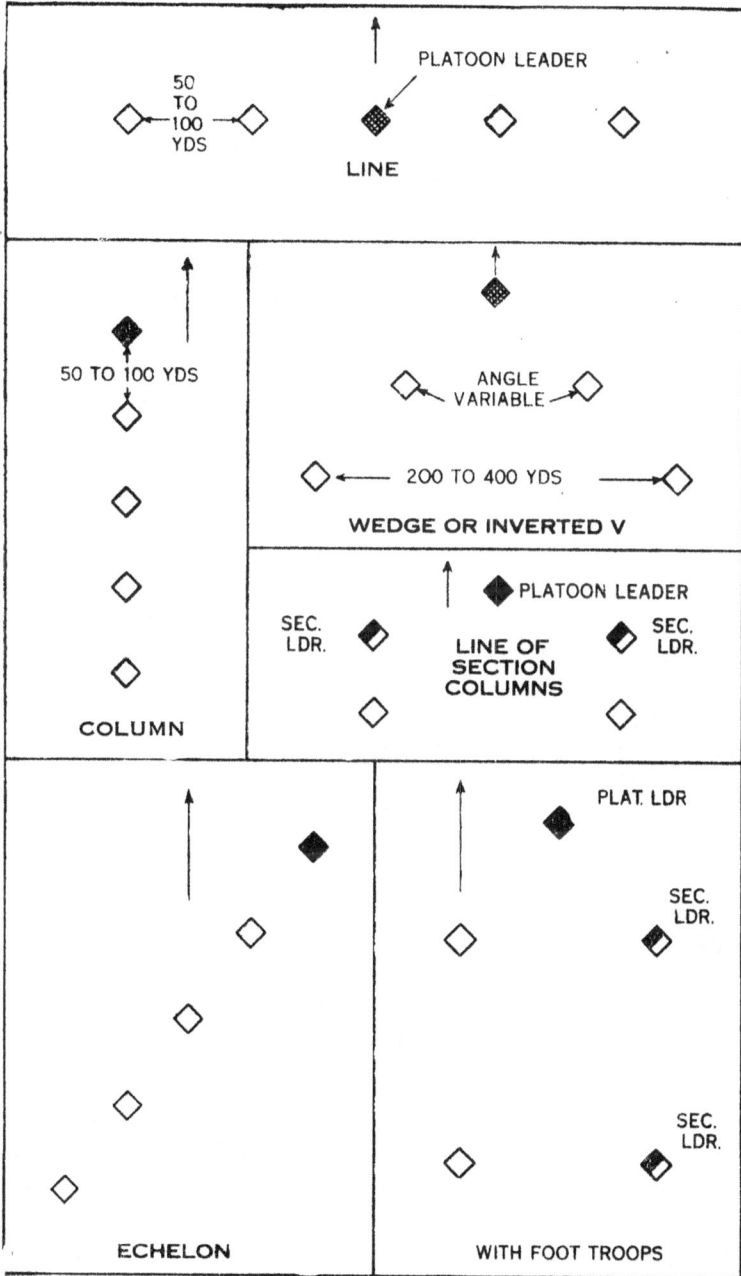

FIGURE 101.—Formations of the 5-tank platoon.

then making recommendations relative to allotment and employment.

b. Reconnaissance.—Tank unit commanders must first gain all available data from such maps and aerial photographs as are obtainable. They reconnoiter by airplane, motor, tank, or on foot. Higher commanders often have airplanes made available to them. Ordinarily, preliminary reconnaissances are made by motor where practicable, and more detailed reconnaissances on foot. Foot troops already in contact with the enemy should be consulted for information on hostile dispositions, obstacles, and terrain features. Daylight hours must be utilized to the utmost for the reconnaissances of all commanders and for pointing out routes, positions, terrain features and hostile dispositions to tank commanders and as much of the crews as can practicably be assembled. Tank units should be moved into position at night, if possible, prior to a daylight attack. Hence familiarity with the terrain must be obtained during the preceding day. Unless commanders of all units down to include individual tanks are thoroughly familiar with the terrain, at least as far forward as the nearest hostile positions, confusion and possible failure are likely to result.

c. Recommendations.—After completing his reconnaissance and considering the various methods of employment, ways to attack, and positions to be occupied, the tank unit commander again reports to the foot troops commander with whom he is to operate and submits his recommendations. Unless these are in conflict with the scheme of maneuver or other plans, they are usually incorporated into the attack order of the infantry unit.

d. Issue of tank attack order.—After receiving the attack order from the foot troops commander, the tank order is issued to subordinate leaders. This should be done, if possible, where the terrain to be passed over in the attack can be seen. When time permits, it is done before the tank unit completes its deployment and movement to the assault positions, inasmuch as the least possible time should be spent in the assault positions. Higher commanders must allow sufficient time for all subordinates, including individual tank commanders, to learn the plans and see the ground over which they are to attack. It may be impossible, at times—as in open warfare—, for tank crews (or even commanders) to see the terrain until the attack is launched. In such cases leaders must give a word picture and point out on maps or oblique (or vertical) aerial photographs where the tanks are to go and what they are to do. Every effort should be made to allow time for orders to be issued, and routes and locations to be described, where and when terrain features can actually be seen.

e. Coordination with other units.—Commanders of tank units indicate to the appropriate officers the assistance they desire from artillery, air corps, smoke, and infantry weapons, and ascertain what support can be expected. They familiarize themselves with the scheme of fire and time schedules planned for the attack. They inform the artillery of their plans to include the probable path, probable hours of reaching certain points, areas to be avoided, and assembly points. Requests for aerial photographs, particularly obliques, are submitted; and any engineer assistance needed for ferrying, bridge repair, or removal of obstacles, is asked for. From the chemical service information is obtained of areas that must be avoided owing to persistent gas; and also its plans for laying smoke or gas.

f. Deployment before attack.—(1) *Contact imminent.*—When tank units are attached to a force that is moving to meet the enemy, they generally move in one or more groups near the rear of the column (or columns), when contact is not imminent. When contact becomes imminent, they may partially deploy along with the larger foot troop units. However, the subordinate elements of a tank unit remain sufficiently close to each other so that they can quickly rally in case they are to be committed as a unit. The formations used are governed by the considerations set forth in paragraph 12. These partly deployed tank units may move near the head of the leading elements of the main body, or farther to the rear between the columns, or near the rear of one of the columns. They should preferably be in or near the column in which the force commander is marching. The commander of the tank unit, with several members of his staff, should be in his command tank or a motor car near the commander of the force. The latter can thus inform him of his plans, direct the allotment or employment of the tanks; and with little delay, these instructions can be transmitted by radio or through a staff officer to the executive marching with the tanks, who, in turn, transmits the orders received to the subordinate commanders.

(2) *After contact.*—After contact with the enemy has been gained, the commander of the force prescribes the disposition of the tanks in conformity with his contingent plans. His decision regarding the employment of tanks is based on the factors already discussed in paragraphs 9 and 10. If the tanks are allotted to one or more subordinate units, they move by whatever covered routes are available to join their designated units. If retained as supporting tanks, they ordinarily move to a suitable assembly area (reserve position, park, or bivouac), from which they can be committed most readily to further the general scheme of maneuver.

(3) *Advance to assault positions.*—Tank units advance to their assault positions so as to occupy them only long enough before the attack to make a final check of vehicles and equipment and obtain additional instructions and orientation. From 15 to 45 minutes should suffice. As previously stated, this advance should be concealed by using covered routes, or preferably by moving at night, in order to protect the tanks from hostile artillery or aircraft fire and to gain the advantages of surprise.

(4) *Advance from assault positions to cross the line of departure.*—Tanks leave their assault positions in time to cross the line of departure at the hour specified for the tank attack. (This is generally when, or a few minutes before, the foot troops advance.) Here again covered routes should be utilized, if available, to protect the tanks

and to keep the direction and point of attack undisclosed until the tanks debouch. This is vital to the life of the tanks for it delays the laying of hostile antitank guns and thereby impairs the effectiveness of their fire. Moreover, the sudden appearance of the tanks at short range is more demoralizing than when their advance can be observed farther away, and in time to take cover or prepare to oppose them. The formations adopted for the final advance and the attack have already been discussed. They should facilitate control but control may have to be sacrificed to increase fire power and make the unit less vulnerable to hostile fire. The platoon sergeant and messengers report to the command post of the supported unit when their platoon advances to attack.

(5) *Deployment of a tank unit joining a force already engaged.*—It is probable that in any major operation during the early part of a future war, the limited number of tanks available will preclude their attachment to corps, divisions, or smaller units, until these units are already engaged with the enemy and the need for tanks becomes apparent. Consequently they will probably, at first, not be found with such units on their march to meet the enemy, except for a few organic divisional tank companies. When a tank regiment, battalion, or company is allotted to a force already engaged—whatever its size— it will usually move by rail (except modern tanks moving less than 200 miles) to a detraining point in rear of the unit it is to join. From there it will advance to assault positions by the same procedure indicated above, ordinarily moving forward only at night. The initial move into the zone will take it to a park or bivouac, from where deployment and advance to assault positions will start, if the unit is divided by attachment to subordinate foot troop units. Here also, the assault positions should not be occupied until just prior to the attack.

14. *THE ATTACK OF THE TANK PLATOON.*—*a. General.*—(1) The discussion in this paragraph is limited to the formation, maneuver, engagement of different tar-

gets, technique of crews, cooperation with foot troops, occupation of an objective, action at assembly points, and resumption of the attack, that constitute the technique of attack of a platoon of tanks attached to an infantry battalion. In other words we are here considering the procedure of accompanying tanks.

(2) A supporting platoon, in contrast to an attached or accompanying platoon, can maneuver over the entire zone of the supported unit (not limited by battalion boundaries), may attack farther into hostile territory (instead of limited objective attacks), and is directly controlled by the platoon leader, who is subject to the orders of the commander of the supported unit. In all other respects a supporting platoon follows a technique identical to that of an attached (accompanying) platoon.

(3) The technique of a tank company, or larger unit, supporting an attack varies from that of a supporting platoon only to the extent that its greater mass affects depth, width, and reserves; and consequently the formations, scheme of maneuver, and missions, which the unit can accomplish. Moreover, the commander of the unit so employed exercises his command function (lost when his unit is distributed), as well as his administrative and staff duties.

(4) In view of the similarity of procedure in the attack of all tank units, irrespective of size or role, this discussion is limited, as already indicated, to that of the tank platoon operating as accompanying tanks attached to an infantry battalion.

b. *Formation.*—As stated in subparagraph *f*(4) of paragraph 13, a tank platoon leaves the assault position in the formation most suitable for crossing the particular terrain that lies in front of it. This formation may be altered at various predetermined points or upon the signal of the platoon leader, as the changing terrain, cover, obstacles, or hostile fire necessitate. The line of departure is crossed in whatever formation is expedient. In the earlier stages of the attack the range to hostile positions may be too great to warrant any firing except occasional bursts where level ground gives a stable gun platform. Consequently tanks depend on their speed, maneuverability, formations, accidents of the terrain, and concealed routes to elude hostile fire. Once the tanks are within effective range of enemy antitank guns, some type of extended formation is necessary to reduce vulnerability to enfilade fire. If the defilade afforded by a draw is utilized to approach the hostile position, every effort must be made to avoid exposure to enfilade fire. Tanks must cling, wherever possible, to the sides of the draw, and take advantage of the cover offered by the smaller transverse ridges and noses. Once a point is reached beyond which it is necessary for the tanks to cover further advance by their own fire, a line, wedge, or other extended formation, must be adopted in order to develop the maximum fire power. (See Figure 102.)

c. *Maneuver.*—Tanks also utilizes their ease of maneuver, in combination with appropriate formations, to take advantage of whatever cover is afforded by accidents of the terrain and vegetation. Where cover is lacking, they must vary their speed, extend their intervals, and follow zig-zag courses with legs of unequal length, in order to make a poor target for hostile antitank guns. (See Figure 103.) Tanks also maneuver, individually and collectively, to avoid obstacles and to obtain favorable firing conditions. Tank commanders must continually observe the movements and actions of the other tanks, particularly the platoon commander's, in order to conform to the maneuver of the platoon. Individual tanks may leave the platoon formation temporarily to engage suitable targets or carry out individual missions, but should return to the platoon as soon as the job is accomplished.

d. *Attacking hostile resistance.*—(1) *Targets.*—The targets engaged by tanks are classified as follows:

(a) Targets dangerous to foot troops but less dangerous to tanks. (These are machine guns, trench mortars, and combat groups of riflemen and automatic riflemen. These may become more dangerous to tanks as the armor-penetrating power of small arms is developed.)

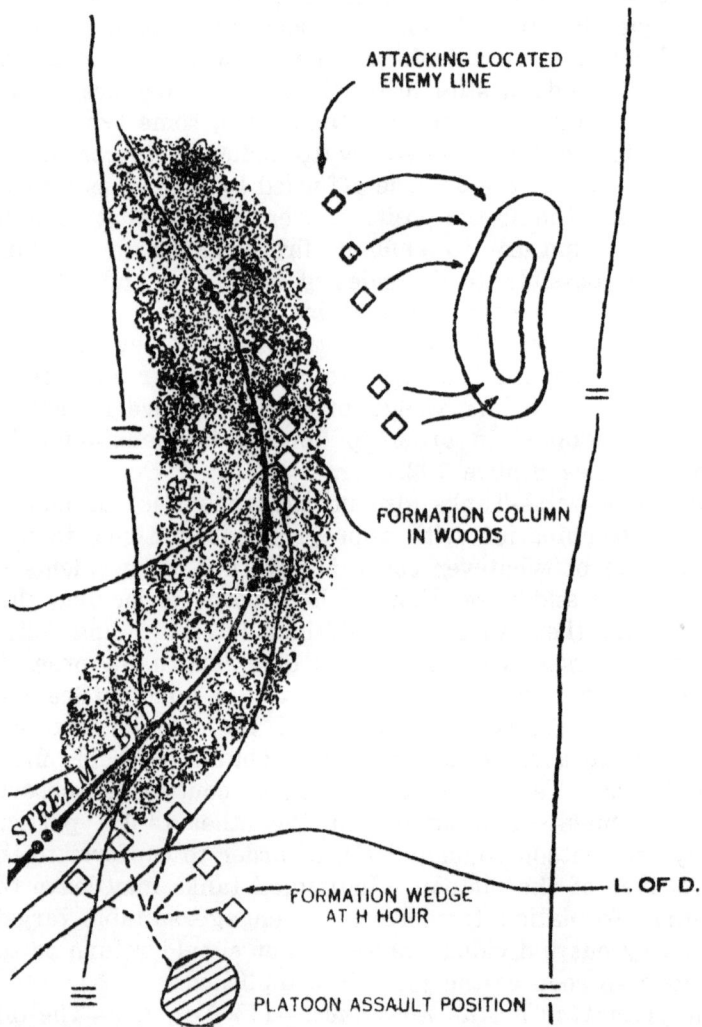

ATTACKING LOCATED
ENEMY LINE

FORMATION COLUMN
IN WOODS

STREAM BED

FORMATION WEDGE
AT H HOUR

L. OF D.

PLATOON ASSAULT POSITION

FIGURE 102.—A method of attack with a tank platoon attached to a battalion of foot troops, when the enemy is active and hidden, and none of his elements have been located, or when time is pressing and reconnaissance is lacking.

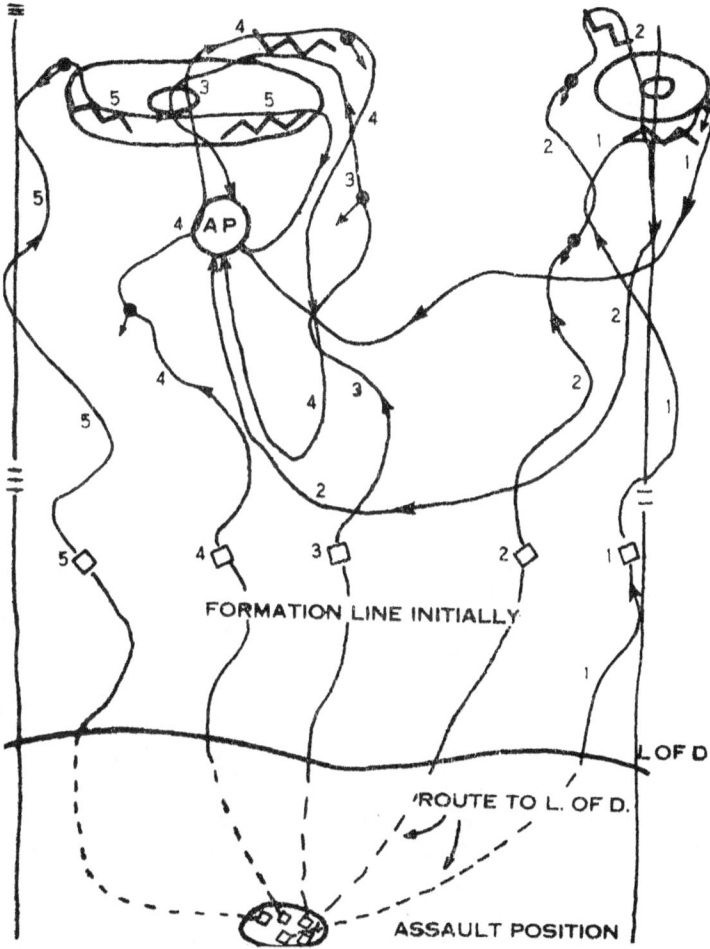

FIGURE 103.—A method of attack by a tank platoon when the enemy is accurately located and time is ample for preparation. The route of each tank is shown to the assembly (rallying) point.

(b) Targets especially dangerous to tanks but not particularly dangerous to foot troops.   (These are antitank guns and field artillery pieces used for antitank defense.)

(c) Targets dangerous to both—supporting artillery and tanks.

(d) Targets offering no threat to either—engineers, signal and supply personnel, transport, supply dumps and bases.

(e) Wire entanglements—destroyed by crushing power.

(2) *Relative importance of tank targets.*—Using as a basis the foregoing classification of tank targets, we can best determine the relative importance of these targets to tank units as they attack.   The mission of tanks is to help the foot troops advance.   Hence, it follows that, in general, hostile elements most threatening to the foot troops —machine guns—are the most important.   Tanks should turn from the destruction of machine guns to engage antitank guns (or artillery so used) only when the tanks' immediate safety requires it.   When tanks receive hostile antitank fire at such range that the danger of disablement is great, they must give their whole attention to their own special enemy until it is defeated.   When tanks are not occupied with machine guns or immediately threatening antitank weapons, they attack other active elements of the hostile defense such as combat groups and mortars.   Seeing none of these, they fire at any suitable targets such as transport or groups of troops.

(3) *Engaging targets.*—(a) *Action of the platoon commander.*—The platoon commander may be able to assign special missions to individual tanks, in some situations, prior to the attack.   More often, the information of hostile dispositions is insufficient, and resistance must be engaged, when it is seen or encountered, in the most practicable manner at the moment.   By means of his radio, or, to a limited degree, by visual signals, the platoon commander can prescribe the action to be taken against targets that appear.   Or he may merely direct the movement and fire of his own tank against targets and rely upon his tank commanders to conform.   This is unquestionably the simplest procedure; and the most likely in combat.

(b) *Action of the tank commander.*—The tank commander listens for radio instructions from the platoon commander and observes the latter's tank for visual signals or changes of movement. In addition he must constantly look for targets to his front (and flanks). He keeps in mind the mission of the platoon and the orders of the platoon leader. His method of approach upon any specific target depends upon the kind of target, the terrain, and the range. Antitank guns are avoided, if possible. If they must be engaged, a defiladed route is sought toward their flank or rear in order to engage them with greater safety, and more effectively. Where this is impossible, a position to fire from chassis defilade may be at hand. As a last resort, antitank guns are attacked directly, using maximum volume of fire and rapid irregular movement. Enemy machine guns are engaged by fire, and when feasible and necessary, are crushed. The tank moves directly and rapidly upon such targets to destroy them. The decision when to avoid a target and when to attack it is based upon the mission, the range of the target, and whether the importance of the target warrants engaging it or another target in view.

(c) *Action of the gunner.*—The gunner (or gunners) looks for suitable targets and engages them as soon as seen, provided they are within effective range and the gun can be held to a fair laying. Because the gun platform of a moving tank is unstable, extreme accuracy of fire is difficult to attain. (Fire from a moving tank has its greatest accuracy at ranges under 300 yards.) As a consequence, the gunner substitutes, to a considerable degree, volume for accuracy. Since even an approximately correct laying can only be maintained for a second or two at a time, it is essential that the maximum volume of fire be developed during favorable periods. (Development of this volume of fire is greatly facilitated by the adoption of the caliber 0.30 machine gun M-2.) Based upon observation of strike or trace, minor adjustments are made while firing. The gunner also responds to signals or oral commands from the tank commander; and when the latter fires his gun, he looks for the trace or strike of the bullets to ascertain the location of the target being engaged. He then joins in this fire with his gun. (The use of tracer bullets mixed with ball, in the ratio of one to five, is considered essential for tank machine guns.)

(d) *Action of the driver.*—The driver aids the gunners when fire is to be delivered, by seeking favorable

FIGURE 104.—Halted tank firing or observing from concealment in woods.

ground. If a zig-zagging advance is used, the driver moves the tank directly towards the target during a brief period while fire is being delivered. He then changes his course (at least 15 degrees) for a time before again moving directly upon the target. The driver also avoids the skyline; if he must cross it, he does so at high speed. An exposed tank not in motion is extremely vulnerable to the fire of antitank guns. Consequently exposed tanks must keep moving. They should only halt to fire from partial defilade (exposing only the turret); or in good cover to deliver fire for brief periods; or in the open, momentarily, to enfilade an occupied trench. Suitable positions for temporary halts to deliver effective fire should be sought, particularly when engaging antitank guns, tanks, or artillery. (See Figures 104 to 106.)

(4) *Hostile tanks.*—Attacking tanks may meet enemy tanks that are advancing in support of a hostile counter-

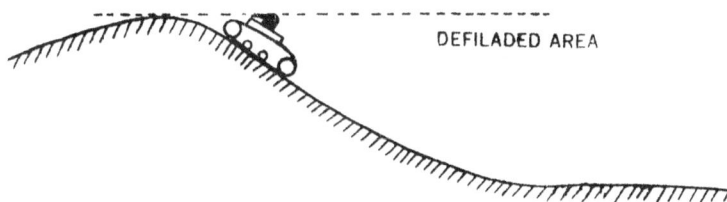

DEFILADED AREA

FIGURE 105.—Halted tank firing or observing from defilade at hilltop.

attack or in delaying action. Provisions for such a contingency should be included in the attack order. If friendly antitank guns are not available or inadequate in number, it may be necessary, for the existence of the tank platoon as well as the safety of the foot troops, to attack the hostile tanks. Maneuver, and the utilization of positions in chassis defilade to obtain effective fire, are the two most effective means for successfully combatting hostile tanks with tanks. The platoon commander must make his decision rapidly, and immediately put it into execution. In general, he should attack hostile tanks whenever they are met.

e. Occupation of an objective.—(1) When tanks reach an objective they destroy or neutralize the enemy resistance by fire and crushing power. They patrol the area, searching out hostile machine guns, but avoid exposed places subject to the fire of antitank guns or the direct laying of artillery. When enemy resistance in the area

FIGURE 106.—Tank taking advantage of the partial defilade of a large shell hole to fire from a halt.

of the objective has been driven out or destroyed, tanks should seek positions in defilade. Tank commanders look for any remaining resistance to the front or flanks, and for opportunities to fire upon hostile support positions or local counterattacks. Tanks stay on the objective or near it long enough to permit the assault rifle units to occupy the position and effect any needed reorganization. If the foot troops are not coming forward rapidly to occupy the position, but seem to be stopped in their advance, a part of the platoon should go back to ascertain the cause. If the advance has been stopped by hostile resistance, the platoon or part of it, should immediately attack such opposition. Then they proceed to their assembly point, or, if their mission has only been accomplished in part, proceed upon their mission, as the assault rifle units take up the advance.

(2) During an attack, tanks attached to infantry battalions ordinarily stay within the zone of action of the unit to which they are attached. However, when the advance within that zone is being delayed by resistance from the zone of an adjacent battalion and there appears to be no other means of reducing it, tanks may be sent temporarily outside of their own zone to reduce it. They should return promptly to their zone as soon as they have completed such a task.

f. Action at assembly points.—(1) At the assembly point, tanks remain in a formation suitable for a resumption of the attack or for defense in case of a counterattack; engines idle and crews remain in close proximity to the tanks. Tanks and weapons are checked and minor repairs made. Necessary reorganization and replacements are made—to the extent permitted by time, materiel, and personnel. Local security is posted, and the platoon leader endeavors to learn the plan or orders for the next phase.

(2) Periods of waiting on local objectives must be as brief as possible. The use of radio communication with the battalion commander will speed up the resumption of the attack. When radio communication is not possible, the foot troops commander, through his staff or the tank liaison personnel, should contact the platoon leader and acquaint him with the maneuver desired in the next phase of the attack.

(3) In many cases, recovery of control can be facilitated if the platoon commander goes in his tank, or on foot, to report to the battalion commander for new instructions.

(4) At assembly points, tank units may have to coordinate with other arms just as was done before the initial attack, in order that a mutual understanding may be had of what the supporting weapons are going to do and how it harmonizes with the maneuver of the tanks. Moreover, the tanks may again need special assistance and should

seek to obtain it if imperative for their protection and success.

g. Continuing the attack.—The attack is resumed at the earliest practicable moment after sufficient fire support is available to cover the advance to the next objective. The tanks advance to the next local objective after learning what this objective is, the location of the next assembly point, what fires, if any, are to support them, and whether any special tank maneuver is to be made. Whether disabled tanks have been replaced or not, the attack continues with those present. The disabled tanks whose locations are known should be reported to the nearest tank maintenance or replacement personnel.

h. Reservicing.—After several hours of action, combat tanks still operating must be relieved for replenishing ammunition, oil, gas, water, etc. This contingency is usually outlined in orders preliminary to the attack.

15. *PURSUIT AND EXPLOITATION.*—*a. General.*— The tank is particularly well fitted to exploit a successful attack, on account of its characteristics of speed and shock action combined with fire power. Tanks can completely demoralize and put to rout a retreating enemy that is already dispirited and exhausted.

d. Direct pressure.—Tanks accompanying infantry battalions against the main body of the enemy assist them as in the attack. Speed, and taking advantage of all weaknesses developed in the enemy position, are essential. When the advance is rapid, foot troop commanders must guard against losing control of tanks.

c. Encircling force.—In order to take full advantage of a defeated enemy, speed of movement by the encircling force is necessary in order to strike the retreating enemy on his flank or rear, or to block roads and delay the enemy so that the direct pressure attack may complete his destruction. Tanks are capable of achieving the speed necessary. They usually operate independently in such groups as are available, or in cooperation with forces moving by motor. When acting independently, control must be held by the tank unit commander. Usually, such a movement and attack is controlled by designating localities or lines not to be passed until a stipulated time. Assembly points and alternate assembly points are designated. Between these points tanks must operate as the situation dictates against any enemy found, particular attention being given to blocking roads and defiles, scattering reserves and transport, and attacking covering forces.

d. *Exploitation.*—Exploitation by tanks is similarly executed. General control is retained only with assembly and alternate assembly points designated where new instructions are issued. In an exploitation, owing to the disorganization of the enemy, the conduct of tanks should be marked by audacious thrusts at places that would ordinarily be avoided. Greater chances must be taken to reap greater benefits.

16. *SUMMARY.*—We have examined the characteristics of tanks, their organization, combat establishments, means of communication, and methods of control. We have studied the principles governing the employment of tanks in which we saw that they should not be committed at random on reconnaissance missions or against ill-defined resistance, but rather conserved for definite simple missions where their mobility, fire power, and shock action can be most profitable to the attack, and turned to best account by the foot troops. The need of secrecy to effect surprise, and the importance of smoke and supporting fires as aids to tank actions, have been shown. The desirability of mass and depth in tank attacks has been emphasized. The relative merits of reallotment (for accompanying tank missions) and retention for general support have been discussed, with preference indicated for the latter. Finally we have traced the tank attack from preliminary preparations, through deployment and assault, to the occupation of assembly positions and resumption of the advance or pursuit. The impression to be retained is that the tank is not all-powerful, but needs the

close cooperation of foot troops and other arms to be of utmost value; and, in addition, that it is a costly weapon, available in limited quantity, which must be held back for the critical moment when it will insure success.

CHAPTER 13

NIGHT ATTACKS

1. *INTRODUCTION.*—An attack launched under cover of darkness with the intention of seizing a prescribed objective before daylight is known as a night attack. Although it is one of the most difficult and hazardous operations resorted to in war, its effectiveness has been repeatedly demonstrated. Furthermore, the continually increasing range and power of modern weapons, and the growing efficiency of aerial observation and attack, point to a much wider use of the night attack in the future than in the past.

2. *CHARACTERISTICS OF NIGHT ATTACKS.*—There are certain general characteristics common to all night attacks:

a. Infantry fire has little or no effect at night owing to lack of observation. Therefore the attacking troops must close with the enemy and seek victory with the bayonet.

b. Troop movement becomes extremely difficult at night. Only the simplest maneuver—a short advance straight to the front with a limited objective—has any reasonable chance of success. Even then, detailed and elaborate plans must be made to insure maintenance of direction, control, and contact.

c. Troops easily lose their morale at night. Men may see enemy infantrymen behind every tree, hostile machine guns in every shadow, an ambush in every clump of trees. Many a night attack has failed because of these phantoms of the dark.

3. *PRELIMINARY CONSIDERATIONS.*—Before deciding upon a night attack a commander should consider:

a. The state of training of his troops.—A night attack demands well-disciplined, thoroughly trained veterans, if the full degree of success is to be expected. Even seasoned veterans have sometimes failed to overcome the dangers of the dark. Situations may be readily conceived, however, in which commanders must attempt a night attack regardless of the state of training of their troops.

b. The weather.—(1) *Visibility.*—Although at first thought it might appear that the darker the night, the greater the chance of success, the reverse is actually true. Some degree of visibility is necessary to solve the problems of direction, control, and contact. It is better to be able to deliver a relatively coordinated blow than to launch an attack in pitch black night and have it disintegrate before it even gets within assaulting distance. A night attack of any import has more chance of success when the moon or the stars furnish some degree of visibility for the attacker.

(2) *The wind.*—A wind blowing from the attack to the defense carries the sounds of the advance to the enemy and is likely to result in its premature disclosure. A wind from the opposite direction helps to preserve secrecy.

c. The terrain.—Open terrain favors control and, whenever possible, should be selected for the attack. Woods or badly cut-up ground heighten the difficulties of maintaining direction, control and contact, and should be avoided. The terrain should also be studied to see whether or not it offers a position where the attack can be formed up at a short distance from, and directly in front of, the objective.

4. *ORDERS.*—*a. Warning order.*—(1) After the responsible officer has determined to make a night attack, where time is a big element, he should immediately issue a warning order to the unit or units involved. If the formal attack order can be written and distributed to the units in time for the preliminary arrangements to be made, there is no need for a warning order. It is essen-

tial that this order and all information of the attack immediately available be sent to the participating units at the earliest possible moment, in order to allow them the maximum time to make preparations.

(2) The warning order should always state the units involved, the date and probable hour of attack, the line of departure, the boundaries, and the objective. Other information may be included.

b. Formal attack order.—(1) The formal order for a night attack must go into far greater detail than a similar order for an attack by day. Nothing can be left to chance; provision must be made for every eventuality and contingency that can reasonably be foreseen. The order must clearly and unmistakably specify assembly positions, the line of departure, formations, routes of approach, the magnetic azimuth of the attack, the objective, means of identification of friendly troops, means of preserving secrecy, method of maintaining contact, action of artillery and other supporting weapons, action to be taken upon capture of objective, a rallying line or rallying points, location of reserve, and its action, if any, in the event the attack is checked or repulsed; action of assault elements in case the attack is checked; prearranged signals calling for fire of supporting weapons, designation, when practicable, of points at which secondary deployments will be made; action to be taken by subordinate units if control is lost; and all other pertinent details.

(2) It is obvious that many of these details would be unnecessary or out of place in an order for a daylight attack. At night, however, a unit must grope its way forward in darkness. Once it has jumped off, it is too late to make corrections or amendments. From that point on it has only the detailed provisions of the attack order to go by. If that order is not clear or has overlooked important details, the attack will suffer for it.

5. *ACTION OF SUBORDINATE UNITS UPON RECEIPT OF WARNING ORDER.*—*a. General.*—As soon as the commander of a subordinate unit receives a warn-

ing order for a night attack he, in turn, should at once warn his unit commanders of the impending operation. Every minute from then on to the time of attack must be utilized in completing the thousand and one essential preliminaries. The situation must be studied, plans made, orders formulated, reconnaissance performed, guides instructed, troops fed, and ammunition issued.

b. Reconnaissance.—The hours of daylight must be utilized to the utmost for reconnaissance of the terrain to be crossed at night. Assembly positions must be selected and marked. The line of departure must be located and identified, and the sector carefully reconnoitered for the most suitable avenues of approach. Reconnaissance groups must bear in mind that maintenance of direction and control is paramount for a successful night attack. Accordingly they must seek every possible means of solving this double and difficult problem. They should select direction points for the approach march and take compass bearings. They should note such valuable directional aids as streams, fences, ravines, ridges, roads, trails, telegraph lines, and similar natural and artificial features that run in the desired direction. Obstacles of any sort that may impede, block, or confuse the advancing troops should similarly be noted; and plans should be made to avoid them during night movement, or to remove them.

c. Guides.—(1) Guides should usually be designated to conduct units to their assembly positions and thence to the line of departure, and finally to lead the attacking columns along previously reconnoitered routes of approach to the hostile position. Whenever possible, commissioned officers should be selected for this important duty.

(2) Men selected to act as guides should accompany the reconnaissance parties. They should thoroughly familiarize themselves with the terrain to be crossed, the routes of approach and all prominent artificial and natural features in the area. They should study the various identifying terrain features just at dusk, thereby familiarizing themselves with the characteristic outlines of these

features at night. This is highly advisable, for familiar objects in daylight become the unfamiliar and unrecognized in darkness.

d. Troop considerations.—(1) Upon receipt of a warning order for a night attack, a unit commander must not become so involved in his preliminary arrangements that he forgets the tool with which he has to work; namely, his troops. There are two important factors that will always boost the morale of the soldier—food and rest. The thoughtful commander, then, will see that his men are fed a hot supper; and if the attack is to be launched a few hours before dawn, he will also provide a hot breakfast and have a cold lunch issued. If the attack is to jump off in the middle of the night, a canteen cup of hot coffee just before moving out will have a buoyant effect on the spirit of the command.

(2) Troops should be afforded as much rest as possible before the attack. They should not be needlessly disturbed, nor should they be formed up hours before the attack and kept waiting. Few things are more detrimental to morale. An irritable, sleepy, hungry soldier is likely to fall out of the night advance at the first opportunity.

6. *TIME OF ATTACK.*—*a.* The exact hour at which a night attack is to be launched may be prescribed by the officer ordering the operation or may be left to the discretion of the commander of the attacking force. If the objective is to be captured and consolidated, the attack should jump off in time to complete its mission and consolidate the captured position before daylight. Strong hostile counterattacks can be expected at dawn. If the position is not consolidated by daylight it is liable to fall before the first well-organized counterattack. An attack for this purpose, then, will usually be launched during the early hours of the night, and almost always before midnight.

b. If the operation is merely a prelude to a daylight attack, it should usually be launched some time between

midnight and a few hours before dawn. The exact hour
will depend upon the amount of time believed necessary
to capture the objective before daylight. Since the attack
is to be continued at dawn, no time is allowed for consolidation of the captured position. In this case it is to
the obvious advantage of the attacker to seize the objective at an hour as close to dawn as possible because
the enemy will thus be deprived of any opportunity to
organize fresh resistance under cover of darkness.

c. The basic consideration, then, for determining the
hour of attack turns on the question of whether daylight
or darkness is desired immediately after the hostile position has fallen.

7. COMPOSITION OF THE ATTACKING FORCE.—
a. *General.*—It is evident that the bulk of the attacking
force will consist of infantry. The questions to be determined, then, are two: first, how large should this infantry force be and second, what auxiliary arms, if any,
should assist it?

b. *Strength.*—There is one principle that applies to a
night attack with special emphasis: the larger the force
the more difficult it is to achieve control and surprise.
Since control and surprise are vital factors in any operation, we see that—from one viewpoint—the smaller the
force, the greater the chance of success. From another
viewpoint we see that a company could hardly be expected to rout a brigade although the control and surprise elements might be perfect. There must be a balance struck between the two.

c. *Auxiliary arms.*—(1) Whether or not the artillery,
engineers, chemical warfare service, air corps, and other
auxiliary arms are called upon to assist the infantry in
a night operation depends entirely upon the special circumstances of the situation. For instance, engineers may
be needed to clear the ground and demolish obstacles in
order to facilitate the advance of the infantry; or they
might be called upon to assist in organizing a captured
position.

(2) Supporting artillery is a valuable aid and can be used in many ways. Particular pains must be taken, however, to insure that when artillery is employed in support of a night attack, it does not reveal our intentions prematurely. The enemy's suspicions must not be aroused by an increased rate of fire or by a concentration on the area to be assaulted. Artillery can avoid arousing suspicion by maintaining a regular rate of fire before the assault and over a wider front than is to be attacked. Artillery fire can also be employed to deceive the enemy. For instance, a heavy concentration may be laid on one area while the infantry attack is launched against another. If the assault is repulsed, the artillery should be prepared to lay down heavy fire on the enemy position to cover the withdrawal of the infantry. The number of ways in which supporting artillery may be used to assist a night attack is limited only by the ingenuity of the leaders of the infantry-artillery team. Infantry commanders must remember, however, that whenever artillery is called upon to support a night attack, its fire must be prearranged. The firing data must be computed before dark and, whenever possible, verified by daylight registration. Therefore the infantry must get in its requests to the artillery at the earliest possible moment.

(3) The air corps may be called upon, especially if the attack is successful, to illuminate the enemy's terrain with flares and to attack his retreating columns and assembly positions.

(4) The chemical warfare service can materially assist the infantry in holding the captured position by interdicting likely avenues of approach with persistent gas. If the attack is to be renewed, however, persistent gas cannot be used; for the attacking troops themselves would suffer by it.

(5) Although enemy tanks will probably not be used in counterattacks during darkness, their presence is always a threat which must be countered. In any night attack, then, antitank weapons should be pushed up close behind the assaulting units, to reach selected positions in

ample time for completed emplacement by daybreak, prepared to meet any counterattacks supported by tanks. Enemy tanks may be used to block defiles by their firepower, or to present a physical obstacle by their bulk; in this case, antitank weapons should be well forward where they can be used against this threat.

(6) As the element of surprise in night attacks is vital to success, the use of tanks in the attack is not favored. Furthermore, tanks cannot see at night, and hence would prove more of a menace to the attackers than a help.

(7) In organizing a force for a night attack, the responsible officer must consider all these agencies. And he must weigh the great factors of surprise and control against the particular mission he wishes some arm to accomplish.

8. *SCHEME OF MANEUVER.*—*a.* The scheme of maneuver for a night attack must be reduced to its simplest terms. The ideal situation occurs when it is possible to form for the attack directly opposite to, and at a short distance from, the objective. The maneuver, in this case, entails no more than a short advance straight to the front culminating in the final rush with the bayonet.

b. Only the simplest movements have any chance in a night attack. The elaborate and complicated will invariably fail, through the loss of one or more of the three decisive elements of night combat—direction, control, and surprise. To attempt wide turning movements and complicated envelopments, both of which necessarily entail changes of direction, is to invite disaster. The maintenance of direction and control, even in a short advance straight to the front, presents major difficulties that many an excellent unit has failed to overcome. But in a long advance involving changes of direction the difficulties are insurmountable. Units lose their way in the dark in spite of the most elaborate precautions, and then quickly disintegrate. If any of the attacking elements come within assaulting distance, the best they can do is deliver a piecemeal, disjointed effort, thrown, more likely than not, against an enemy alerted by the blundering confusion of the advance.

c. The surprise element is most vital. Military history bears witness to the fact that a surprised enemy is usually a defeated enemy. On the other hand, many a well-planned, well-conceived night attack has failed because it lacked that element.

d. Many night attacks have failed disastrously because leaders forgot that direction and control are also vital, and that the only chance of maintaining them is through the utmost simplicity of maneuver.

9. *THE NIGHT ADVANCE.*—a. *General.*—At night, considerations of direction and control require the maintenance of close contact and relatively compact columnar formations until within assaulting distance. The decisive element of surprise demands a stealthy advance characterized by the absence of rifle fire.

b. *Formations.*—(1) The formations employed in the night advance are adopted with a view to facilitating maintenance of direction and control. Obviously, line formations are at once ruled out; for they are impossible to control and immediately lose direction in the dark. Column formations of some sort are mandatory even though the advance be no more than four or five hundred yards straight to the front. A unit going into its assembly position under cover of darkness usually moves in route column. The move from the assembly position to the line of departure can also be made in route column if the circumstances warrant. Guides are generally furnished to lead the unit to the proper place on the line of departure.

(2) The degree of deployment at the jump off depends upon the distance from the enemy, the character of the intervening terrain, the degree of visibility and the activities of hostile aviation. If the distance is short, the terrain open, and visibility relatively good, the advance elements may immediately go into line of squad columns with reduced intervals. Conditions that warrant this

much dispersion at the jump off will, however, seldom obtain.

(3) Instead, it is usually found best to deploy the leading elements in line of section or platoon columns with greatly reduced intervals. Officers and guides equipped with compasses lead their units forward by continually checking the direction of the advance. An officer or noncommissioned officer marches at the rear of each column to keep the men closed up, prevent straggling and enforce silence. Local reserves follow in line or column of platoon columns, with reduced intervals and distances.

(4) It is evident, of course, that this formation cannot be maintained until the leading elements come within assaulting distance of the hostile position. Such a procedure might all too easily end in disaster if the enemy caught the attackers in such a dense formation. Obviously, then, the formation must be further deployed before it reaches the point where premature discovery would be costly. Again, no rule or approximation can be given. The previous reconnaissance and the terrain will determine the time to further extend the formation. Occasionally, it may be possible for the reconnaissance parties to mark this point of secondary deployment; but more often sheer impracticability and considerations of secrecy will prevent it. The ideal situation occurs when some prominent and unmistakable feature of the terrain can be used as a guide. This should be borne in mind during the daylight reconnaissance. If no such feature exists, and the line cannot be indicated, each unit commander must act on his own initiative or conform to the neighboring units.

(5) When the leading elements make this secondary deployment, they should go into line of squad columns with reduced intervals and distances. Reduced distances in the columns facilitate control of individual squads by their leaders. The reduced intervals result in a dense irregular skirmish line when final deployment is made for the assault. This density is necessary to assault the enemy with the bayonet and carry the position on the first rush.

(6) Machine guns belonging to the assault elements should generally move with the local reserve echelons in order to be immediately available to cover the troops during reorganization after they capture the objective, and to help in the consolidation of the position.

(7) The dispositions of the local and general reserves vary little from those of daylight operations. The only material difference lies in the fact that the local reserve is usually closer to the leading elements whereas the general reserve is held well to the rear and well off the axis of advance. This disposition is a precautionary measure taken because the attack may fail. If it were to fail, with the general reserve on the axis of advance, the reserve would in all probability be swept away by the rush of troops to the rear. In any event, confusion would result and morale would be seriously undermined. Finally, the reserve would be unable to cover the withdrawal by fire because the retreating troops would be directly in the way.

(8) The formations suggested in this text are not to be construed as rigidly applicable. Many other equally effective columnar formations for the night advance have been used successfully. The situation itself is, as always, the best guide. It is not likely, however, that any situation will warrant jumping off in a line of skirmishers. The criterion of any formation for a night advance to the attack is this: Does it facilitate direction and control? If the answer is in the affirmative, the formation is probably sound.

c. *Rate of advance.*—A definite rate of advance may be prescribed in case of a long advance after leaving assembly positions. No exact figures can be given for this, but certainly it cannot exceed the night rate for troops marching across country. Taking into consideration the difficulties of control and contact between units and the necessity for secrecy, it will actually be much slower. It is believed that 100 yards in from six to ten minutes will be about the rate actually used. It is well to prescribe the rate of advance in definite time tables, giving the time the unit will cross certain terrain features. When contact has been gained or is imminent, a definite rate should not be used. Each unit should then press forward rapidly within its zone of action to capture its objective.

d. Contact.—(1) At night, close contact must be maintained between all elements of the attacking force; for otherwise units become confused in the dark and lose their way, and the attack disintegrates before it comes within assaulting distance. All leaders must give this difficulty serious consideration. The formal order of the higher commander should take the first step in the right direction by prescribing that contact be maintained from right to left, left to right, or both. The adoption of a column formation with reduced intervals and distances further simplifies the matter. Leaders must not think, however, that these two provisions end the difficulty. Far from it. Many times in war units that have jumped off with only 20-yard intervals between elements have frequently lost contact before going 500 yards. Darkness is deceiving; the quietness and the ease of advance must not lull commanders into the belief that everything is clicking in parade ground fashion. Leaders of each unit must continually check their guides by compass bearing, constantly assure themselves that positive contact exists with adjacent units, and unendingly verify and maintain the cohesion of the subordinate elements of their own command. Such checking must never cease.

(2) Many devices have been used in the endeavor to insure close contact at night. In one night attack in the Boer War the attacking force inclosed itself with ropes. A continuous rope, stretched across the entire front of the attacking force, was grasped and carried along by the leading men of the advance elements. A rope running down each flank was held by the right and left columns. Such elaborate precautions speak eloquently of the difficulty of maintaining contact. Many other methods of facilitating contact can be devised. But it must be re-

membered that these are merely supplemental and should never replace the positive contact afforded by patrols and connecting groups or files.

(3) In the darkness it is impossible to tell friend from foe unless friendly troops wear some unmistakable badge of identification. White brassards on both arms, or undershirts worn over the outer clothing, are the most common means used for this purpose. At night neither the arm bands nor the undershirts can be seen for more than 15 or 20 yards. Thus they serve the double purpose of identifying friendly troops, and promoting good contact between connecting files and between columns marching at reduced intervals. At the same time, they do not prematurely reveal the advance to the enemy.

(4) Leaders of subordinate elements must also be impressed with the fact that if contact is lost they must push on to the objective on their own and not mill about in the darkness looking for other units.

e. *Surprise.*—(1) As stated several other times in this text, the supremely decisive element of a night attack is surprise. History records dozens of night actions in which a unit unexpectedly driving out of the darkness has carried a position manned by a force three or four times its own strength. It also records many a failure which resulted solely from a premature disclosure of the attack. In fact, it may be stated as a general truth that a night attack which does not achieve surprise will fail. Obviously, then, every effort must be bent toward the preservation of secrecy to the end that the assaulting force may fall unexpectedly upon an unprepared and unsuspecting enemy.

(2) The advance to the assault must be characterized by silence and stealth. Talking, smoking, and the display of lights are prohibited. Special precautions are taken to prevent equipment from rattling. Horses should not be included in the advance. They are likely to neigh and thus warn the enemy of the approach. Connecting files should be taught to rely upon identification badges for maintaining contact and should be strictly enjoined against shouting, calling, talking from a distance, or otherwise making noise.

(3) The most important provision of all prohibits rifle fire. Many a night attack that promised fruitful results has been completely spoiled by some nervous scout or connecting file's blazing away at an imaginary enemy, or even at a real one. If patrols or scouts covering the advance encounter the enemy, they close swiftly with the bayonet seeking to silence him before the alarm can be given. Under no circumstances, except a direct order from an officer, is rifle fire employed. It has no place in the night attack. In practically every case where attacking units have cut loose with their rifles they have inflicted more losses among their own troops than upon the enemy. In the darkness a fire fight results in the utmost confusion. Men fire at anything that moves, for they see the enemy everywhere. Badges of identification mean nothing in such cases; for excitement, fear, and instantaneous reaction rule the fight.

(4) The bayonet is the weapon of the night attack. It is silent and it is effective; and before a man can kill with it he must be close enough to his opponent to determine whether he is friend or enemy, without the possibility of making a mistake.

10. *THE ASSAULT.*—*a.* The assault is the culmination of the attack. The enemy usually determines at what point it is to be launched because he generaly discovers and opens fire upon the advancing troops when they are 100 or 200 yards away. This is the signal for the final rush with the bayonet. There is no longer any need for secrecy; the enemy has no time now to prepare or mount his defense. Officers and noncommissioned officers urge the men forward. The compact squad columns deploy into a dense skirmish line and drive forward with the bayonet. Every effort must be made to carry the enemy's front line in the initial rush. Delay, hesitancy, or repulse in front of the position will afford the hostile machine guns time to get into action and inflict heavy casualties. The safest place is inside the hostile position.

b. When the enemy discovers the attack and opens fire, the long awaited mental and physical shock usually brings the advance elements of the attacking troops to an abrupt halt. It has been found that this reaction occurs in almost every night attack. And this is the crisis of the whole operation. If aggressive, vigorous leadership is lacking at this moment, the entire line is liable to break in panic. Officers and noncommissioned officers must overcome that moment of inertia. Personal example will count for little in the darkness, for officers, if seen at all, cannot be distinguished from enlisted men. They must resort to orders and shouts of encouragement such as "BAYONETS! GO GET 'EM! WE'VE GOT 'EM NOW! THEY CAN'T GET AWAY!" Each officer knows his troops and should suit his actions to their temperament. The main thing is to get the assault under way. Once it has started, the excitement and spirit of battle will carry it along until the position has fallen or the attackers have literally been physically repulsed in hand-to-hand combat.

c. If the enemy is thoroughly asleep on his job, a thing that seldom happens, the attacking force itself must decide when it is near enough to launch the assault. This, too, will usually be altogether automatic. Some one unit will generally see the enemy's position first, owing to the irregularities of both the advance and the terrain. As a rule that unit will initiate the assault. Its rush forward will serve as the signal for the rest of the command to drive forward to the assault. However, as previously stated, it is almost always the enemy who discovers the attack and gives the signal for the assault by opening fire.

11. *ACTION FOLLOWING THE ASSAULT.*—*a. General.*—Once the objective has been carried immediate preparations must be made to resist the counterattack which can be definitely expected. To this end a hasty reorganization is effected, and steps are taken to defend the position against the expected enemy blow as soon as possible.

b. Reorganization.—(1) Although it is a general principle that reorganization will never be made in any attack unless necessary, it is inconceivable that a night attack can be launched and carry a hostile position without the utmost confusion and intermingling of units. A reorganization of some sort therefore has to be undertaken. It is usually impossible to extricate and reform units in the darkness. However, officers and noncommissioned officers should seek to form groups of men and place them in the more decisive tactical localities. The defense of the position during the hours of darkness and at dawn will depend almost altogether upon the individual leaders and these mixed groups of men that they have formed, placed, and temporarily command.

(2) The commander of the operation should make provision in his attack order for a force, usually detailed from the local reserves, to mop up the position as soon as it is definitely taken.

c. Consolidation.—(1) The degree to which a position is consolidated depends upon the purpose for which it was taken. If an objective is carried with the intention of continuing the attack at daylight there is little need or reason for much consolidation, the time being more profitably spent in a thorough reorganization. We have seen in an earlier paragraph that an attack of this nature is launched at a time computed to take the objective shortly before dawn, merely allowing a half hour or so of darkness as a safety factor for unforeseen delays. If the hour of attack has been correctly computed, there will scarcely be time for the enemy to organize a counterattack in the brief period before dawn. Therefore, although security provisions are immediately taken, every effort is bent toward accomplishing an effective reorganization in time to renew the attack at daylight.

(2) If the position is to be held, emphasis must be placed on consolidation. Automatic rifles and machine guns must be placed to cover likely avenues of approach. Security and observation groups must be established to the front and flanks. Local reserves must be held in a central

locality ready to intervene to the front or to either flank. In short, every provision that can be made under cover of darkness to prepare the position against counterattack must be accomplished.

12. *ACTION FOLLOWING A REPULSE.—a.* If the assault elements of a night attack are repulsed, one of two courses is open to the commanding officer. He can either commit his reserve or order a withdrawal to the rallying points or line prescribed in his order. The enemy will be thoroughly alerted, the surprise element will be completely missing, the terrain in front of the position will be illuminated with flares, and the hostile machine guns will play havoc with a fresh effort.

b. There can be little order in the withdrawal of a unit defeated in a night attack. Rallying points are prescribed in the attack order for just such a contingency. It is largely a matter of individuals and small groups working their way back to these points. Confusion, casualties, intermingled units, and bewildered men will be inevitable. It would be desirable, of course, to withdraw in an orderly fashion by successive units from the right or left but this is seldom, if ever, possible. Leaders are not only unable to find the left or the right but unable to find their own units. Isolated men and groups are scattered over the field at varying distances from the assaulted position. Many men will have long since been on their way to the rear.

c. If a withdrawal is ordered, the general reserve located off the axis of advance covers the movement. A prearranged signal should be fired to bring down the supporting artillery on the enemy and further shield the retreating troops. Little else can be done.

13. *PSYCHOLOGICAL ASPECT OF NIGHT OPERA-TIONS.—a.* Although a detailed treatment of the peculiar psychological aspects of a night operation is beyond the scope of this text, a brief discussion of several of the

more important characteristics should lead to a keener appreciation of the extreme difficulties that attend such operations.

b. The entire psychological structure of the night attack revolves about the simple truth that man has never outgrown his fear of the dark. At night that inherent fear sharpens all the soldier's faculties. The suggestion of a shadow becomes an enemy scout; the imaginary scout miraculously grows to a combat patrol; and in a few seconds the patrol has expanded into a formidable night attack. The soldier has seen nothing but a shadow. His imagination has seen the rest. It is a strange truth that the imagination never paints an encouraging picture at night; it always conjures up the thing that is most feared. It is equally strange and equally true that the greater the inaction at night, the greater the fear. The attacking force moving through the darkness to the assault has its fears and its apprehensions, but these are negligible compared to those of the defending force. In the night advance, men see and hear and feel other men moving all about them. They feel a certain degree of security and companionship in that simple fact. On the other hand, those same men on guard in their front positions at night feel isolated and abandoned. They strain to see the dark terrain before them, and their fear and imagination accomplish the rest.

c. It is owing largely to this psychology of the dark, and the wide variation of morale between action and inaction, that a small force driving unexpectedly out of the night has so often routed a force many times its size.

d. The strict prohibition against rifle fire is rooted not only in its ineffectiveness at night but in the further fact that scouts, patrols, and the foremost men in the advance invariably see the enemy in many an innocent shadow and at once open fire. It has been shown again and again that one rifle shot by a scout almost invariably leads to general firing. The leading elements visualize the enemy on every side and blaze away into their own units. The rear elements, not to be outdone, cut loose against the assault echelons, usually with deadly effect. As a positive provision against such a disastrous contingency many commanders have supplemented the mandatory prohibition against rifle fire with the further provision that rifles will not be loaded.

e. Is it possible to completely allay these phantoms of the dark? Probably not. However, they can be materially lessened by adequate and intelligent training in night operations. The battlefield is no place for this training. If it is to be accomplished at all it must be accomplished in time of peace. Individual soldiers must be trained to move from point to point in the dark utilizing the stars and previously determined landmarks to maintain direction. Units must be practiced in advancing over all kinds of terrain at night, still maintaining direction, and cohesion. They must be trained to change formations and deploy quietly, rapidly, and without confusion. It is best for such maneuvers to be conducted against real troops, and not an outlined or simulated enemy. Organizations accustomed to operating in the dark against a friendly enemy lose many of their imaginary terrors when actually engaged in real night combat.

14. *HISTORICAL EXAMPLES.*—The foregoing text has dealt exclusively with the underlying tactical principles of a night attack from a theoretical point of view. Six examples of night operations that occurred during the World War are now presented in order that the student may study the practical applications of those principles in actual warfare. Of the actions recorded, some resulted in complete failures, some achieved a partial success, and some won through to striking victories. In every case the operation is analyzed in a discussion immediately following the narrative. It is to the student's advantage, however, to analyze the operations himself before turning to the discussion.

EXAMPLE I

In 1914, during the first days of the Battle of the
Marne, the German Fifth Army suffered so heavily from
French artillery fire that its infantry was unable to close
with the enemy. In order to come to grips, the army
ordered an attack for the night 9/10 September on a
front of 20 kilometers. Portions of four army corps
participated. One of the units engaged in this action was
the 30th Infantry of the 34th German Division.

On September 9 this regiment, which had just received a number of replacements, was in reserve near
Bulainville. (See Figure 107.) That afternoon the
colonel received the division attack order. In this order
the 30th Infantry was directed to launch its attack from
the vicinity of Amblaincourt, which was believed to be
occupied by the French. The small hill about 1800 meters
southwest of Issoncourt was assigned as the regimental
objective.

Realizing that the Bunet stream would have to be
crossed, the colonel promptly dispatched an officer's patrol to reconnoiter for crossings. Before dark this patrol
returned with the necessary information.

At nightfall the regimental commander assembled his
officers and issued his order. The regiment would attack
with the 2d and 3d Battalions abreast, the 2d on the right.
The 1st Battalion would be in reserve. The 2d Battalion was to move forward along the west edge of the
Chanel Wood and then turn eastward following the south
edge toward the objective. The 3d Battalion would move
on the objective by guiding on the north edge of the
wood. Weapons would not be loaded and there would
be no firing. Silence was mandatory. What commands
had to be given would be transmitted in whispers. The
progression of first-line battalions would be made "with
units well in hand, preceded by a thick line of skirmishers".

By midnight, the 30th Infantry and adjacent troops
had reached attack positions north and west of Amblaincourt without alarming the French. Rain was falling.

The advance was started. As the leading elements
neared Amblaincourt there was a sudden burst of firing.
Immediately every one rushed toward the town. There
were no French there. In the confusion some straw

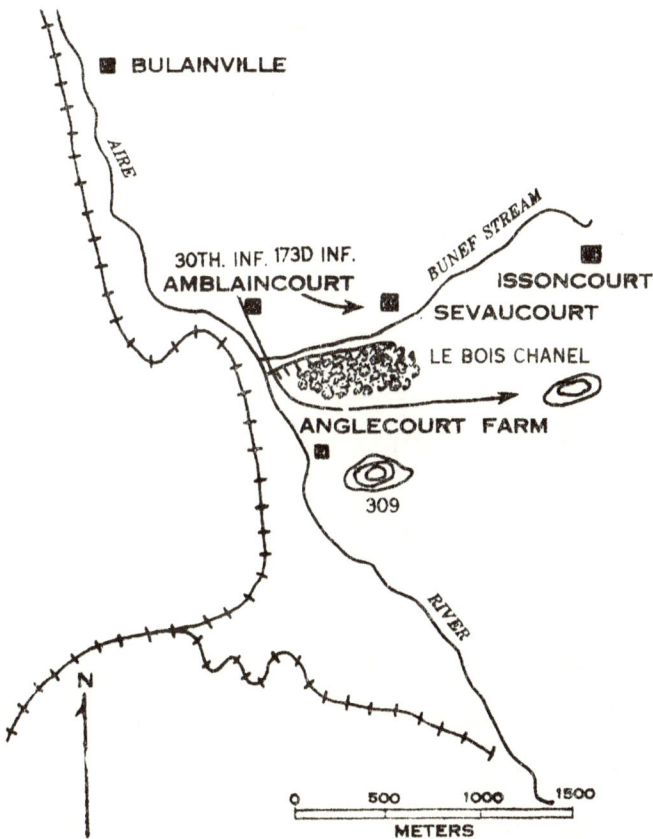

FIGURE 107.

piles nearby caught fire, revealing the milling Germans to
the French who actually occupied the Chanel Wood, and
who promptly opened a heavy fire. The German assaulting units forthwith fell into the greatest disorder, and

the 30th Infantry became intermingled with the 173d on its left.

In spite of the confusion and the heavy fire, most of the men of the 30th and some of the 173d pushed on toward the dark outline of Chanel Wood. They crossed the Bunet Stream, the water reaching to their breasts and sometimes to their necks. Emerging from the stream they charged the wood in one confused mass. They reached the edge and hand-to-hand fighting followed. German accounts state that an irregular fire came from all sides; that no one knew friend from foe. Neighboring units had lost direction; and there were even men from other army corps mingled with the troops of the 30th. About 2:30 AM the Germans were in possession of the Chanel Wood, but their losses had been terrific. The history of the 30th Regiment says:

The most complete disorder reigned in the units after the incidents of the Chanel Wood. Near Anglecourt were moving units of the 30th, 173d, 37th, 155th and even Wurtemburgers (XIII Corps). Officers strove to organize at least squads or half platoons, but the smallest group, as soon as formed, became lost in the obscurity. It was only at the south edge of the Chanel Wood that sufficient order was reestablished to continue the advance.

The 8th Company managed to push on and capture several cannon after a hand-to-hand fight with the gunners. Unfortunately they had to withdraw soon afterward, having come under an intense fire from their own comrades.

At daybreak the 30th Infantry, completely intermingled with the 173d, held the line: northeast corner of the Chanel Wood—hill *309.* Although ground had been gained, the attack was considered a failure.

On September 12, a German colonel who commanded a unit in the same division with the 30th, met the German Crown Prince, who commanded the Fifth Army, and asked permission to speak frankly regarding this attack. This being granted he said: "Imperial Highness, one more night attack like that one, and the army will be forever demoralized."

(From the article in *La Revue d'Infanterie,* August, 1927, by Colonel Étienne, French Army.)

DISCUSSION

The history of the 30th Infantry refers to this night as "St. Bartholomew's Eve". In the memory of the survivors, it was the most terrible of the entire war.

Although the German troops fought with incontestable bravery, and although they achieved miracles, the action was doomed before it began. Subordinate commanders were not given the opportunity for daylight reconnaissance. The infantry was not opposite its objective at the jump-off. The objective itself was more than 6000 meters away with the intervening terrain unknown. As a crowning touch the regiment was ordered to execute an abrupt change of direction—this in the dead of night and at the height of the attack. To demand that the 30th Infantry, in a night attack, take Amblaincourt, capture the Chanel Wood, then change direction and push on to a distant objective, meanwhile maintaining control, was more than demanding the impossible; it was presenting an unequivocal invitation to disaster.

Night attacks must have limited objectives. The results can be fully exploited only by day.

EXAMPLE II

On the night of 6/7 October, 1914, the French 2d Battalion of Chasseurs moved by truck to Vrely, where it arrived at 7:00 AM. (See Figure 108.) There it was attached to the 138th Brigade. At 2:30 PM it was ordered to march via Warvilliers on Rouvroy-en-Sauterre in order to participate in a night attack by the 138th Brigade. Another attack, coordinated with this, was to be launched from Bouchoir toward Le Quesnoy.

The 138th Brigade planned to attack with the 254th Infantry on the right, the 2d Chasseurs on the left and the 251st Infantry in reserve. The dirt road between Rouvroy and hill *101* was designated as the boundary between the 254th and the 2d Battallion of Chasseurs.

The terrain between Rouvroy and Parvillers was flat and presented no difficulty to movement at night.

The 2d Chasseurs was an elite organization. However, as a result of previous fighting, it was reduced to an effective strength of approximately 150 men per company. Most of the battalion's six companies were commanded by noncommissioned officers.

At 5:45 PM, with dark closing in, the battalion moved forward through Rouvroy. Not more than an hour had been available for reconnaissance. Information was vague. It was believed that Parvillers was held by the Germans.

The 2d Chasseurs formed for attack as follows: Two companies were deployed in one long line of skirmishers, preceded by patrols. Four companies followed in second line. These four companies were abreast, each having two platoons leading and two following. Platoons were deployed in line of skirmishers. The distances ordered were:

150 yards from the patrol to the first line;

200 yards from the first line to the second line;

50 yards between leading elements of second-line companies and their supports.

The machine-gun platoon was placed 50 yards behind the left of the third line.

Shortly after dark the battalion advanced on Parvillers. As the advance neared hill *101,* one of the patrols ran into an enemy outguard which promptly opened fire. Many of the French answered this fire without knowing what they were shooting at or why. Soon firing became general.

The two leading companies halted. A cry of "Forward!" ringing through the darkness was caught up and echoed by hundreds of voices. Abruptly the second-line companies rushed forward, racing toward Parvillers. Pellmell they charged through the leading companies, one of which followed. A terrific uproar ensued, punctured with shouting and cheering.

The rush reached a trench 250 yards northwest of Par-villers. The defenders had fled, leaving weapons and equipment, but the enemy farther in rear had been

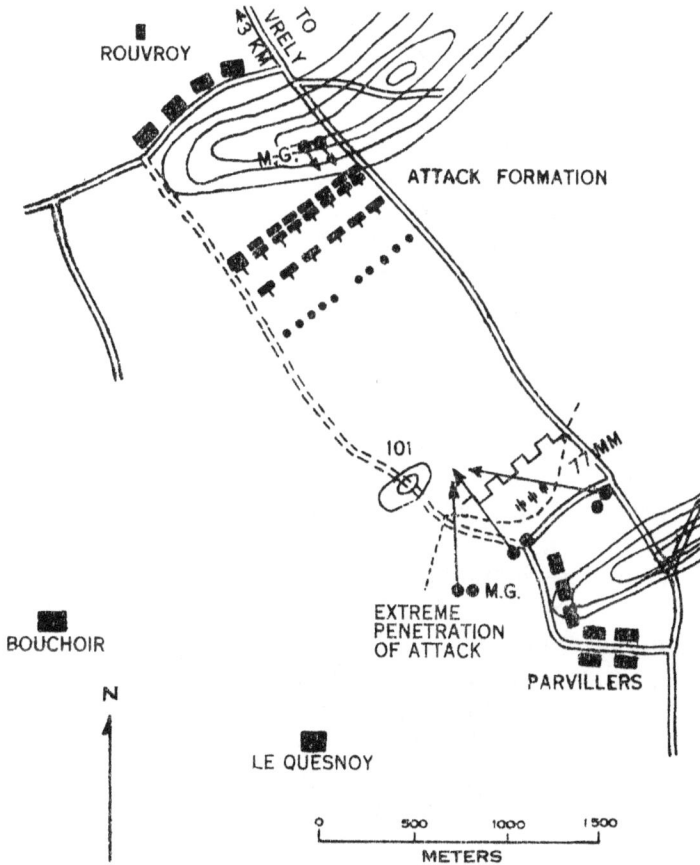

FIGURE 108.

warned. Suddenly three 77-mm cannon, 150 yards behind the trench, opened at point-blank range on the French. By the flashes artillerymen could be seen serving the guns. The French in front of the battery stopped, but those on the right closed in and captured the three pieces.

In great disorder the advance continued toward the village. As they moved forward, their left flank came under fire of hostile machine guns located on the road 600 meters northeast of Parvillers. By this time all French units were hopelessly intermingled, many company and platoon commanders had become casualties, and in many places, confused by the dark, the French were firing on their own troops. The attack wavered and stopped.

It was 11:00 PM. With much difficulty, noncommissioned officers rallied a few scattered groups and occupied the conquered trench. It was realized that further concerted action by the battalion was impossible.

Meanwhile the right assault company, which had not followed the general movement, was still under partial control. The battalion commander ordered it to a central position 600 yards in rear of the trench to cover the withdrawal of the battalion. When the order to withdraw was given, voices, whistles, and bugle calls were heard. Firing continued during the entire movement; but finally the battalion managed to extricate itself and reform just in rear of Rouvroy. It had suffered in the neighborhood of 300 casualties.

The entire French attack failed.

(From an article in *La Revue d'Infanterie*, June, 1924, by Lt. Col. Jeze, French Army.)

DISCUSSION

In this engagement the French solved the problem of direction but failed completely in the co-existing problems of control and surprise.

As a matter of fact the direction phase practically solved itself, for the road paralleling their advance on Parvillers made any great loss of direction virtually impossible. Unfortunately, no kindly terrain feature could eliminate the remaining difficulties.

It was inevitable that the widely scattered formation they adopted should result in loss of control. At night, distances and intervals must be diminished and formations kept compact. In this instance, section columns or even larger groupings would unquestionably have gone a long way toward keeping the battalion in hand. Particularly was a compact formation mandatory here where most of the company and platoon leaders were noncommissioned officers with little or no experience.

The patrol which encountered the hostile outguards on hill *101* should have closed with the bayonet without firing. It failed to do this, and firing soon became general. The usual results followed with clock-like precision; once started, the firing could not be stopped; officers were unable to get the leading elements to continue the advance; and the attacking units fired into their own troops.

The second-line companies, with due courage, but with undue cheering and firing, charged. The tumult, the firing, and the onrush of hundreds of men from a distance, gave the Germans ample warning. It was an attack—an assault that had started too far off. The French lines, revealed by their cheering, were swept by hostile machine-gun fire. In the utmost confusion the assault wavered to a halt, and shortly afterward the battalion withdrew.

Loss of control through a vicious formation, and loss of surprise through yelling and firing, had wrecked one more night attack.

EXAMPLE III

On 12 January, 1915, the French were attacking northward near Soissons. (General Map, Figure 109.) At 7:00 AM, the 1st Battalion 60th French Infantry, which was in reserve, marched from Villeblain to Maast-el-Violaine where it arrived at 10:45 AM. There it received an order to move back to Courmelles which it reached at eight o'clock that night.

At Courmelles the battalion commander was informed that his battalion, would make a night attack, without delay, to retake hill *132* which had just been captured by the Germans.

No large-scale maps were available, no one in the battalion knew the terrain, and there was no information as to the exact location of the hostile positions. The order received by the battalion commander more than met the requirement of brevity—"attack the enemy when you get close to hill *132*. He will shoot at you." A guide had been provided, however, to conduct the battalion to the French front line.

The two battalions cleared Courmelles at 8:30 PM and arrived at the Vauxrot Glass Factory two hours later, where they dropped packs. They now marched along the road in single file. At Point A (Special Map, Figure 109) the guide turned off into a communication trench that was knee-deep in mud and blocked in several places by fallen trees. At these blocks the column was broken, companies became disorganized and considerable time had to be spent in reorganizing platoons when the front line was reached. As a result it was 3:30 AM before the attack formation could be taken.

The battalion was directed to form with two companies in assault and two in support, each company in column of platoons, and each platoon deployed in line of skirmishers. On forming up, company commanders found that they had near them only squads without leaders and portions of platoons. Entire units were missing. Someone lit a match to check the compass. German flares became increasingly frequent.

At 4:00 AM the attack jumped off—but in places only. The troops were poorly oriented. They did not know where to go or where to stop. There was no liaison. One assault company lost direction. The company behind it pushed on and the two became hopelessly intermingled.

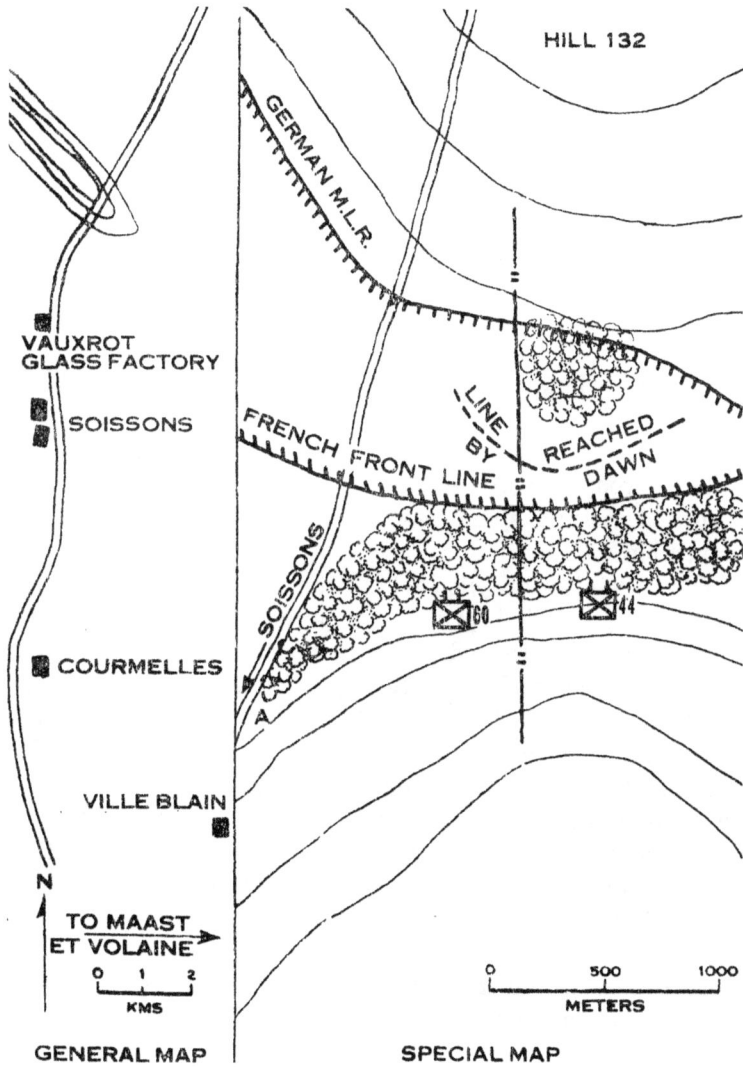

HILL 132

GERMAN M.L.R.

VAUXROT
GLASS FACTORY

SOISSONS

FRENCH FRONT LINE

LINE REACHED BY DAWN

SOISSONS

COURMELLES

60

44

VILLE BLAIN

N

TO MAAST
ET VOLAINE

0   1   2
KMS

0   500   1000
METERS

GENERAL MAP

SPECIAL MAP

FIGURE 109.

German artillery and machine guns opened a withering
fire on the disorganized units, forcing them to halt, take
cover and wait for daylight.

At daybreak it was learned that the battalion of the
44th had attacked on the right of the 60th at a slightly
later hour.   Both battalions had failed, virtually no
ground having been gained.   The losses in the 60th were
exceptionally heavy.

(From the article in the *La Revue d'Infanterie*, June,
1924, by Lt. Col. Jeze, French Army.)

DISCUSSION

This attack is a conspicuous tragedy of error.   A deliberate effort at failure could not have been more
thorough.

The troops were exhausted when the attack was
launched, having spent the day in marching and counter-
marching.

The precipitation with which the battalion was engaged
definitely precluded proper preparation, particularly re-
connaisssance.   Indeed, the troops were in the dark
figuratively as well as literally, not even knowing the
exact location of the hostile position.   Add to this, the
additional handicaps that the battalion was not under
control at the jump-off, that the formation was entirely
unsuitable, that lighting matches, firing and yelling removed any chance of surprise, and we have a situation
that not even a Bonaparte could retrieve.

As Lt. Col. Jeze concludes: "In doing exactly the opposite of what was done, they would not have been far
from realizing the most favorable condition for the success of the operation."

EXAMPLE IV

On 10 October, 1918, the 2d Battalion 30th U. S.
Infantry, was in reserve in the Bois de Cunel.   (See

Figure 110.)   On the previous day, as an assault unit, it had reached the north edge of the wood and was therefore somewhat familiar with the terrain beyond.

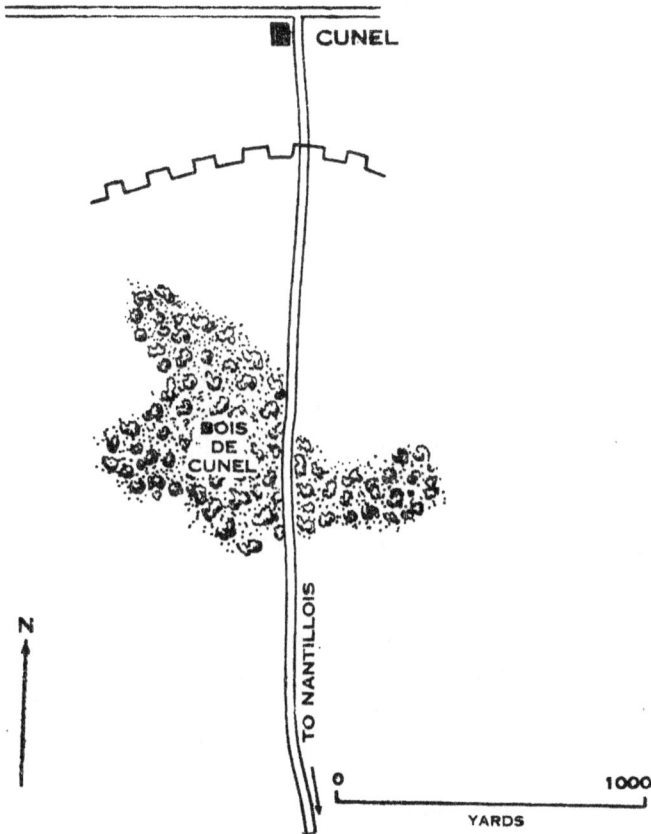

FIGURE 110.

Early on the 10th, the 1st Battalion 30th Infantry, had attacked to the north, but after progressing a short distance had been pinned to the ground in front of a German trench located north of the Bois de Cunel.   It was ordered to withdraw to the Bois, reorganize and resume the attack at 7:30 PM, assisted by a new artillery preparation. The withdrawal began shortly after dark, but in the process the battalion became so disorganized that it was unable to launch the attack at the designated hour. However, the division commander had ordered that the trench 800 yards north of the Bois de Cunel be taken on the 10th. Since the 1st Battalion had failed to accomplish this mission and was unable to make a second effort in time, the 2d Battalion 30th Infantry and one company of the 7th Infantry were directed to attack the hostile position at 10:00 PM. The northwestern edge of the wood was designated as the line of departure for the 2d Battalion, and the northeastern edge for the company. There was to be no artillery support.

After all units were in place, the battalion commander assembled his company commanders and thoroughly explained the details of the attack. The formation was three companies in assault and one in reserve. As the frontage was large, and since all organizations had been depleted some forty per cent in previous fighting, each company, in turn, employed three platoons in assault and one in reserve. The assault platoons were deployed as skirmishers with intervals of two to five yards. The reserve company, formed in line of squad columns, was directed to follow the center assault company at 100 yards. The machine-gun company attached to the battalion was ordered to remain in place until the enemy had been driven from the trench, then to displace forward and assist in the organization of the captured position as rapidly as possible.

The attack was launched on time. Exactly two and a half hours had elapsed since the Germans had been subjected to a heavy artillery preparation, following which the attack of the 1st Battalion had failed to materialize. When no attack had followed this 7:30 PM bombardment, the Germans apparently concluded that there would be no further American advance into their territory that night.

The movement forward of the 2d Battalion was slow and cautious. Secrecy had been stressed. German flares went up frequently. Each time one began to illuminate an area, all men remained motionless, resuming their movement only when the illumination disappeared. This method of advance was continued until the assault units were close to the hostile position. Finally the movement was discovered and machine-gun and rifle fire ripped into the assaulting units from front and flanks. But the Americans were now too close to be stopped. In a swift charge, they closed with the enemy, overcame a determined resistance, and captured part of the disputed trench. The Germans, however, were still holding portions of the trench on the flanks.

By this time every vestige of organization had disappeared. Many company, platoon, and section leaders were casualties. The reserve company was completely intermingled with the assault companies. All was confusion. Immediate steps were taken to reorganize the battalion, while a message requesting reinforcements was sent to the regimental commander.

At 2:30 AM the battalion commander reported to the regimental CP. There he informed the colonel that the 2d Battalion was now occupying the trench in the zone of the 30th Infantry and had established contact with the company from the 7th Infantry on the right, but that reinforcements were necessary on the left where the enemy still held the trench in considerable force. One company was promptly dispatched to this dangerous flank where, after severe fighting, it succeeded in driving the enemy from his position.

At 6:00 AM the strength report of the troops that had made this attack showed the following effectives: Company E—1 officer, 30 men; Company F—1 officer, 40 men; Company G—1 officer, 20 men; Company H—1 officer, 27 men; Company G 7th Infantry—1 officer, 10 men. Not all of the missing were casualties. Many men who could not be accounted for had merely become lost in the darkness.

(From the personal experience monograph of Major Turner M. Chambliss, Infantry.)

DISCUSSION

Here most of the conditions essential to the success of a night operation are evident.

The battalion knew the terrain.

It was close to its clearly-defined line of departure.

It was placed opposite its objective.

The objective was limited and was unmistakable even in the dark.

The troops had not been engaged during the day and were therefore comparatively fresh.

Details of the attack were carefully explained by the battalion commander.

The movement was made in silence, great care being taken to avoid alarming the enemy.

The attack was made at a time when the Germans had concluded that no further effort would be made that night.

All of these factors made for success.

On the other hand the extended formation contributed to the loss of control, and the subsequent confusion and intermingling of the reserve company with the assault units necessitated a call for help to clear up the situation on the left flank.

The figures giving the effective strength of units are indications of the disorder which may attend even a successful night attack. True, the companies were depleted at the start; but, even so, the small effective strength at the conclusion of the operation is striking.

Example V

On November 5, 1918, the 123d French Division was attacking to the east. Late in the afternoon it had been stopped west of Esqueheries. (See Figure 111.) The troops were almost exhausted but despite this the 12th Infantry, with the 1st Battalion 6th Infantry attached, was directed to prepare an attack to take Esqueheries without delay.

The colonel of the 12th Infantry issued orders for an attack at dark. The 2d Battalion 12th Infantry, was directed to attack the town from the west, encircle it on the north and seize the exits toward Petit Foucomme and La Voire. The 3d Battalion was ordered to pass north of the 2d Battalion, and hold the exits leading to RJ 191 and Le Nouvion. The 1st Battalion 12th Infantry was told to push forward to the Le Nouvion Road and occupy a position where this road entered the forest. The 1st Battalion 6th Infantry was directed to seize the southern exits of Esqueheries and those leading toward Sarrois. After securing these exits, the attacking battalions were then to strike toward the center of the town.

The commander of the 1st Battalion 12th Infantry received this order at 8:00 PM and immediately sent for his company commanders. At 9:00 PM none of them had yet reached the battalion CP. In the interim, the battalion commander learned that the 1st Battalion 411th Infantry, on his left, had received no order to attack. The night was pitch dark and rain was falling in torrents. There was no road or trail to guide the commander of the 1st Battalion 12th Infantry to his objective; he would have to advance three kilometers across country over terrain that bristled with thick hedges. Considering his men incapable of such an effort, he requested authority to remain in position until daybreak. This was granted.

The 2d Battalion 12th Infantry, also received this order about 8:00 PM. Its commander at once endeavored to get in touch with adjacent battalions to arrange details. He was finally informed that the 1st Battalion had received authority to delay its advance until daylight. He was unable to get in touch with any other unit.

Undeterred by this, the battalion commander of the 2d Battalion issued his order. The 5th Company was directed to move forward until it reached the road: Petit Foucomme—Esqueheries, which it was to follow into the town. The 6th Company was ordered to advance until it reached the road leading from LaVoire to Esquheries, then follow that road to the town. The 7th Company, which had been in reserve, was directed to send one platoon to attack astride the road entering Esqueheries from the west and capture the western part of the town. The rest of the 7th Company was directed to remain in reserve.

The company commanders protested that their men were extremely fatigued, the rain torrential, the night dark, and the terrain unknown. The battalion commander was obdurate. He stated that the operation would be carried out according to his order and that the movement would start as soon as the 3d Battalion 12th Infantry arrived.

At 11:00 PM the 10th Company of the 3d Battalion arrived abreast of the 2d Battalion. Its commander stated that, at the start of the movement, the 10th had been the rear company in the 3d Battalion column, but that now he had no idea where the remainder of the battalion was. After a further fruitless delay, the commander of the 2d Battalion directed his companies to move to the attack without awaiting for the 3d Battalion.

Shortly after the attack jumped off, the remaining units of the 3d Battalion arrived. The men were exhausted, the companies disorganized and the officers

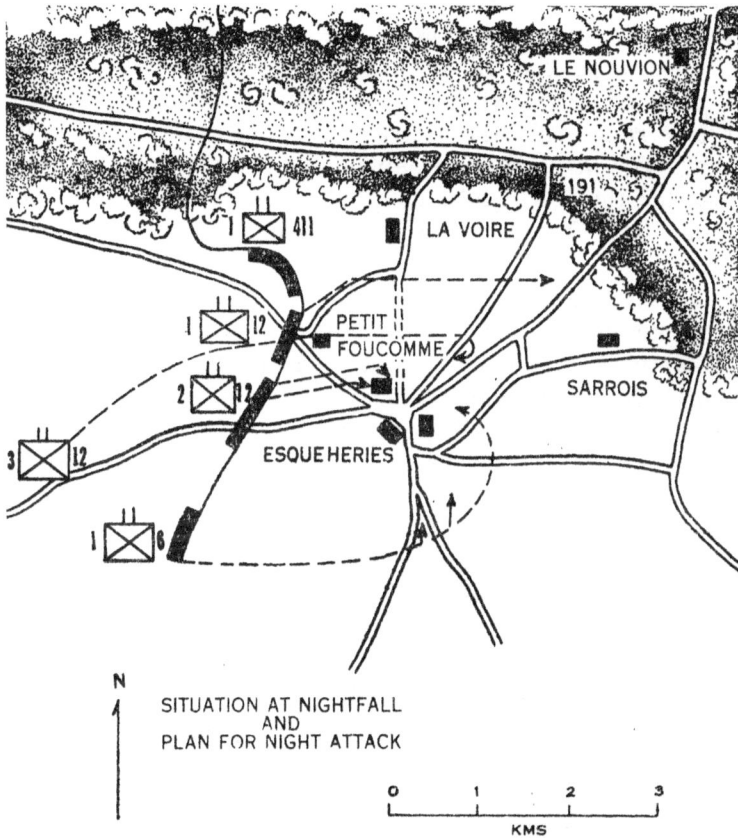

LE NOUVION

191

1 ⊠ 411

LA VOIRE

1 ⊠ 12

PETIT
FOUCOMME

2 ⊠ 2

SARROIS

3 ⊠ 12

ESQUEHERIES

1 ⊠ 6

N

SITUATION AT NIGHTFALL
AND
PLAN FOR NIGHT ATTACK

0   1   2   3

KMS

FIGURE 111.

unoriented. The battalion commander thereupon decided to remain in position until dawn.

Meanwhile the 2d Battalion had moved out at 1:00 AM, much later than had been expected. Darkness and the heavy rain made the forward movement slow and difficult. The 5th and 6th Companies did not reach the north edge of the town until daylight. Similarly the 1st Battalion 6th Infantry did not reach the southern exits until 5:00 AM. However, the one platoon of the 7th Company ordered to attack Esqueheries from the west, with an improved road to guide on, progressed rapidly. It captured the western part of Esqueheries, whereupon the Germans evacuated the entire town, leaving this one platoon in undisputed possession.

(From the article in *La Revue d'Infanterie*, April, 1918, by Major Jaunet, French Army, on the advance of the 123d French Division from the Sambre Canal to the vicinity of Chimay.)

DISCUSSION

The failure of the attempted encirclement of Esqueheries is instructive. Four battalions were ordered to participate in the operation. So serious were the difficulties that two did not even make a start; and the other two, with the exception of one platoon, did not arrive within striking distance of the objective until daylight.

This one platoon had a positive means by which it was enabled to maintain direction; namely, the road that ran past its initial position straight into the town. Other units, lacking points on their route which could be readily identified, spent the greater part of the night in a disheartening game of blind-man's buff.

Again it is pointed out, that of four battalions ordered to the attack, but one platoon closed with the enemy; and yet this single platoon was entirely successful in capturing the objective.

This operation graphically demonstrates the following facts: troops that are to take part in a night attack should be familiar with the terrain; the ground should not present too many obstacles to movement; the troops should be close to, and opposite, their objective; the axis of advance should be clearly marked and unmistakable; and finally the troops involved should be in good physical condition and imbued with high morale.

EXAMPLE VI

On 11 November, 1914, the 121st French Infantry was entrucked and moved to the north where a great battle was in progress along the Yser River.

In three months of war the 121st had been both lucky and successful. Morale was excellent. As an added touch, many officers and noncommissioned officers, wounded in earlier fights, had recently returned to the regiment.

About noon the 121st arrived at Oostvleteren. (See Figure 112.) Here it was directed to march to Reninghe where further orders would be issued.

At Reninghe at 4:00 PM, a division commander informed the colonel:

Your regiment is attached to my division. The Germans have crossed the Yser Canal between the Drie Grachten bridge and a point 800 meters south of it. There is nothing in front of Noordschate to prevent their rapid progress toward Reninghe.

At 1:00 AM tonight the 121st will attack and drive the enemy over the Yser.

The XX Corps will be on your right and a regiment of Zouaves on your left. There is a gap between them. Their flanks are near the canal.

You will find the Colonel of Zouaves at Noordschate. Carry on without further orders.

The regimental commander designated the 2d Battalion, supported by a machine-gun platoon, to make the attack

Noting from his map that the terrain between the Yser and Yperlee Canals appeared extremely difficult, the battalion commander determined to make a personal reconnaissance. Saddle horses having just arrived, he mounted all his company commanders and moved rapidly toward Noordschate. Finding no one there, the party climbed to the second story of a house and studied the terrain. Although dusk was closing in, sufficient light remained to show that the problem of reaching the Yser with troops at night would present grave difficulties. The intervening terrain was a quagmire, interlaced with small canals and large ditches, which would obviously make the maintenance of direction and control extremely difficult. No French units could be seen. Apparently there were some Germans near the Yser Canal.

Following the reconnaissance, the battalion commander issued an oral order for the attack. He directed the battalion to move forward without delay to Noordschate and to form by 11:30 PM along the Yperlee Canal with three companies abreast, their right 400 meters south of the Noordschate Bridge, and their left just north of the Yperlee bend. Patrols would be sent out to seek liaison with units on the flanks. Reconnaissance of the canals directly to the front was limited to 200 yards in order not to alarm the enemy. Two companies were directed to search Nordschate for light material such as ladders and planks, with which to cross the canals.

At 12:15 AM the 8th Company would move out along the ditch 400 meters southeast of and parallel to the road: Noordschate—Drie Grachten, and follow this ditch to the Yser.

The 6th Company, starting at 12:30 AM, would at first follow the ditch just south of the road: then incline to the right and march on the junction of the Yser and the Martie Vaart.

The 5th Company, at 12:40 AM, would follow the road or the ditch just north of the road and attack the Drie Grachten bridge.

The 7th Company and the machine guns were to remain east of Noordschate in reserve.

The battalion commander further directed that there be no firing, that leading elements wear a white brassard, and that particular attention be paid to control, each company moving in a single column, preceded by an officer's patrol.

About 6:30 PM, the battalion started its march on

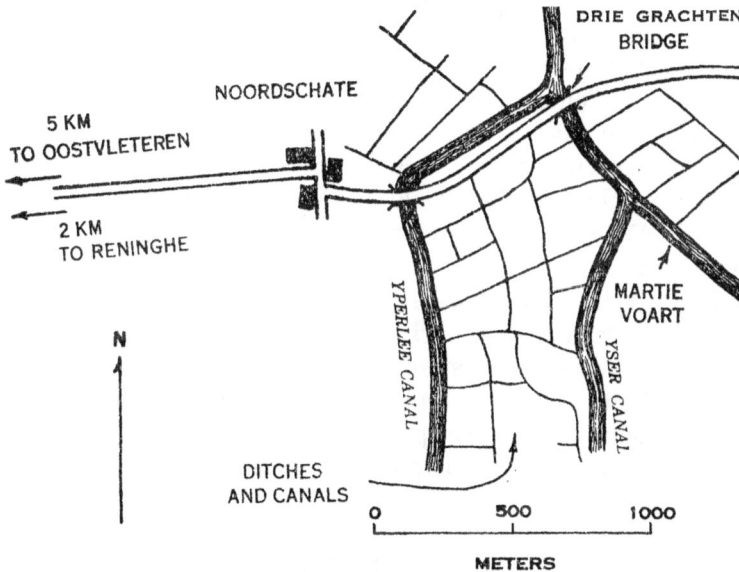

FIGURE 112.

Noordschate. In the meantime the battalion commander had reported the results of his reconnaissance and his plan for the attack to the regimental commander, who approved his dispositions, but informed him that he was going to try to have the attack postponed 24 hours.

At 8:00 PM the battalion commander met the colonel of the Zouave regiment at Noordscate, which still appeared entirely deserted. The Zouave commander said that he knew reinforcements were coming, but not that a night attack was contemplated. He added that he could not furnish any guides who knew the terrain in question. There were no evidences of the XX Corps to the south.

The battalion reached Noordschate at 8:30 PM. Efforts to find the commander of the front-line battalion of Zouaves on the left failed. However, the few Zouaves in the vicinity were notified of the proposed action of the 121st and told not to fire. Likewise a few tired soldiers of another unit were found just north of the road: Noordschate—Drie Grachten; and their commander, a noncommissioned officer, was informed of the plan to attack. To questioning he replied that he knew nothing of the terrain south of the road but believed that the water in the ditches would be about a meter deep. At 11:00 PM patrols reported that water in these ditches was breast high.

Just at this time an order was received countermanding the attack and directing the battalion merely to hold its ground. All companies were immediately notified.

At 11:00 PM a patrol reported that it had gained contact with the XX Corps to the south and found it in a state of complete confusion; no one there knew the location of even the larger elements.

At 12:30 AM came a new counterorder directing the attack to be launched at 3:00 AM. The battalion maintained all its previous arrangements with the exception of the times at which companies were to move.

The 8th Company moved forward at 2:15 AM. At 2:30 AM the captain of this company reported that it was almost impossible to cross the canals. Several men had fallen in and were unable to climb out of the sticky mud. He added that under such conditions movement to the Yser would require several hours, that many men would be lost en route, and that there would be no surprise. Having implicit confidence in this company commander and feeling that he would not exaggerate difficulties, the battalion commander immediately ordered:

"The 8th Company will follow the 6th, and on reaching the Yser, move south to its objective."

The 6th Company moved out on the road: Noordschate—Drie Grachten, and followed it almost to the Yser before turning south. A few minutes later the 8th Company followed the 6th. The 5th Company then moved by the same route to the Drie Grachten Bridge.

The attacks of all three companies succeeded.

The Germans, completely surprised, were thrown back over the Yser without more than a shot or two being fired. The battalion captured 25 prisoners and suffered no losses.

(From the article by Lt. Col. Baranger, French Army, in *La Revue d'Infanterie*, April 1929.)

DISCUSSION

This attack succeeded despite conditions that might easily have led to failure, such as fatigue of the troops, the almost impassable state of the ground, vagueness as to the situation of adjacent units, and the fact that the troops arrived on the scene after dark.

Why did it succeed? DIRECTION! CONTROL! SURPRISE!

The column formation in which the advance was made facilitated control. Each company was preceded by an officer's patrol; thus, when contact was first made, it was made by a group under a responsible leader.

The road and the Yser guided the troops to their destination. In the original order, these companies were to advance abreast, each in a column and each following a specified ditch. When this was found to be impracticable, all used the road.

Extreme precautions were taken to obtain surprise. Despite the obvious desirability of ascertaining the state of the terrain, the battalion commander limited reconnaissance to 200 yards to the front in order to avoid alarming the enemy. In the advance he insisted on silence and prohibited firing.

Finally, the battalion consisted of good troops and determined leaders, and as a result of success in three months of war, a feeling of mutual trust and confidence had been established.

"The symphony in black was not known to this battalion," says Colonel Baranger.

15. *CONCLUSION.—a.* Experience in the past has led to certain definite conclusions in regard to the undertaking and preparation of night attacks. For the sake of emphasis they are repeated.

(1) Night attacks should be undertaken preferably by troops that are fresh, well-trained, in good physical condition. These troops must be under excellent control at the start.

(2) The objective should be clearly defined and easily recognized in the dark, or already known to the attacking troops.

(3) The units making the attack should be able to form facing the objective and at no great distance from it.

(4) Generally speaking, there can be no maneuver.

(5) Simplicity in night attacks is vital in order to maintain direction and control. Routes of approach should be clearly defined and unmistakable in the dark. Trails, railroads, edges of woods, ridges, valleys or other prominent features leading toward the objective, which can be readily located and followed by the troops, are most desirable.

(6) Subordinate leaders should be afforded an opportunity for daylight reconnaissance.

(7) Detailed preparation is usually necessary if the attack is to succeed.

b. For the troops carrying out the attack, the following points are important:

(1) The formation must facilitate the maintenance of direction and control. This means a column formation in the early stages, and, as the enemy is approached,

a line of small columns preceded by patrols, in preference to a line of skirmishers.

(2) A strong leader with a few determined men should be at the head of each column. An officer or a reliable noncommissioned officer should be placed at the rear. Thus emergencies can be handled silently and decisively.

(3) Instructions must be explicit. All men should know the objective, the compass direction of attack, the formation to be taken at the place of deployment, the exact mission of the unit, the signal for the final rush to the assault, action in case the enemy is not surprised, locations of rallying points in the event the attack is definitely checked, action upon capturing the hostile position, and the means of identifying friendly troops.

(4) Secrecy and silence are essential. There must be no firing, no yelling, no smoking, no striking of matches. Absolute silence must be maintained until the troops are among the enemy.

c. Night attacks are highly difficult operations. They are frequently the expression of a vigorous leadership, which, regardless of difficulties, is determined to carry through to a successful conclusion. But despite vigor of leadership, these attacks will fail unless extreme attention is accorded to that military trinity of the night—*direction, control, and surprise.*

CHAPTER 14

ATTACK IN WOODS

1. *GENERAL CHARACTERISTICS.*—*a.* The following general characteristics are true of all combat in woods:

(1) Movements of troops are concealed from hostile air and ground observation.

(2) The effectiveness of all fire is less and the importance of close combat is greater.

(3) Direction, control, and communication are difficult to maintain.

(4) The possibilities of surprise and ambush are greater.

(5) The effectiveness of gas concentration, especially those of persistent gases, such as mustard, is increased.

(6) Action by small forces is favored.

b. These factors vary, of course, with the density of the growth, the size of the woods, the conformation of the ground, and the weather. Every woods presents a distinct problem which must be met and solved on its own merits.

2. *PHASES.*—The attack on a woods may be divided into three general phases:

a. Advance over open ground to the edge of the woods.

b. Actual fight and advance through the woods.

c. The debouchment from the woods.

3. *ADVANCE TO EDGE OF WOODS.*—During the advance over open ground to the edge of the woods, the attacker is confronted by the manifest disadvantage that the enemy can see all of his movements whereas the

385

enemy's positions in the woods are concealed. To offset this disadvantage as much as possible, the attacker may adopt any one of the three following methods:

a. He can have his scouts reconnoiter to the front and flanks of the advancing column. Visual contact must be maintained between the scouts and the advancing units or between the scouts and connecting files. As soon as the scouts indicate "all clear", the platoons and sections advance on the woods, and the scouts move on to continue their reconnaissance.

b. The attacker may employ smoke screens to conceal his advance over the open ground, providing wind and atmospheric conditions are favorable.

c. The approach to the edge of the woods may be made under cover of darkness.

4. *ACTUAL ASSAULT.*—*a.* The disposition of troops for attacking the edge of a defended woods is similar to that for attacking any other strong position. The strength and position of the defending force is not likely to be fully uncovered in the first stages of the fight. Nevertheless, the attacker can usually adhere to his initial dispositions until he has captured the edge of the woods. Seizure of the edge of the woods is of paramount importance, for once this is accomplished, the attacker is on an equal footing with the defender, in so far as natural cover is concerned. After the edge of the woods is taken, it should be used as a line of departure on which a reorganization is made and from which the advance through the woods is launched. Unless the attacking troops halt at the edge to reorganize, the advance immediately becomes sporadic and unequal, and confusion and loss of direction result.

b. During the halt at the edge, patrols advance and maintain contact with the enemy. Combat patrols protect the flanks, and measures are taken to keep contact with the troops to the flanks and rear. A commander must bear in mind, however, that the edge of a woods is a distinct terrain feature which can readily be registered upon by hostile artillery, and swept by hostile attack aviation. Hence, the period of reorganization for further assault at the wood edge should be brief.

c. Moreover, a wood should never be attacked unless its possession is indispensable. Of course, when tactical or strategical considerations render it necessary, and it is unsafe to pass by hostile defenses in a woods, the decision is unhesitatingly made to attack, woods or no woods. The penetration is the usual method employed in dislodging the enemy from the edge of a large wood. Against a small wood an envelopment (following artillery preparation) is the more common method.

d. The same thoroughness in reconnaissance, care in formulation and transmittal of orders, provision for maintaining direction and control, and aggressiveness in execution of plan, that are indispensable to night combat, are equally essential in woods fighting. Reconnaissance agencies must thoroughly reconnoiter the ground to the front and flanks. Intelligence agencies must seek information concerning the interior of the woods, such as roads, trails, marshes, and hostile strength and dispositions. Considerable information can be obtained from local inhabitants and from maps or aerial photographs. If the information is not complete, it is probably best to uncover the hostile position by feints.

e. When the time arrives for a large force to make the true attack, it is usually more effective to launch two or three distinct attacks simultaneously, rather than trust to a single main effort. The principal reason for this lies in the difficulty of intercommunication in woods. Thus several attacks, all delivered at the same moment, lessen the chance of the defender's properly employing his rear echelons. This obviously affords to the attacking units a greater opportunity to discover and break through the weakest link in the defense chain. It must be borne in mind, however, that when several attacks are launched, they must be thoroughly coordinated, and not delivered piece-meal. The plan of attack is, of course, governed by the nature of the ground cover in front of the wood.

5. *ADVANCE THROUGH THE WOODS.*—*a.* Let us assume that the edge of the woods has been carried and that the attacking troops have been reorganized. They now push vigorously into the woods, but the advance should nevertheless be methodically organized, with a view to maintaining cohesion among the attacking units. Besides depending largely upon reconnaissance and other security, the rate of advance is influenced by the character of the growth. The advance is made under protection of patrols to the front and flanks. Patrols must maintain visual contact with the troops they are protecting, and adjacent units also, if possible. In view of these facts, the advance through the woods is better accomplished, in most situations, by making a series of bounds.

b. Direction is maintained by means of compass bearing. Roads, trails, streams, clearings and the like are also used as guides for the maintenance of direction. Short halts, to fix direction, and to restore order, communication and control, are made on predetermined, well-defined lines, such as roads running perpendicular to the direction of advance. If such features do not exist, halts for this purpose should occur on a prescribed time schedule.

c. Too many halts of this kind are inadvisable. It is desirable to get some of the leading elements through the woods, established in positions at the farther edge as soon as practicable. This is not, however, a rule applicable to every situation. The emphasis that should be placed upon communication and control varies according to the extent of the woods, the opposition encountered, and the nature of the operation.

6. *FORMATIONS.*—*a.* Since two of the most difficult problems to be solved in the advance through woods are maintenance of direction and control, it follows that the

formation adopted is of great importance. Tactically,
it should prevent surprise by ambush, permit rapid deployment, furnish opportunity for the use of assault fire,
and allow a vigorous advance with fixed bayonets. Technically the formation depends upon the type of woods
encountered. In dense woods a compact formation is
essential to control. In open woods, however, the leading
elements may often be deployed in an irregular line, and
control still maintained. Line formations of this type in
thick woods are virtually impossible; for men get lost
or slip away in the underbrush, and the formation soon
degenerates into leaderless groups incapable of concerted
action. On the other hand, section columns, or columns
of section columns, increase the difficulty of deployment
and retard the development of the available firepower.
Therefore squad columns is often the best formation for
the leading elements of an attack driving through thick
woods. The squads work forward rapidly and easily;
moveover, they can rapidly deploy and maneuver within
themselves as necessary, and can be controlled without undue difficulty. Squads can best advance in patrol formation. The leading sections of assault platoons, then,
should be deployed in line of squad columns or groups,
in visual contact, and the rear sections and platoons in
section columns, or columns of section columns.

b. The rear platoons comprise the local reserves; they
should follow the assault units closely. The reason for
this can be readily seen if we visualize the leading elements
moving forward against an enemy they cannot see, and
whose strength they therefore greatly magnify. In such
a situation close support is essential to morale and confidence. The adoption of deep formations in wooded
terrain makes for poor cohesion of units and may result in
the absence of support at a critical time.

c. It is nevertheless true that in extremely dense woods,
reserve units may have to move forward in single file,
if a reasonable rate of progress is to be made. With

leading units advancing in this manner, rear units in
column of twos may find it impossible to keep up in passing through dense underbrush.

d. The local reserve of a company or platoon generally
follows the assaulting troops at between 100 and 200
yards, depending on the density of the woods and the
roads, paths, and clearings available. Contact with the
assault units is maintained by patrols or connecting files
sent forward from the reserve. In a similar manner
contact is preserved with adjacent units.

e. It is as useless to outline formations to be rigidly
maintained by troops attacking through a wood, as to do
this for units advancing over open terrain. Unless the
advance is weakly resisted, it will be difficult, if not impossible, to preserve any regular formation, or even to
guide the fighting to any great extent. The commanding
officer will come nearer success if he is prepared to face
and meet a constantly shifting situation.

f. This breaking up of so-called "regular" formations
in woods is inevitable. There is no magic formation by
means of which a commander can lead a body of troops
through a defended forest and maintain continuous communication and unbroken battle lines. In fact, the evidence
from past operations implies that troops engaged in woods
fighting usually become so intermingled, after a short
period of close combat, that any semblance of a controlled
action by the senior commanding officer ceases, and that
bush fighting, by isolated groups, is the usual result. Add
to the poor visibility, the "hide-and-seek" nature of the
combat, the continual whine of ricochets, the exaggerated
noise of bursting shells, and the dropping of large branches
from trees struck by shells, and disorganization is not
surprising. This is by no means an argument against
adherence to easily controlled formations for as long as
the situation permits. It is merely intended to indicate
the necessity for common sense tactical adaptation and
to point out what has almost invariably occurred in the
past. On the other hand, a commander must not accept

disorganization and loss of control as factors that he can do nothing about. On the contrary, all leaders must be forewarned of the difficulties and must take every precaution to preserve cohesion and liaison, and avoid confusion.

7. *FIGHTING WITHIN WOODS.*—*a.* As soon as the near edge of a woods is captured, vantage points are seized, reorganization is effected if necessary, and the advance is pressed before the defenders can rally to counterattack. As long as it is possible for assault units to advance, reserves should be held out. If the scouts encounter resistance too strong for them to overcome, the leading rifle elements rapidly form skirmish lines facing this resistance and, utilizing all cover, close with the enemy. Units in rear move to one side of the line of advance with a view to striking the resistance in a flank, or checking a counterattack against the flanks of the leading elements.

b. For the same reason brought out above with regard to our own occupation of the edge of a woods—the fact that it is a precise target for artillery and attack aviation—, attacking troops are likely to find the hostile defenses disposed with considerable depth, and not mainly along the actual fringe. In fact, the edge may be thinly held, or chiefly protected by a series of separated machine-gun posts, with the strongly held defenses farther in. But here, as in all warfare, we must remember that an enemy may have a different conception of defense from our own, and hence, be ready for the unexpected.

c. If the forward move is temporarily checked, the local reserve of the attackers will probably become involved. Commanders should then bend every effort to form scattered troops into small bodies for employment as a fresh reserve. Groups of this nature can be thrown in at key points with telling effect. In command of such groups noncommissioned officers and junior officers find opportunity for real leadership. The group leader should try to establish and retain some semblance of formation and control and at the same time avoid cramping the "Indian-fighting" style of the individual soldier. Higher commanders quickly lose control of the fight in the face of stubborn resistance. The leaders of smaller units, however, may effectively control their respective groups, and still make an effort to preserve in some degree the general connection of the whole.

d. The longer troops can fight in woods as a unit, the better the chance of success. To obtain this result, a commander must realize that small units (platoons, sections, and even squads) are more or less independent in their fighting, once the advance encounters stiff resistance. Battalion commanders almost invariably, and company commanders usually, will find that an attempt to exercise direct control for too long a time will cause the assault lines to dissolve into individuals; whereas, if the situation is frankly faced and actual control of subunits relinquished to subordinates at the proper tactical moment, effective group fighting will ensue, rather than fruitless individual "duelling".

e. In the event that only one part of the assault line is checked, units on the right and left continue to push forward, increasing their flank security, but, in the absence of definite orders, leaving the job of assisting the checked unit to rear elements. This is done because uniformity in the advance is usually less important than getting as many troops as possible through the forest and establishing them on the far edge (or, when attacking through a deep forest, at a predetermined line), in the least possible time.

f. As previously mentioned, one of the main characteristics of woods fighting is the reduction of the effectiveness of fire and the increase in the importance of close combat. As in night combat, this brings the bayonet to the fore as the decisive weapon. Unlike night fighting, however, rifle fire is important in woods fighting, especially in cleared spots and in areas devoid of thick undergrowth.

g. During the advance through woods, progress may be signalled by flares, which can be readily seen by airplanes. Rocket signals are used to indicate the arrival of a unit at the objective or on important terrain features. Pre-arranged signals of this type are especially important when the advance is being made in conjunction with artillery support. There should be due caution, however, to see that such signals are not displayed to enemy observers instead of our own.

h. In any advance through woods, troops must be particularly watchful for gassed areas. Woods are well-nigh ideal localities for the use of peristent gases in defense. A withdrawing enemy may, indeed, drench a strip, either by aircraft or ground gas weapons, across the entire width of the woods to prevent, or at least greatly hinder, our advance. The enemy may also fill certain areas with mustard or similar persistent vesicants in order to force us to follow lanes of approach into his best fields of fire.

i. The enemy is also likely to use obstacles for this same purpose. Hence, extra caution and forethought is requisite whenever troops advancing through woods find obstacles in their path.

j. When units have been successful in passing through a wood, provision must be made by the commander for a careful mopping up of that area. This duty is given to reserve units.

k. As we have previously seen, direction is maintained by compass. This applies, of course, more to a march through woods or an advance against feeble resistance than to hard fighting in woods. Practically, the direction of the enemy's retreat usually determines the direction of advance of the leading assault groups; for the fire of the enemy acts as a magnet to draw the attackers on, representing as it does the immediate and obvious opposition to be overcome. This tendency for the enemy to dictate the direction of advance is counteracted by deploying the reserve as a new assault wave facing in the right direction, and, when possible, withdrawing the old leading units, and reforming them into a new reserve. In any case, it is likely to be futile to seek to correct the direction of the straying units, in view of the well-established principle that deployed troops under fire can change front only at the risk of heavy losses, and also because, in woods particularly, control of troops once committed to the fight is extremely difficult to retain.

8. *DEBOUCHING FROM THE WOODS.*—*a.* Before debouching from a wood to follow up a success, the various echelons should be reorganized, if necessary.

b. If the situation permits, the terrain over which the advance out of the woods is to be made should first be covered by reconnaissance patrols. Machine guns, mortars, and 37-mm guns should be used to cover the debouchment.

c. Here, again, it must be remembered that the actual edge of the woods is an excellent target for hostile artillery and aircraft. If the enemy is expected to place heavy artillery fire on the edge of the woods as soon as his own troops are clear—and he will if he has any artillery—the attacking force should move out of the woods by a single rush, utilizing all cover and taking advantage of surprise. Likewise, when heavy fire from close-combat weapons is possible, the advance from the shelter of the trees should generally be effected by rushes of small groups.

9. *MACHINE GUNS.*—*a.* Machine guns, correctly used, are of great assistance to troops atacking a wood. During an advance to capture the near edge, they may support the attack by overhead fire, using either direct or indirect laying. This fire is often delivered frontally, penetrating the woods to a considerable depth. Their assistance is usually greater if the guns can be emplaced to deliver oblique fire, since this creates a greater danger zone just within the edge of the woods.

b. When the enemy holds a small woods not wholly included in the zone of action of an attacking unit (See

Figure 113), machine-gun fire along the side *AB* prevents his debouch from, or entry into, the woods from other points. After the near edge *DA* of a small woods (less than 1,000 yards in depth) has been taken, machine guns may be advantageously moved to positions on the flank. From these positions fire can be delivered on enemy troops approaching or leaving the far edge of the wood *BC*.

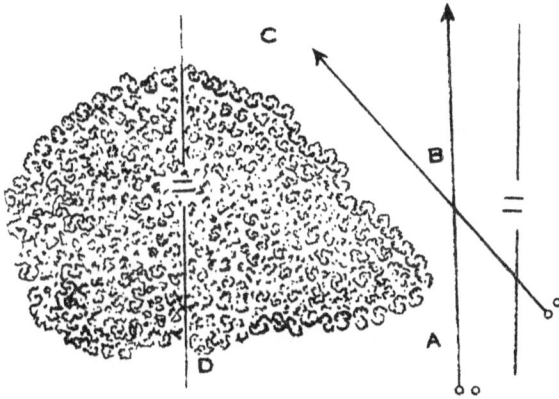

FIGURE 113.—Machine guns used in attacking a small woods.

This may require that guns be moved into the zone of action of an adjacent unit.

c. If the woods extend over the entire width of the zone of action and are more than 1,000 yards in depth, machine-gun units, after helping the attack gain the near edge, should close up on the rifle companies, usually with one platoon supporting or attached to each. Within the woods, machine guns are placed to fire along roads and paths, and cover clearings. When positions of sufficient elevation are available, plunging fire delivered over the tops of trees may materially assist the advance.

10. *OTHER SUPPORTING WEAPONS.*—*a.* The infantry mortar and the 37-mm gun can be employed to

advantage. Their most effective use, however, usually comes before a wood is actually entered. These weapons are much less effective, as soon as it becomes impossible to observe the strike of the projectile.

b. In thick woods, hand and rifle grenades may inflict damage on friendly troops by striking tree trunks and branches and exploding within a radius dangerous to friendly troops; hence they must be used with care. In relatively open woods, troops that are thoroughly trained may employ these weapons with excellent effect.

11. *TANKS.—a.* Woods limit the operation of tanks in actual mechanical movement and in control. They cannot pass through close set trees of 8 inches or more in thickness. When tanks are used in woods fighting, they should be kept well forward in order to make them immediately available to the unit to which they are attached. Tanks move in column, and from cover to cover, until such time as their profitable employment is possible. When committed to action in woods they are governed by the same tank tactics as in any other form of combat, except that intervals between tanks are necessarily reduced, for purposes of control and coordination.

b. In the assault of a woods, tanks should avoid well-defined lanes and roads leading toward the enemy's position, because these are liable to be covered by hostile antitank weapons or mines.

c. Tanks are valuable in breaking through abatis, or wire, or other obstacles.

d. If the woods are of such character that tanks can operate in them, antitank weapons should be employed with the assaulting units. If the woods are too dense, then antitank weapons should be held back to cover the intervals between woods.

12. *ARTILLERY.—*Artillery support furnished to a unit attacking through woods usually takes the form of successive concentrations on sensitive points (hills, cross-

roads, machine-gun nests, command posts, dumps, and the like). Although the physical effect of 75-mm shells bursting on targets in wooded areas is limited, the effect on morale is great. Artillery fire must, of course, be placed safely in advance of the assault line. This can be accomplished only by maintaining the closest liaison between artillery and infantry.

13. *GENERAL RULE OF SUPPORT.*—Each phase of a battle in woods determines the use to be made of each supporting weapon. It should not be supposed, moreover, that in the heat of battle—when the fighting is by groups rather than by controlled units—, machine guns can be emplaced, sited, and then fired so as to contribute effective support. Woods combat presents in rapid variation an ordinary attack, close combat, independent group clashes, and finally pursuit. All this combines to make the opportunities for supporting weapons fleeting and irregular, opportunities to be exploited immediately or not at all.

CHAPTER 15

PURSUIT

1. *INTRODUCTION.—a.* A primary object of all warfare is the destruction of the armed forces of the enemy. It is obvious, then, that the commander who defeats the enemy but who pauses to draw breath or who is content to rest on his momentary laurels, and thereby allows his opponent to escape, is abandoning his mission. Furthermore he is indulging in the most hazardous form of gambling. Too often has the defeated force of one day been the victor in a decisive action on another. Witness the unhindered withdrawal of the German Eighth Army following its defeat by the Russian First Army at Gumbinnen on the 20th of August, 1914. Nine days later that same Eighth Army completely destroyed the Russian Second Army in the great battle of Tannenberg.

b. A defeated army is not a vanquished army until its complete destruction has been accomplished. This is not achieved through the mere process of driving the enemy from his position. That is but the first step in the right direction, an initial advantage that may be quickly lost unless the enemy is relentlessly pursued. No time must be given him to reform his shattered battalions. Continual and unending pressure must be applied to the end that his demoralized columns must eventually halt, give battle, and so be destroyed.

2. *THE DECISION TO PURSUE.—a.* It is a fundamental principle of war that a victorious force, unless prohibited by a specific mission or by considerations of the situation as a whole, will pursue its defeated opponent

with the object of destroying him in decisive combat. Bearing in mind that the term pursuit is used in this text in the broad sense of aggressive, offensive action against a withdrawal of the defeated enemy, the first question that arises is: When should the decision be taken to pursue? The answer is definite and unequivocal —at the first indication of the enemy's *intention* to withdraw.

b. The real problem is to determine whether the enemy contemplates or has already begun a withdrawal. This is far from obvious. The movement will almost invariably be initiated under cover of darkness and will be extremely difficult to detect during its early stages. There are no simple and precise rules by which it can be instantly determined. The initiative and aggressiveness of the commander and his subordinate leaders, and the persistent effort of all reconnaissance agencies down to include those of the smallest unit, largely determine whether or not the enemy slips away undetected.

c. There are many circumstances that increase the likelihood of the enemy's attempting a withdrawal; for example, if he occupies a poor position, if his flanks, line of communication, or base are seriously threatened, if he is heavily outnumbered, if he knows his opponent is about to be strongly reinforced, if he has just been defeated, or if the morale of his troops has crumpled as a result of previous defeats and excessive casualties. The situation should be constantly studied for these influencing factors. And when they apparently become of sufficient weight to render a hostile withdrawal likely, every possible precaution should be taken to detect the movement at its inception. Observation airplanes if available should scan the roads for troop or train movements to the rear. At night they should illuminate the hostile position and rear areas with flares. The infantry must be increasingly vigilant and aggressive. At nightfall strong combat patrols should feel out the hostile line to determine the presence or absence of the enemy. They must make every effort to find out whether the resistance they encounter is offered by the hostile force as a whole or by weak covering detachments acting more or less independently. Prisoners should be taken; and, if possible, patrols should push through to small-unit command posts to determine what activities are under way.

d. All information, whether positive or negative, must be rushed back to the responsible headquarters. There it is pieced together bit by bit. If the final picture indicates a withdrawal, and neither the mission of the command nor the general situation prohibits it, the responsible commander at once makes the decision and immediately prepares to launch the pursuit.

3. *PRELIMINARY CONSIDERATIONS.—a.* Before organizing and planning the pursuit, the responsible commander and his staff should strive to discover the enemy's intentions. Answers should be sought to four main questions:

(1) First, Why is he withdrawing? Is it voluntary as for instance, in the preconceived plan of a delaying action? Or has he been forced to it? This is an important consideration. A commander would not be justified, for example, in taking the same chances against a well-organized force known to be fighting a delaying action that he would against a badly beaten enemy whose morale is shaken.

(2) Second, What is the enemy's destination and what routes are open to him? There is seldom any definite information on this point. Usually all that can be done is to determine the destination he is most likely to seek and study the road net by which he must travel. When his destination can be reasonably predicted this assists materially in formulating the plan of pursuit.

(3) Third, What is the enemy's scheme of withdrawal with particular reference to the size, composition, and use of his covering force? This information, as will be seen later, may be used to great advantage by that por-

tion of the pursuing force used to apply direct pressure.

4. The fourth consideration, and one that largely determines the risks a pursuing force may take, is the effectiveness of the hostile troops. Is their morale shaken? Are they fatigued? Has their fighting efficiency been seriously impaired? What degree of control remains?

b. A commander will do well to consider these four major points and all their implications before launching his pursuit. A few minutes devoted to a study of these varying factors prior to issuing the order for the operation, may often save many an hour of wasted effort once the action is under way. It must be remembered, however, that more often than not only one or two, and frequently none, of these questions can be answered. Again and again there will be no information of the enemy other than the fact that he is withdrawing; and even that will arrive late. Often, too, little or nothing is known of the terrain over which he is moving. His entire situation will be vague to us, and ours, particularly after a pitched battle, will usually be far from clear. These difficulties have to be taken in stride if the enemy is to be destroyed. Risks there will be, but as John Paul Jones once said: "It seems to be a law, inflexible and inexorable, that he who cannot risk, cannot win."

4. *PLAN OF PURSUIT.*—*a*. It will be recalled that pursuit was previously defined as aggressive, offensive action *against* a withdrawal of a defeated enemy. Thus, it is seen, that in the broadest sense of the word, pursuit may occur even when the enemy merely seeks to abandon his position but is prevented from carrying out his intention. Such a situation may arise when the intelligence agencies of an attacking force are fortunate enough to ascertain and recognize the first indication of a retrograde movement by the enemy. With this information it may be possible for the high command to force the enemy to abandon his intention before his withdrawal has even

begun to get under way. A determined attack immediately launched against his entire front may serve to hold him in place whether he wishes it or not. Occasionally an energetic commander who has a highly mobile force at his disposal may be able to throw it across the hostile line of retreat, and thus definitely prohibit any rearward movement. In the past this latter expedient has rarely been possible, and in those relatively few instances when it could be accomplished, other circumstances have frequently rendered it inadvisable. However, the tremendous mobility now afforded by trucks, armored cars, and the new fast tanks, indicates that far more opportunities for such action will occur in the future.

b. A situation such as that just described represents the ideal. Unfortunately it seldom obtains. Even the most painstaking precaution rarely ever discloses a hostile withdrawal in its early stages. In consequence, the plan of pursuit with which we are most concerned deals with an enemy who has successfully disengaged all or the major portion of his troops from combat and is withdrawing under the protection of covering forces. How can he be stopped and engaged in decisive combat? Experience has shown that this is most effectively accomplished by dispatching *encircling forces* (Figure 114) around one or both flanks to cut him off or slow him down, while the remainder of the command, *the direct pressure,* (Figure 115) follows up ready to strike a decisive blow as opportunity affords. Let us consider, then, the encircling maneuver and the conduct of the direct pressure under separate headings and in detail.

5. *THE ENCIRCLING MANEUVER.*—*a. Mission.*—A force dispatched on an encircling maneuver (Figure 114) is assigned one of two missions; to block or cut off the enemy's retreat in whole or in part, or to slow him down enough to permit the remainder of the command to come up and strike the decisive blow. The destination of the encircling force is prescribed by the

commander sending it out. The route thereto is also
prescribed by the same commander up to the time the
encircling force will have cleared known resistance but

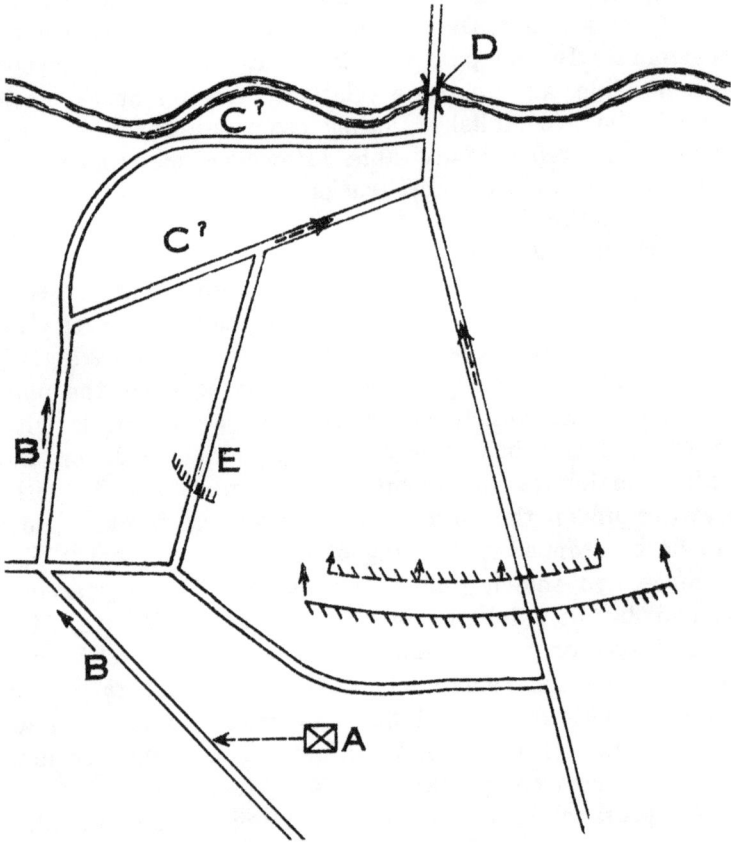

FIGURE 114.

thereafter it is left to the discretion of the encircling
force commander.

(1) *Blocking the enemy's retreat.*—(a) In an earlier
paragraph it was pointed out that the high command
should seek to discover the enemy's destination and the

route or routes he is using—in other words, his line of
retreat. The importance of this at once becomes evident
when considered in relation to the mission of the encircling force. Without this information—or a reason-

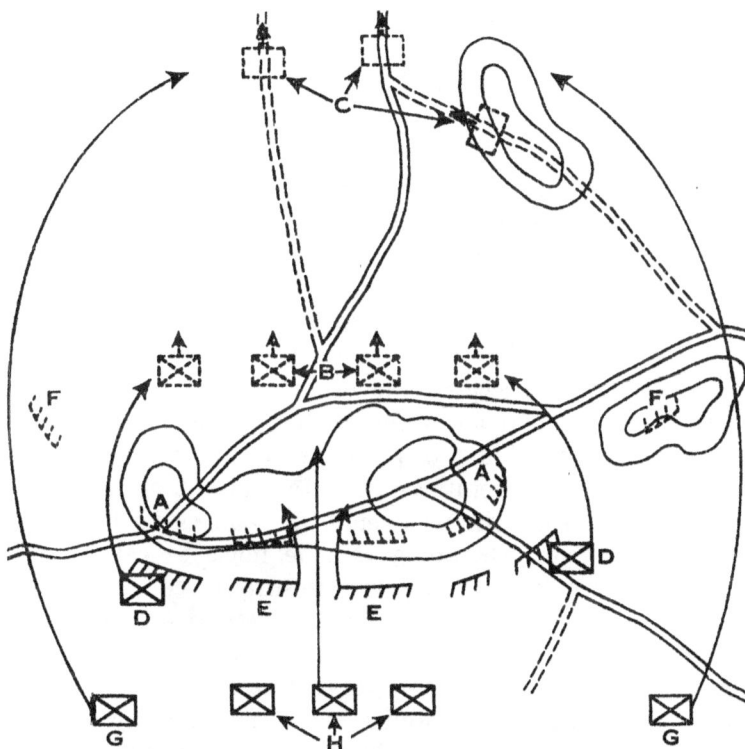

FIGURE 115.

ably accurate prediction—, this force will be constrained
to act without any definitely preconceived plan. The blind
groping of such action obviously militates against success
and materially increases the always considerable risks
that this force runs.

(b) Let us assume, then, that the line of retreat is
known, or that the lay of the ground and the situation

as a whole permit a reasonably accurate forecast. The objective of the encircling force must now be selected. Three considerations should be borne in mind in this selection. First, the position should be vital to the success of the enemy's retirement. In other words, unless the enemy holds it his withdrawal is blocked. Second, the encircling force should be able to reach it before the enemy. Third, this force must be strong enough to hold the enemy at the point selected until the remainder of the friendly troops—the direct pressure—comes up. It must be remembered that the encircling force will initially be beyond supporting distance. Therefore, unless it is reasonably strong and soon followed by other troops it runs grave risk of decisive defeat.

(c) A little thought will show that an encircling force used in this manner runs serious risks despite any precaution that may be taken. On the other hand, however, its action when successful is often decisive.

(d) The commander ordering the pursuit should back up the actions of his original encircling force by dispatching others to reinforce it or to assist it by attacking the flanks of the retiring enemy, as soon as he has other troops available. On doing this he must be careful not to unduly weaken his direct pressure forces.

(2) *To slow down the enemy's retreat.*—It may be neither advisable nor practicable to send out an encircling force to cut the hostile line of retreat. Troops in sufficient strength to accomplish this mission may not be available or the enemy may have too great a start. When such conditions occur, a compromise has to be made. Frequently no more can be done than to hastily organize those troops at hand and send them out to attack the retreating enemy in flank with a view to delaying him as much as possible. Obviously, the nearer this encircling force strikes the head of the retreating column, the greater the chance of decisive action.

b. Composition of encircling forces.—(1) *General.—*

(a) A victorious force must launch its pursuit of the
defeated enemy at the earliest possible moment. Every
minute of delay increases the enemy's chance of escape
by just that much. It should be remembered that an
encircling force must move farther and faster than its
opponent. It has to swing wide and at the same time
outdistance the retreating columns. Obviously the
greater the delay in starting that force, the less chance
it has to intervene at a decisive point. The rule, then,
is this: launch the pursuit in the least time possible
with those forces that are available and nearest at hand.
Reserves, or those troops that can be withdrawn from
action quickly and with the least difficulty, should be
used. The important thing is to get some force under
way at once. Other forces can follow as circumstances
permit.

(b) An encircling force must be able to move more
rapidly than the enemy no matter how fast he moves;
or it cannot accomplish its mission. For this all-important reason the responsible commander should, whenever
practicable, employ the most mobile units of his command on encircling missions.

(c) The strength and composition of the encircling
force must be such as to accomplish the desired end.
If the means at hand are insufficient to carry out this
purpose, the mission must obviously be modified.

(d) With the foregoing considerations in mind let us
proceed to a brief discussion of the various arms with
reference to their employment in encircling forces.

(2) *Cavalry.*—This arm is particularly valuable for pursuit because of its relatively great mobility. The encircling
force may be composed entirely of cavalry if there is a
sufficient amount available. Ordinarily this condition
will not obtain, and the encircling force must be made up
of two or more of the combined arms. When this occurs
the cavalry is employed in one of two ways. It is either
attached to the infantry encircling force or sent out to
operate as an independent unit. If a relatively strong

body of cavalry is available, it can usually operate to better advantage sent out as an independent force, particularly if the hostile cavalry is comparatively weak. If, on the other hand, but little cavalry is available or the hostile cavalry is strong by comparison, it is wiser to attach the cavalry to the infantry force making the encircling maneuver.

(3) *Tanks.*—(a) *Old types.*—Although World War types have long since been obsolete, a considerable number remain on hand. Only a few test models of the new fast types have been issued to the service. Therefore it is highly probable that the old models will be pressed into service in the initial stages of any conflict that occurs in the near future.

If the old type light tanks are available and the terrain in front of the direct pressure does not favor their employment, they may be considered for attachment to the encircling forces. Such use, however, would be exceptional in view of the fact that these tanks lack the mobility vital to the success of the encircling force. If they are used in this manner it will almost always be necessary to carry them on trucks. They would ordinarily be of far more value to the direct pressure.

(b) *New types.*—The new, fast tanks are most suitable for the pursuit in that they combine the highest degree of mobility with fire and shock action. They could either be attached to the encircling force or used as an independent unit. Employed as an independent force to deliver surprise attacks against the enemy's flanks and rear, thereby delaying, disorganizing, and demoralizing his retreating columns, their value is apparent. Their demoralizing power and shock action can also be employed to great advantage with the encircling force by assisting it to win through to its objective when the enemy offers a stubborn resistance.

(4) *Artillery.*—Distances at which encircling forces operate and difficulties of maintaining communication with the main body make it impracticable for the ar-

tillery accompanying the direct pressure to support the
encircling maneuver. Therefore artillery should be attached to the encircling force. The range and firepower
of artillery render it particularly valuable on encircling
missions. Accordingly the forces charged with these
missions should have a large proportion of artillery
attached to them. It should be borne in mind, however,
that only light artillery has the requisite mobility for
encircling missions. Students should also keep in mind
the still greater mobility of modern motorized artillery
units.

(5) *Infantry.*—(a) Ordinarily the encircling force's
mission, and the absence or shortage of cavalry and fast
tanks, will force the use of infantry. When this contingency arises—and it almost always will—every effort
must be bent to increase the mobility of the infantry.
Obviously the ideal situation occurs when trucks are
available in sufficient quantity to transport the encircling
force. All present indications point to a large percentage
of motorized infantry in the future. Even relatively
small forces of such infantry would be of great value
to the pursuit. It is also worth while to consider the
possibilities of using commandeered automobiles. In this
day of motor cars, almost every community in a civilized
country is well enough equipped to furnish sufficient
cars, trucks, and busses for the transportation of infantry
on these and other missions. The value of civilian transportation for military purposes was graphically demonstrated twenty years ago in the historical dash of the
Paris taxicabs to the Marne.

(b) When no form of transportation is available and
none can be commandeered, leaders must resort to other
means of increasing the infantry's mobility. Care should
be taken to select those units that are the freshest. If
no choice exists in this respect, it may be possible to
form detachments of men especially selected for their
strength, health, and general endurance. Only equipment
that is absolutely essential should be carried. The lighter

the load of the infantryman, the faster and farther he can march. Similarly, all unnecessary trains should be left behind.

(c) The German pursuit of the defeated Russians in the Winter Battle of The Mazurian Lakes graphically illustrates the extremes that are sometimes adopted to heighten the mobility of an infantry force sent out on an encircling mission. In this action the Germans formed picked detachments of sturdy, robust men. They left all supply trains, even the kitchens, behind. With rations on their backs, they started out wide on the flanks. In spite of the intense cold of the Russian winter and the heavy snow that averaged more than a foot in depth, the encircling force drove forward at a rate of 21 to 31 miles a day for three successive days. More than 100,000 Russian prisoners were taken as a result of this operation. Without such heroic measures the full fruits of the German victory could not have been realized.

c. Route and action of encircling forces.—(1) The forces used on the encircling maneuver must move rapidly around the known or probable enemy covering forces and gain the hostile flank or rear. To provide for this the commander of the main force usually prescribes the route to be followed around these covering forces. In view of the uncertainty of future developments, the route thereafter is left to the initiative of the commander of the encircling forces.

(2) Although every effort is made to conceal the movement of the encircling forces, it should be kept in mind that an alert enemy will discover them sooner or later and detail forces to oppose them. To engage such forces would be to play into the enemy's hand. They should be avoided if practicable. If this cannot be done they should be contained by a fraction of the encircling force while the remainder continues on, around their flank, to the objective. If these hostile forces can neither be avoided nor contained, no alternative remains but to

attack. This attack must be prompt and the action so decisive that it will definitely eliminate this hostile barrier from that point on.

(3) The encircling forces must often move under conditions that are obscure and possibly hazardous. Boldness, without foolhardiness, should characterize their action. Special attention must therefore be given to local security, particularly if the enemy has cavalry or fast tanks.

(4) No situations call for greater exercise of initiative than those that usually confront the commanders of encircling forces. The rapidly changing situation and the difficulties of communication call for prompt decisions by the commander on the spot. Fortunate is that higher commander who has a Sheridan to keep in mind the mission of the whole force and act accordingly.

6. *DIRECT PRESSURE.*—*a. Mission.*—The mission of of those forces directly pressing the rear of the enemy (Figure 115) is to keep as many of the enemy's troops engaged as possible, to keep close on his heels, and be ready to strike with full strength at a decisive moment.

b. Conduct of direct pressure.—(1) *Initially.*—The pursuit should begin when the enemy starts to give way and attempts to disengage himself from action. Often the actual order to pursue will not arrive until hours after the enemy has begun his effort to disengage and withdraw. Therefore all subordinate commanders should press forward on their own initiative and without orders, spur on their troops, and clinch any advantage gained by them for the prompt employment of reserves. Direct pressure is maintained by the troops already in contact with the enemy and is a continuation of the preceding action characterized by an aggressive advance and the maximum employment of all available firepower. Troops, before whom the enemy is giving way, attack him in front, break through his covering forces, and strike his main body.

(2) *Later.*—(a) Front line troops cannot advance indefinitely against even relatively little resistance without disorganization and loss of control. Furthermore small hostile groups armed with automatic weapons can effectively delay large forces. Beyond a certain point no increase in the rate of advance can be reasonably expected by any increase of the troops in direct contact. Efforts to do this only result in heavy losses. These factors indicate that troops initially engaged in the direct pressure should be reformed at designated places preparatory to taking up the later phases of pursuit. At the same time, halts for reorganization must not result in losing contact or even in any diminution of the pressure by those in direct contact. Smaller bodies are pressed forward on a broad front to maintain this contact and pressure. Others are quickly reorganized and continued forward as local reserves. General reserves are reconstituted and employed as directed by the higher commander. All must continue forward. The primary object of overtaking and destroying the enemy's main body must be kept ever in mind and the pursuit so conducted as to permit the bulk of the troops to be promptly launched in a concerted attack against the enemy's main body when opportunity arises.

(b) During pursuit the situation is constantly changing, local opportunities come and go rapidly, and communication becomes increasingly difficult. Decentralization of command is therefore essential. It is impracticable for the higher commander to prescribe details. He must tell his subordinate commanders in general terms what he wants done, provide them with the means of doing it, start them off, and then, under such general supervision as he can exercise, depend upon their energy and skill for results. Hence zones of advance will usually be prescribed and supporting weapons frequently attached.

(c) Commanders of flank units should endeavor to push around the flanks of the enemy's covering forces and strike his main body. Interior commanders should endeavor to push through gaps in the enemy's covering forces (i.e., around the flanks of local units of these covering forces).

(d) The rapidity of progress and the unavoidable disruption of communications that it causes make it necessary for commanders of all units down to include platoons to keep up constant liaison with troops to the right and left. They must also send back frequent reports for the information of higher headquarters. Neglect to do so inevitably results in complete disorganization, loss of control, and consequent breakdown of the operation, and may lead to defeat.

7. *PURSUIT AT NIGHT.*—*a.* The general principles governing night operations apply in pursuit as in other forms of action. The attendant difficulties of maintaining direction and control are likewise present. There are certain special features, however, that should be kept in mind.

b. Night offers the enemy the best opportunity to effect a withdrawal. When the situation indicates a possible withdrawal, reconnaissance measures must be doubled. Patrols and raiding parties must push through or around the hostile covering forces and gain contact with the main body, taking prisoners, and observing what is happening in the rear of the hostile covering forces.

c. When the enemy withdrawal has begun, or better yet when it can be anticipated, pursuit must start. At night, the encircling forces should swing wider around the flanks in order to avoid the delay that even a very small enemy covering force may effect. Unable, at night, to effect much delay of the enemy's main body by the firepower of artillery or other weapons, an encircling force will more frequently endeavor to place itself across the enemy's line of retreat. A small force, so located at night, can seriously delay, or even effectually stop, a force many times its size. History affords many examples of the delay, confusion, and congestion of traffic resulting from such action. The consequent break-down in the enemy morale is of tremendous value.

d. If the encircling force is unable to place itself across the enemy's line of retreat, it should endeavor to reach a position on the flank from which, when day breaks, it may effectively stop or delay the enemy withdrawal. Harassing attacks at night on the enemy flanks, except by cavalry and light fast tanks, are difficult and the results only local.

e. During the night, troops in the direct pressure should endeavor to push small forces through the gaps in the hostile covering forces, take them in rear, or contain them, and pass on. The difficulties will be great and the progress slow with a resulting tendency, even greater than in daylight, to push reserves around the flanks in support of the encircling forces.

f. Pressure must be maintained, and as long as a semblance of control can be maintained, the pursuit must be pushed to the limit of the physical endurance of the troops. *He who sleeps at night may wake to find his enemy gone.*

8. *VARIATIONS IN PURSUIT.—a.* Pursuits differ depending upon the situation in each particular case. For example, the mission and situation may limit the distance to which a pursuit is to be carried. Encircling forces may be sent around one or both flanks or not sent out at all. Direct pressure may be effected by a fraction of the command while the bulk of the force follows around a flank, or flanks, in support of the initial encircling forces, or vice versa. Such action involves considerable risk. It may, indeed, be the very action desired by the enemy, who may be only simulating a withdrawal for the purpose of causing his adversary to scatter his forces in pursuit and thus make himself vulnerable to counterattack. Hence a maneuver of this kind should only be carried out when it is definitely known that the enemy cannot or will not take advantage of the

opportunity thus afforded him.

b. In all cases, however, the pursuit should be carried out on a front wider than that of the enemy. Commanders of units on the flanks should endeavor to push forces around the flanks of the hostile covering forces. These forces, though small at first, may be supported later as other troops are made available. The commander of the entire force, using any reserves he may have, or may quickly obtain, should start them off around the enemy flank or flanks. If these forces are weak, or the situation obscure, they may be attached to the flank unit and the commander of the latter directed to delay or intercept the enemy retreat. These forces should be supplemented, or supported, later by others as they can be withdrawn from units in the direct pressure. Thus, that which starts as an envelopment of the enemy flank may grow, becoming an ever increasing threat on the flank (or flanks) of the withdrawing enemy, until his retreat is effectively stopped. When the retreat is thus stopped, the pursuers, with their weight already developed on one or both flanks, are in a favorable position to promptly launch an attack. Action then must be aggressive, and as coordinated as possible. Otherwise the wolf may turn and successively destroy the hounds that have pursued and surrounded him.

c. The conduct of these forces, which operate on the flanks of the retreating enemy, corresponds to that previously described for encircling forces. Although the mission of these forces as a whole may be to cut off the hostile retreat, certain units may have the mission of attacking and delaying the enemy, leaving to the more advanced elements the mission of blocking the hostile retreat. They may be within or beyond supporting distance of each other, or of the main body. They may be under one commander, or under several separate commanders in which case the action is coordinated by the high command. They are all encircling forces, the others are the units in direct pressure.

9. *CONCLUSION.*—*a.* In undertaking a pursuit, leaders must bear in mind the all-important mission that the enemy must be forced to halt and then destroyed. All other missions assigned the pursuing forces are secondary and are designed with the aim of furthering this one basic mission.

b. The situation must be constantly studied for indications of a hostile withdrawal. Unless the utmost vigilance and constant contact are maintained, the enemy will slip away undetected. History abounds in such occurrences. In one instance on the Western front, an entire German corps withdrew, and its absence went undetected for more than twenty-four hours.

c. The fatigue of troops is not a valid excuse for failure to pursue. The enemy is just as exhausted—probably more so. In addition, the morale factor is all on the side of the pursuer. The pursuit must be continued until the enemy is destroyed or until actual physical exhaustion of the troops precludes further movement.

d. To sum up, then:

(1) Start early with the most mobile units that are available.

(2) Pursue on a broad front. Keep pushing forces around the enemy's flanks to cut off his retreat or delay his main body.

(3) Decentralize.

(4) Keep all forces up and well in hand in order to be ready to strike with full strength when the opportunity occurs.

(5) Be relentless. Give the enemy no rest by day or by night. Push the pursuit to the extreme limit of physical endurance.

www.ingramcontent.com/pod-product-compliance
Lightning Source LLC
Chambersburg PA
CBHW021546210326
41599CB00010B/325